TWO-STROKE PERFORMANCE TUNING

TWO-STROKE PERFORMANCE TUNING
Second edition

A. GRAHAM BELL

Haynes Publishing

© A. Graham Bell 1999

All rights reserved. No part of this book may be reproduced or stored in a retrieval system or transmitted, in any form by any means, electronic, mechanical, photocopying, recording or otherwise, without prior permission in writing from the publisher.

First published by G. T. Foulis & Company, June 1983
Reprinted February 1984
Reprinted October 1985
Reprinted November 1987
Reprinted July 1989
Reprinted April 1991
Reprinted August 1992
Reprinted November 1994
Reprinted December 1995
Reprinted March 1997
Reprinted February 1998
Second edition published by Haynes Publishing September 1999
Reprinted September 2000
Reprinted May 2002
Reprinted June 2003
Reprinted January 2006
Reprinted December 2007
Reprinted February and November 2009
Reprinted November 2010
Reprinted January 2012

British Library Cataloguing in Publication Data
A catalogue record for this book is available from the British Library

ISBN 978 1 85960 619 3

Library of Congress catalog card no. 99-73267

Haynes Publishing, Sparkford, Yeovil, Somerset BA22 7JJ, UK

Tel: 01963 442030 Fax: 01963 440001
Int. tel. +44 1963 442030 Fax: +44 1963 440001

E-mail: sales@haynes.co.uk Website: www.haynes.co.uk

Haynes North America, Inc. 861 Lawrence Drive, Newbury Park, California 91320 USA

Typeset by G&M, Raunds, Northamptonshire
Printed and bound in the UK by MPG

Jurisdictions which have strict emission control laws may consider any modification to a vehicle to be an infringement of those laws. You are advised to check with the appropriate body or authority whether your proposed modification complies fully with the law. The publishers accept no liability in this regard.

While every effort is taken to ensure the accuracy of the information given in this book, no liability can be accepted by the author or publishers for any loss, damage or injury caused by errors in, or omissions from, the information given.

Contents

		Preface	6
Chapter	1	An introduction to two-stroke tuning	7
Chapter	2	The cylinder head	12
Chapter	3	Porting and cylinder scavenging	31
Chapter	4	The exhaust system	88
Chapter	5	Carburation	120
Chapter	6	Ignition	164
Chapter	7	The bottom end	192
Chapter	8	Lubrication and cooling	222
Chapter	9	Power measurement and gearing	249
Appendix		Table of useful equivalents	266
		Index	267

Preface

From a very humble beginning, the two-stroke internal combustion engine has now been developed to a degree that was not thought possible only a few years ago. I am sure even the engineers who have stood by the two-stroke principle for so long find it staggering that this mechanically simple device can produce as much power as it does today, with relative reliability.

Originally, I looked upon two-stroke engines with contempt. They made a horrible ring-ding noise, nothing like the beautiful note of four-stroke racing engines. They emitted a blue haze from their tailpipes too, which appeared unsightly, long before any of us heard of the word pollution. On hot days these engines seized with monotonous regularity. Difficult starting, flooding and plug fouling seemed the order of the day.

Consequently I wrote off two-strokes, convinced I would never lower myself to develop one of these unreliable little beasts in my workshop. But that all changed when two of my friends bought themselves 250cc Bultaco Pursang motocross bikes and insisted I prepared them. I took up the challenge and was rewarded with the knowledge that a 'ring-ding' I had developed came home third in the National Motocross Championship with a B grade rider on board.

From then on the challenge has not abated as I have strived to unravel the mystery of what makes a two-stroke fire. Instead of looking on the two-stroke with contempt, I now view this little marvel with fascination, as there is much yet to be learned about the two-stroke power unit.

It is my hope that this book will assist the enthusiast involved in motocross, enduro, desert, road or go-kart racing to develop and tune his two-stroke engine for horsepower and reliability.

A. Graham Bell
Maitland
New South Wales

Chapter 1

An introduction to two-stroke tuning

Mechanically, the two-stroke engine is very simple, and unfortunately on too many occasions this apparent simplicity has fooled would-be tuners into believing that this type of power unit is easy to modify. Just a few hours work with a file in the exhaust and inlet ports can change the entire character of the engine for the better, but if you go just 0.5mm too far, you could end up with a device slower than its stock counterpart.

Therefore modifications must be planned carefully, keeping in mind that seldom, if ever, is the biggest (or most expensive) the best. As you plan your modifications, always tend to be conservative. If necessary, you can go bigger later.

Possibly the worst viewpoint you can start out with is that the manufacturer didn't know what he was doing. I started out thinking that way too; but then I began to realise why the engineers did it that way. Pretty soon I was learning more about what makes a two-stroke fire – and making fewer mistakes.

You must keep in mind that all production engines are a compromise, even highly developed racing engines like the Yamaha TZ250. You can make the TZ churn out more power, but will you be able to ride it with the power band narrowed right down, and do you have the experience to handle a sudden rush of power at the top end on an oily or wet track? Also, think about the added wear caused by more rpm and horsepower; do you have the finances to replace the crankshaft, pistons and cylinder more frequently now that you are running at 14,000rpm instead of 12,500rpm? When you begin to think about things like this, you start to understand a few of the reasons why manufacturers make compromise engines and machines. Remember the TZ250 started out as a road racer, so you can imagine some of the problems you could come up against if you were to modify a single-cylinder 125 motocross engine for use in a road racer.

Obviously the first work you should do is bring the engine up to the manufacturer's specifications. This is termed blueprinting, and involves accurately measuring everything and then correcting any errors made in production. You will be amazed at the gains to be made, particularly in reliability, and to a lesser extent in performance, by correcting manufacturing deficiencies. I am convinced manufacturers bolt their road racers together merely to make shipping all the pieces easier, such are their tolerances.

I have seen engines that have never been started with piston clearances larger than the manufacturer's serviceable limit. Conrods that vary 0.4mm in centre to centre length and 20 grams in weight, on the same crank. Crankwheels which are 0.1mm outside true centre. Cylinder heads with a squish band clearance of 1.7mm, instead of 0.7–1.0mm. Cylinders with port edges so sharp that the side of the piston and rings would have been shaved away in a few minutes' running. New pistons with cracks. New cylinder heads that are porous.

Included in blueprinting is cleaning the rough cast out of the ports, and matching all gaskets so they don't overlap the ports. The transfer ports must be matched to the crankcase. The carburettor should coincide with the mounting flange and inlet port. Anything the manufacturer has not done (presumably to cut costs), you should do.

Blueprinting is slow, tedious work, and it can be expensive when crankshafts have to be separated and then machined and trued, or when cylinder heads have to be machined to close up the squish band without raising the compression ratio. It is not very exciting work because when you have finished the engine is stock standard, and telling your mates all the work you have done won't impress them. But don't let this put you off, the basis for any serious tuning must begin with bringing the engine up to the manufacturer's specifications.

Most people won't believe how close to standard are the motors used by the factory racing teams. Other riders are convinced that, because the factory boys are quicker, they must have more power and lots of trick parts. In truth, the differences are in frame geometry and the ability of the factory rider to ride faster and make the right choice of tyres, suspension settings, gearing, jetting, etc. Plus, of course, they use blueprinted engines.

I think all enthusiasts have to make a choice as to whether they desire to be a competitor or an engine tinkerer. If you want to compete successfully, then use the latest factory stuff which you can afford. Blueprint the engine, but otherwise leave it alone. If you find some weakness in a particular part, then fit a more robust part. On the other hand, if at heart you are really a tinkerer then compete in a class catering for tinkerers. This means running on the circuit in a street bike class which demands stock frame and exhaust and street tyres, but which permits unlimited engine modification.

The greatest potential for tuning exists when an engine is 'hybridised'; when it is modified for a purpose for which it was not originally intended. For example, modifying a motorcross engine for circuit race use in a kart or bike, or modifying a street bike engine for circuit use. If however the engine was originally built for motocross, enduro, trials, kart or circuit and the professional tuner is limited in what he can do to make the engine more competitive, be realistic and consider how little the enthusiast can do to 'improve' the engine.

Factory engineers strive to achieve maximum power with the widest possible power band to suit a particular type of competition. A higher state of tune will mean that any hp gains at peak will be accompanied by an equal or greater loss somewhere else. Simply put this means that the engine will be less flexible and gaps between gears may become wider than the power band. The result will be a machine which is much harder to ride and which is slower than stock.

Table 1.1 shows what was found when a modified Yamaha TZ125 was run on the dyno. The owner had expended a lot of money getting the engine reworked, but could no longer get close to his previous best lap times. Looking at the power curve it is no wonder he was exhausted after only a few laps. The engine made very little power up to 12,000rpm and then in one thousand revs nearly doubled its output. Sure, the engine has good peak power, but it all happens over a very narrow 1,200rpm band. When the engine was 'de-tuned' back closer to Yamaha's factory specifications the engine lost 3.7hp at the peak, about 9 per cent, which sounds a lot but the power band is now 2,500rpm wide.

Table 1.1 Yamaha TZ125 power comparison

rpm	Test 1 hp	Test 1 Torque	Test 2 hp	Test 2 Torque
9,500			23.7	13.1
10,000	14.3	7.5	26.3	13.8
10,500	15.2	7.6	27.7	13.9
11,000	20.5	9.8	31.0	14.8
11,500	19.9	9.1	34.8	15.9
12,000	22.8	10.0	36.8	16.1
12,500	28.1	11.8	37.2	15.6
13,000	37.4	15.1	35.4	14.3
13,500	40.9	15.9	31.4	12.2
14,000	33.1	12.4		

Test 1 – Modified engine with revised porting, new expansion chamber and 40mm flat slide Lectron.

Test 2 – Engine 'de-tuned' with close to stock porting, revised expansion chamber and 38mm flat slide Mikuni.

In reality I find that many riders are more comfortable and faster when engines are 'de-tuned' even from factory specs. It seems some consumer machines of today are too close to the snappy works bikes of last season. Consequently they are more of a handful for riders with lower skill levels than expert professionals. Getting the best out of such machines is hard work, not fun.

To get the fun factor back the engine can be de-tuned in a number of ways. If it has a power valve in the exhaust port with mechanical control it should be adjusted so that it does not open fully, and fit a softer spring in the governor to delay the rpm at which opening commences. You can also lower the compression ratio 10–12 per cent to take away some of the top end 'snap'. Going down one or two sizes in the carburettor (2–4mm) will do wonders to pump up the mid-range. Also, to make a motocrosser more of a play bike you will need some extra flywheel effect to increase traction and smooth the power surge. You can achieve this by bolting a weight to the ignition rotor.

So that there is no misunderstanding of the two-stroke operating cycle I will describe what goes on in each cylinder, every revolution of the crankshaft.

The first example is the piston-ported Bultaco Matador Mk4 which, like most two-strokes, operates on the loop scavenge principle. As the piston goes up, the inlet

A heavy ignition rotor provides additional flywheel effect to increase traction and tame the sudden rush of power on peaky engines.

port is opened by the piston skirt at 75° before TDC (Top Dead Centre) and the atmospheric pressure (14.7psi) forces air/fuel mixture in to fill the crankcase (Figure 1.1). The piston continues to rise to TDC, compressing the fuel/air charge admitted on the previous cycle. At 3.2mm before TDC the spark plug fires, sending the piston down on the power cycle. As the piston continues its descent the inlet port is closed and the fuel/air mixture is partially compressed in the crankcase. At 85° before BDC (Bottom Dead Centre) the exhaust port is opened by the piston crown and the exhaust gases flow out. After another 22° (63° before BDC) the blow-down period finishes and the piston crown exposes the transfer ports to admit the fresh fuel/air charge. This is forced up the transfer passages due to the descending piston reducing the crankcase volume by the equivalent of the cylinder displacement, in this instance 244cc. As the piston begins rising, the mixture continues to flow into the cylinder and the exhaust gas continues flowing out. The piston continues rising, closing off first the transfer and then exhaust ports. Next the inlet port opens, to start the cycle over again.

Rotary valve engines operate on the same loop scavenge principle, but in this case a disc partially cut away and attached to the end of the crankshaft opens and closes an inlet port in the side of the crankcase. The Morbidelli 125 twin road racer is a rotary valve engine. The inlet port opens 30° after BDC and closes 79° after TDC. The piston crown opens and closes the exhaust and transfer ports.

The following pages will provide you with the knowledge necessary to develop a successful two-stroke competition engine, but do keep in mind the principles outlined in this chapter so that you avoid the most basic pitfalls associated with two-stroke tuning.

An introduction to two-stroke tuning

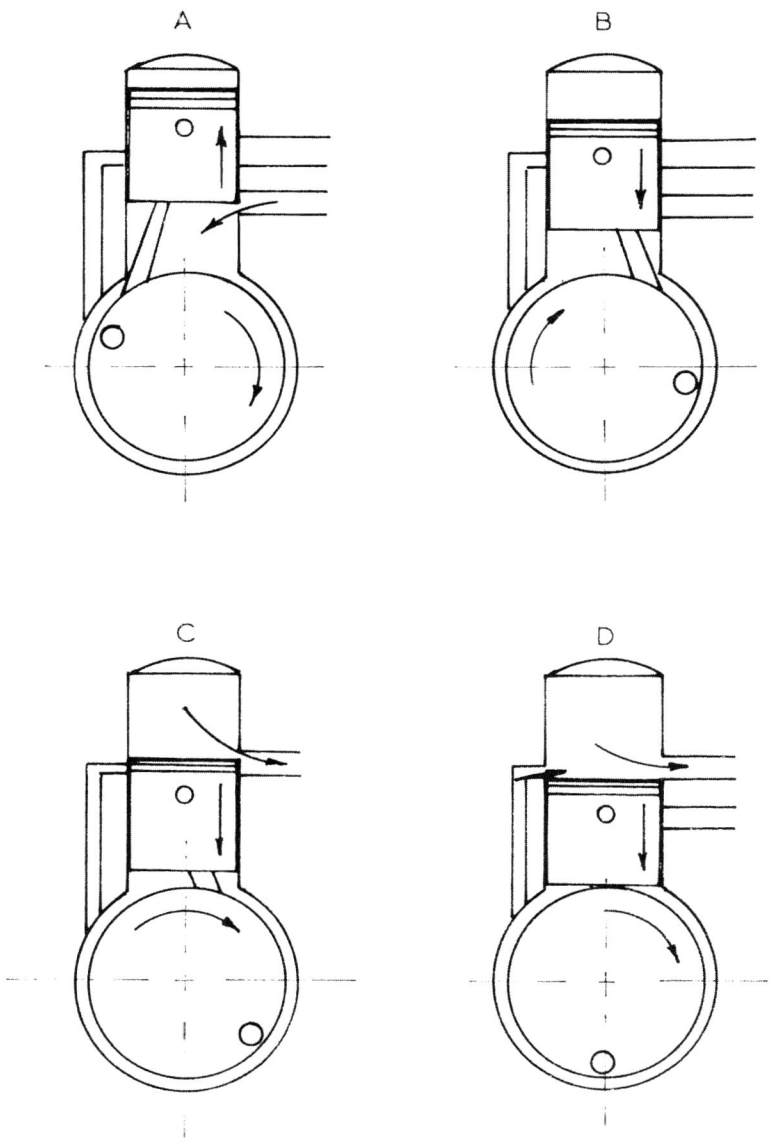

A - mixture in cylinder is compressed & inlet cycle begins.
B - mixture in crankcase is compressed.
C - exhaust cycle begins & primary compression continues.
D - transfer cycle begins & exhaust cycle continues.

Figure 1.1 Two-stroke operating cycle.

Chapter 2

The cylinder head

The two-stroke cylinder head, either air or water-cooled, certainly doesn't look very exciting but its design has a large bearing on how well your engine will run. Manufacturers use various external shapes and cooling fin patterns but the main requirement here is that the cooling area be large enough to adequately cool the engine. Some people feel that an air-cooled head must have radial fins to be any good, but I disagree. Conventional finning is entirely adequate. It is the surface area which counts, not the fin pattern.

What is more important is the shape of the combustion chamber and the location of the spark plug. Over the years many combustion chamber designs have been tried, but only a couple are conducive to a reliable, high horsepower engine. The one thing a powerful two-stroke doesn't need is a combustion chamber that promotes detonation, the killer scourge of all racing two-strokes.

To understand the type of combustion chamber you need it is necessary to appreciate just what detonation is and what can be done to be rid of the problem. Detonation occurs when a portion of the fuel/air change begins to burn spontaneously after normal ignition takes place. The flame front created by this condition ultimately collides with the flame initiated by the spark plug. This causes a rapid and violent pressure build-up, and the resulting explosion hammers the engine's internal components.

Detonation leaves many tell-tale signs for which the two-stroke tuner should have an ever-wary eye. The most obvious sign is a piston crown peppered around the edge as though it has been sand blasted. Bikes with plated aluminium cylinders will usually show the same sand blasted effect around the top lip of the bore. A cracked (not molten) spark plug insulator also indicates detonation. If kept running, a detonating engine will eventually seize and/or have a hole punched right through the top of the piston.

The conditions leading to detonation are high fuel/air mixture density, high compression, high charge temperature and excessive spark advance. A high piston

The cylinder head

With air-cooled engines fin surface area sufficient to provide adequate cooling, rather than the fin pattern, is the more important consideration.

crown or combustion chamber temperature can also lead to this condition. In a racing two-stroke all of these detonation triggers are virtually unavoidable, with the exception of excessive spark lead.

Researchers have found that it is the gases at the very outer limits of the combustion chamber, called the 'end gases', that self-ignite to cause detonation. These end gases are heated by the surrounding metal of the piston crown and combustion chamber, and also by the heat radiating from the advancing spark-ignited flame. If the spark flame reaches the outer edges of the combustion chamber quickly enough, these end gases will not have time to heat up sufficiently to self-ignite and precipitate detonation. Herein lies the key to prevent detonation – keep the end gases cool and reduce the time required for the combustion flame to reach the end gases.

The most obvious step that would satisfy the second requirement is to make the combustion chamber as small as possible, and then place the spark plug in the centre of the chamber. Naturally the combustion flame will reach the end gases in a small combustion space more quickly than if the chamber were twice as wide. Additionally, a central spark plug reduces flame travel to a minimum. (Figure 2.1)

In meeting the second requirement, the need to keep the end gases cool can also be accommodated. If we move the combustion chamber down as close to the piston crown as possible, no combustion will occur around the edges of the chamber until the piston has travelled well past TDC. This large surface area acts as a heat sink and conducts heat away from the end gases, preventing self-ignition.

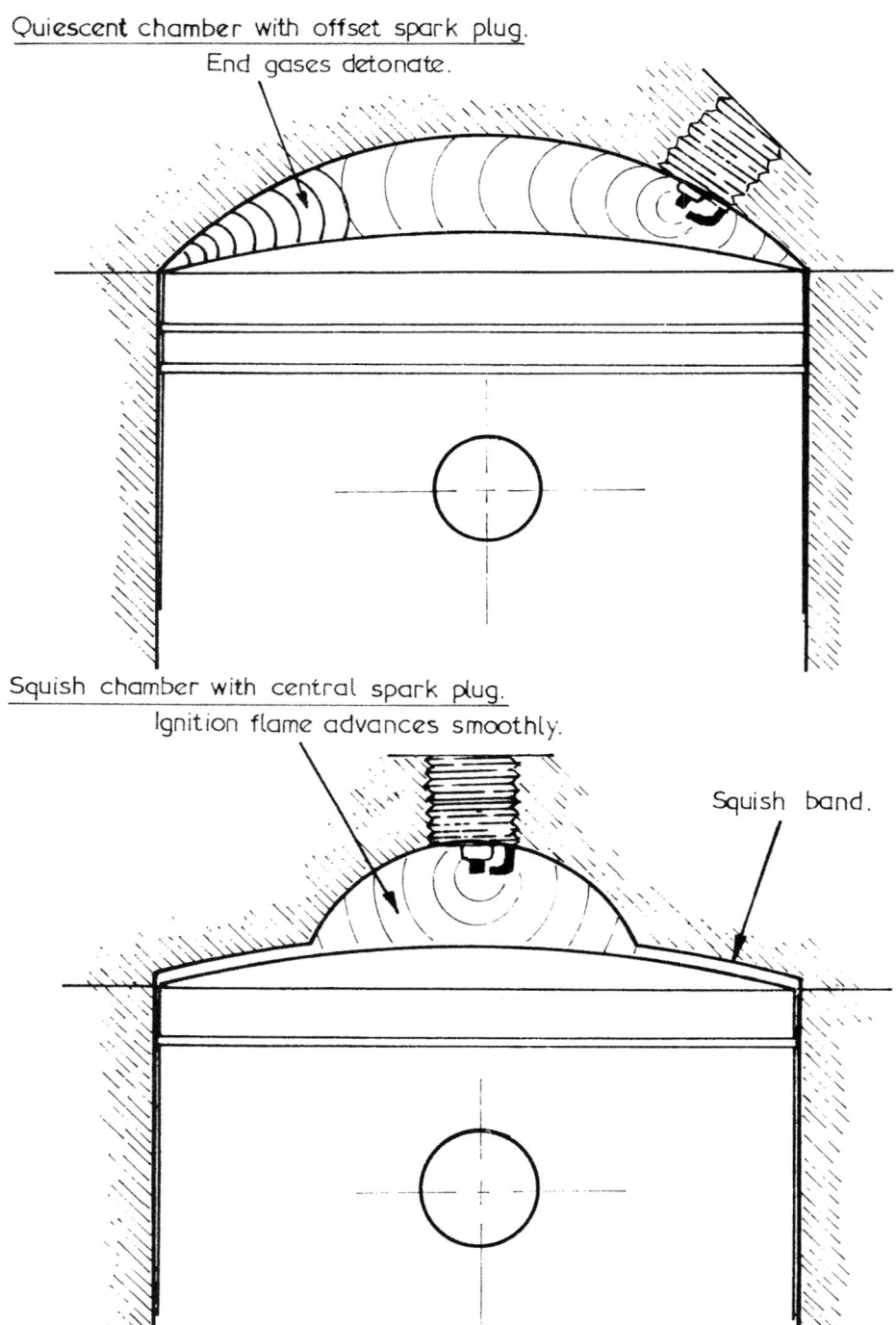

Figure 2.1 Squish-type combustion chamber promotes rapid combustion and reduces detonation.

The chamber just described is called a squish-type combustion chamber because of the squish band around its edge. Originally, the squish band was designed to squish the fuel/air charge from the edges of the cylinder toward the spark plug which, of course, it still does. The fast moving gases meet the spark plug and quickly carry the combustion flame to the extremity of the combustion chamber, thus preventing detonation.

Since that time, more benefits of the squish chamber have come to light. The mixture being purged across the combustion chamber from the squish band homogenises the fuel/air mixture more thoroughly and also mixes any residual exhaust gas still present with the fuel charge. This serves to speed up combustion by preventing stale gas pockets from forming. Such pockets slow down, and in some instances, can prevent flame propagation.

Turbulence caused by the squish band also serves to enhance heat transfer at the spark-initiated flame front. Without proper heat transfer, jets of flame would tend to shoot out toward the edges of the combustion chamber, prematurely heating the surrounding gases to start off the cycle leading to detonation.

Rapid combustion has other advantages besides controlling detonation. With an increase in combustion speed there is, of necessity, a corresponding decrease in spark advance. The closer to TDC we can ignite the charge, the less negative work we have to do compressing a burning charge that is endeavouring to expand. Also there is less energy loss in the form of heat being transferred to the cylinder head and piston crown.

When less heat is conducted to the head and piston, the engine runs cooler and makes more power. A side benefit resulting from the cooler piston also enhances the power output. A cool piston does not heat the charge trapped in the crankcase as much, therefore a cooler, denser fuel/air charge enters the cylinder each cycle, to make more power.

If you think about it, you will see that the compact squish type combustion chamber also contributes to a cool piston by confining the very intense combustion flame to about 50 per cent of the piston crown just before and after TDC.

Engine designers have known about these things for a considerable time. This is why you will find the best racing engines follow the squish design. Also you will notice that these engines have a very small bore in relation to their stroke, as this too cuts down the size of the combustion chamber and reduces the area of piston crown exposed to the combustion flame.

In an effort to minimise cylinder and piston distortion, some manufacturers have chosen to use an offset squish type combustion chamber (Figure 2.2). The exhaust side of a two-stroke cylinder and piston is always the hottest, even though cooling air flow is much better here than on the back (inlet side) of the engine. There are several reasons for this, all associated with the passage of very hot (630°C) exhaust gas through the exhaust port. The escaping gas heats the exhaust port and cylinder wall as well as the side of the piston. This can cause the piston to expand abnormally and in some circumstances to seize. To take care of this possibility, the manufacturer may choose to increase piston to cylinder clearance, but this may not be desirable as extra clearance can increase leakage past the rings and usually results in high piston wear. A safer step is to move the combustion chamber to the rear of the head. If this is done, the front of the piston crown is shielded from the combustion flame by the squish surface. Then, when the front of the piston is heated during the exhaust stroke, it will not expand so far due to it being much cooler initially.

Two-stroke performance tuning

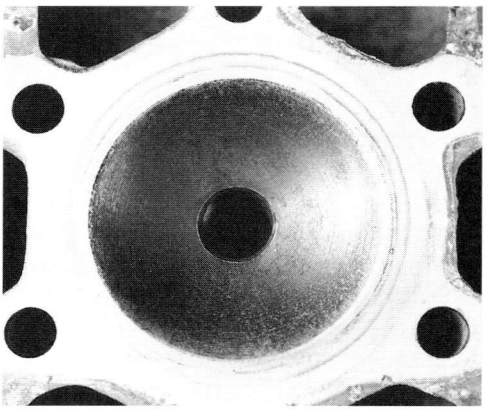 

A quiescent combustion chamber (left) does not have a squish area. Note the corrosion and mineral build-up due to plain town supply water being used as coolant. The offset squish chamber (right) concentrates combustion heat at the rear of the piston away from the hotter exhaust side. This evens out the temperature gradient across the piston crown, but more importantly it moves the actual combustion space to the cooler inlet side of the piston. Hence, prior to ignition, less heat is transferred to the inlet charge, so reducing the risk of detonation.

Several two-stroke engines are produced with squish and offset squish chambers, but unfortunately mass production usually reduces their effectiveness. It is a very difficult task to keep tolerances of closer than about 0.2mm in production. Therefore you find many engines with a squish clearance of 1.3–1.8mm instead of the 0.6–0.8mm clearance that is required.

If your machine is only for play, and that's the use to which many motocross

Figure 2.2 Offset squish-type combustion chamber is offset away from the exhaust port.

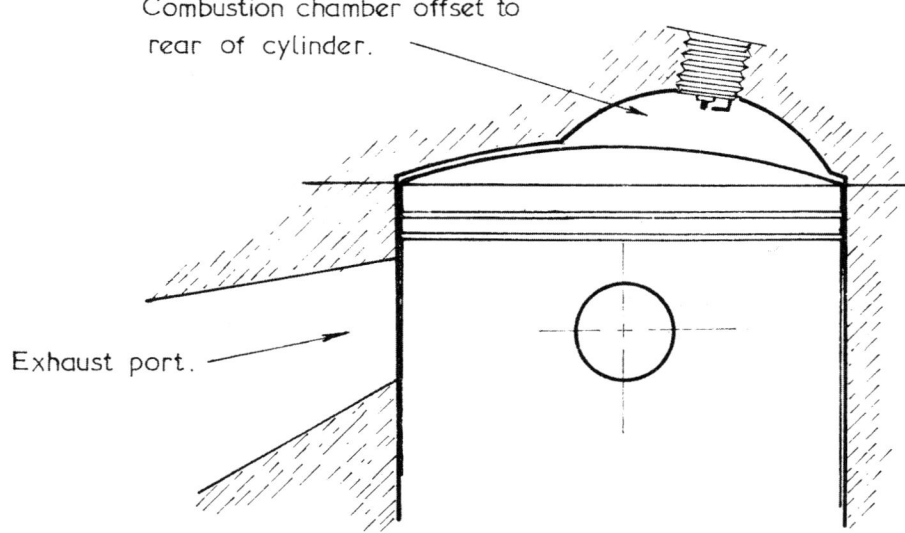

The cylinder head

bikes are put, a wide squish clearance will not matter. You will not get peak power, but you will possibly never know the difference. And you will probably never ride hard enough to experience detonation.

However, if you want top power and no risk of detonation, the squish clearance must be closed up. A squish band that isn't working is worse than no squish band at all as it wastes part of your fuel/air charge. Wasted fuel charge spells less horsepower.

To give you an idea of how much horsepower you could be losing it would be good to consider the example of a TZ250 Yamaha road racer with a bore 54mm in diameter. The compression ratio uncorrected is about 15:1, meaning that the trapped charge is compressed into a space 8.8cc in volume. If the squish clearance is 1.7mm (lots of motors come from the factory like that) 1.94cc of the trapped charge will not be burned until well past TDC, too late to produce any power. That 1.94cc represents 22 per cent of the inlet charge lost, but when the squish clearance is reduced to 0.8mm the charge loss is reduced to 0.92cc, or 10.5 per cent. On paper it would seem an easy way to pick up 11.5 per cent more power, but losses reduce this increase in peak hp to about 5–6 per cent on the dyno. Mid-range power however can rise as much as 10 per cent, so the bike is easier to ride and it doesn't detonate.

Reducing the squish clearance is not easy, you can't just machine 1mm, or whatever, off the head as the compression ratio would end up many numbers too high. Also you must be sure not to reduce the clearance so much that the piston will bang into the head at high rpm. The clearance required will vary from engine to engine, and also on how careful you intend to be each time you replace a piston, rod or barrel.

Pistons usually vary in compression height by up to 0.2mm. Conrods are supposed to be within a 0.2mm range but they can be up to 0.5mm out. Cylinder heights are maintained within 0.4mm. In the worst case you could rebuild the motor with a new piston, rod and barrel. The piston could be 0.2mm taller and the rod 0.2mm longer. Together with a cylinder 0.4mm shorter than before, the new parts could reduce your squish clearance by $0.2 + 0.2 + 0.4 = 0.8$mm which would result in a blown motor if the clearance was set at 0.8mm previously. Manufacturers realise this, so they purposely set the clearance wide to make allowance for the worst possible parts size combination.

If you are willing to measure the squish clearance each time you do a rebuild, and then compensate for inadequate clearance by fitting a thicker barrel base gasket or a thicker head gasket, you can reduce the clearance down to the amount shown in the accompanying table.

Table 2.1   Minimum squish clearances

Cylinder size (cc)	Clearance (mm)
50–80	0.5–0.7
100–125	0.6–0.8
175–250	0.8–1.0
300–500	0.9–1.2

Note – Kart engines running to 20,000rpm require increased squish clearances of 0.8–1.0mm

To find accurately what the squish clearance figure is, the barrel must be tensioned

Two-stroke performance tuning

down on a standard thickness base gasket. Clean all traces of carbon from the head and piston. Place a strip of modelling clay 20mm wide by 3mm thick across the piston crown. Fit the head gasket and head, and turn the crank to move the piston just past TDC. Remove the head and then cut the clay down the middle with a sharp, wet knife. Carefully pull one strip of clay off the piston and then measure the thickness of the clay left on the piston. You have to be accurate, so use the end of your vernier calipers. As a cross-check also measure the clay thickness on the other side of the piston. If the thicknesses vary this would indicate that the head gasket surface has been machined on a different plane to that of the combustion chamber. Also at this time measure and record, for future reference, the compressed thickness of the base gasket and head gasket. (Figure 2.3).

After the clay thickness is measured you can work out how far the head must be machined to give the desired squish clearance. As mentioned previously, the combustion chamber must also be machined deeper into the head to keep the compression ratio at an acceptable level. If you wish to keep the compression ratio the same as standard, the combustion chamber will have to be machined twice as deep as the amount skimmed off to reduce the squish clearance, assuming a 50 per cent squish band. Therefore if 0.9mm is removed, the combustion chamber will have to be made 1.8mm deeper. A 50 per cent squish band is one having an area equal to half the

Figure 2.3 Measuring the squish clearance using the modelling clay method.

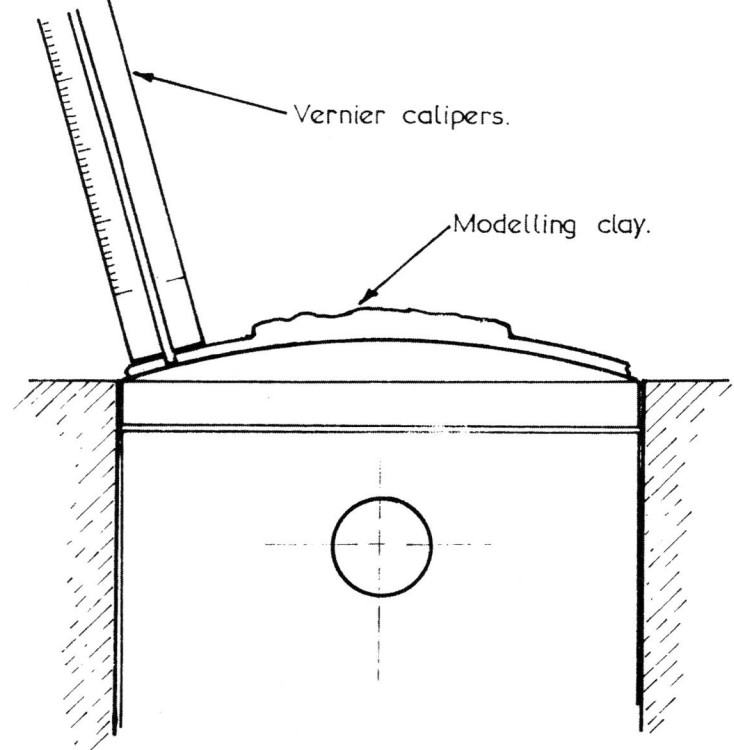

The cylinder head

 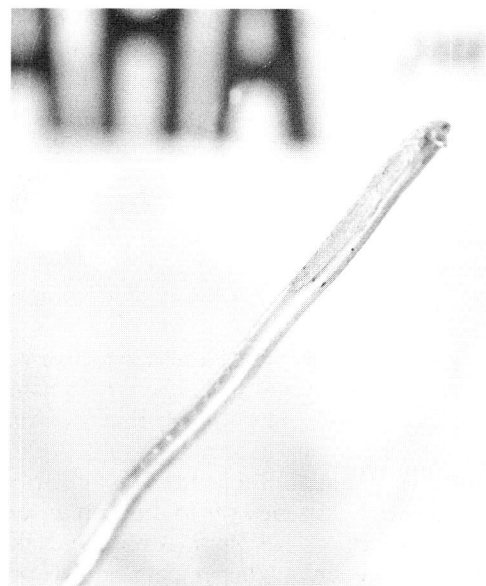

Measuring the squish clearance (above left) using soft 1.2mm resin core solder. With the engine at mid-stroke to ensure that all ports are closed, two pieces of bent solder are inserted down the spark plug hole, one aimed toward the front and the other aimed toward the rear of the cylinder bore. The crank is rotated through 180° to crush the solder in the squish band. (Note only one piece of solder shown for clarity.) The solder is crushed (above right) as the piston passes TDC. When measured with vernier calipers (below) the squish clearance is found to be 0.7mm.

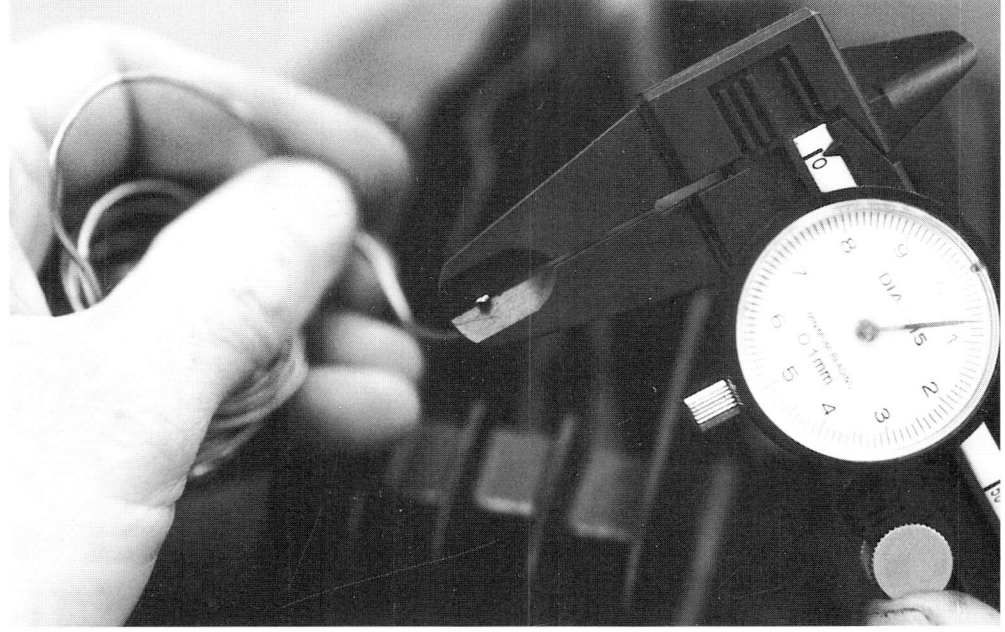

Two-stroke performance tuning

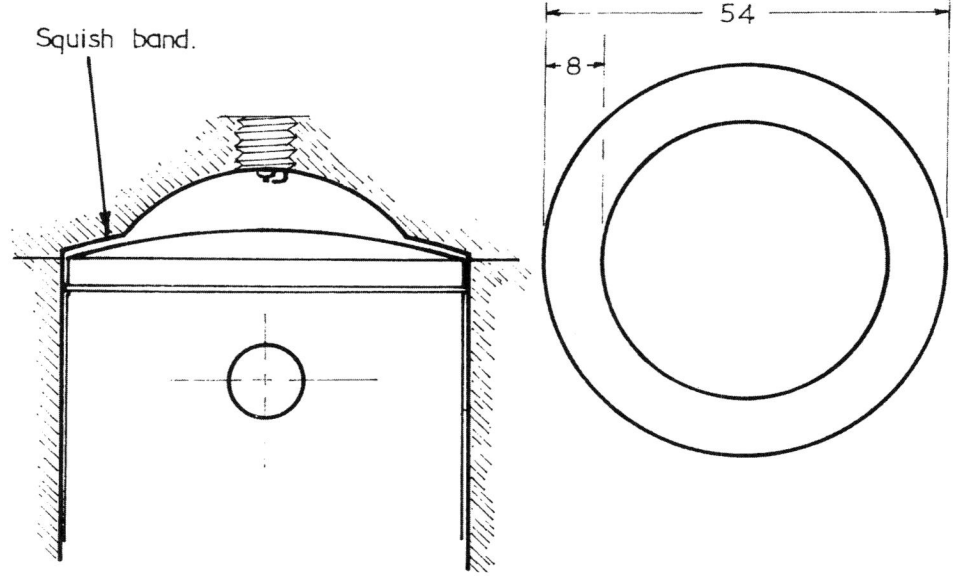

Figure 2.4 A 50 per cent squish band masks half of the piston crown area during the early phase of combustion.

cylinder bore area, ie an engine with a 54mm bore would have a squish band approximately 8mm wide (Figure 2.4).

Not all engines run a 50 per cent squish band, this is basically a middle-of-the-road figure. Fixed gear kart engines typically run a much wider band, about 12mm wide on a 50mm bore or 73 per cent, while road race engines required to make best upper rev range power may at times run a band width down to 4mm on a 54mm bore, or 15 per cent. Why this difference? The fixed gearing kart engine is very highly loaded pulling out of slow corners. This is frequently close to the torque peak when the engine is more prone to detonation. A wide squish band provides better combustion control in this situation, so power is improved even though there is less actual volume of mixture exposed to the ignition flame front. When equipped with a close ratio gearbox an engine will not remain heavily loaded in this 'detonation zone', but will fairly quickly accelerate through it before the heavy engine load raises temperatures and induces detonation. Therefore the squish band can be narrower because not so much area of the piston crown and combustion chamber have to work as a heat sink, drawing combustion heat away during that vital time 'window' while the fixed gear engine dwells in the detonation zone. With reduced danger of detonation the gearbox engine's squish band can be narrowed to expose a larger portion of fuel/air mixture to the combustion flame, thus pushing up high rpm power.

In addition to squish band width, the actual angle at which the band is machined has an influence on detonation control and hp output. Thus piston ported engines which have compromised low speed hp run a shallow squish band angle of about 15–17°, and down to 10°. The Yamaha KT100S combustion chamber in Figure 2.5 is machined with a 90mm radius, close to a 17° angle, to help improve low speed power. Rotary valve engines which can be more easily ported to give a wide power band run a

The cylinder head

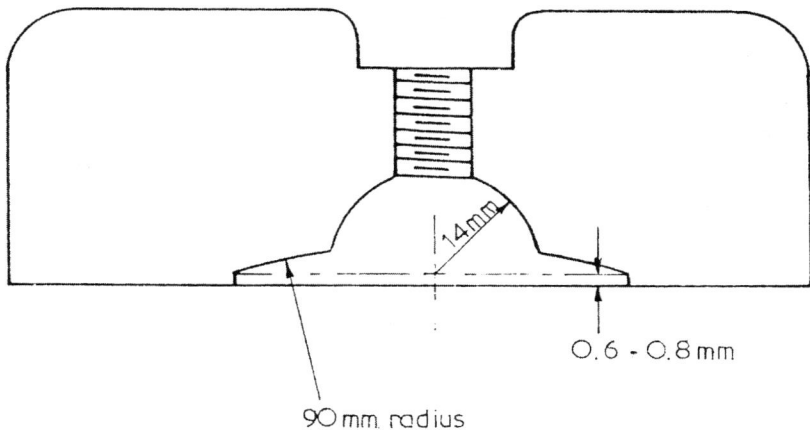

Figure 2.5 Yamaha KT 100S kart combustion chamber.

steeper 19° or 20° squish band angle to assist high rpm power output.

To check that the machine shop recuts the combustion chamber to the original contour when it is deepened, you will have to make a template of the chamber shape before you send the head off. The template can be made out of any light gauge metal or even stiff cardboard. (Figure 2.6).

Most people like to see the compression ratio pushed up as high as possible. High compression has always been equated with high horsepower. I agree that the compression ratio should be made as high as practicable, but often the manufacturer has already found the limit and built his engines accordingly. All you can do in this instance is check that production tolerances have not lowered the ratio significantly below that which the manufacturer intended.

Something you must always remember when dealing with two-stroke engines is that increasing the compression ratio will not give a power gain equivalent to that which you would pick up with a four-stroke engine. Heat is the enemy of two-stroke engines and pushing the compression ratio to give an expected 6 per cent power increase will possibly result in a 1–2 per cent power rise at the most; the rest will be lost in heat energy and pumping losses. However, at lower engine speeds the cylinder will not be completely filled with fuel/air mixture and the power may jump by 3–4 per

Figure 2.6 Combustion chamber contour checking template.

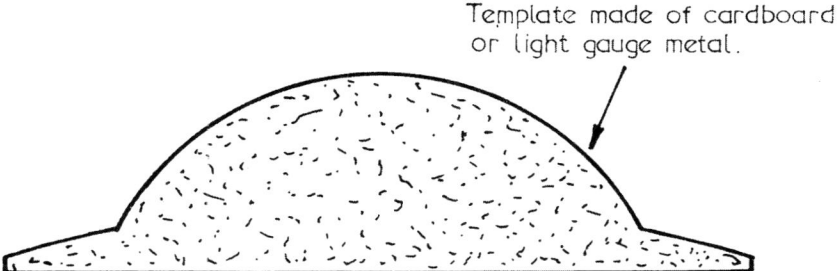

cent because there is not such a heat loss. This is, in fact, the real benefit of raising the compression ratio, not to increase maximum power but to pick up mid-range power and possibly widen the power band.

In times past when porting was less efficient than it is today, and when expansion chambers were less well understood, which compromised their ability to evacuate the cylinder of exhaust gas and recharge it with fuel/air mixture, compression ratios could be pushed quite high without resulting engine damage. This was possible because not only was the cylinder not completely filled, but what fuel mixture was drawn in was diluted by a large volume of exhaust gas which dramatically slowed the rate of combustion. With combustion slowed by inert exhaust gas getting between the oxygen and fuel molecules, combustion temperature and pressure spikes were flattened, so a high compression ratio was safe.

However, as engine breathing improved there was an increase in the volume of fuel mixture drawn in. At the same time advances in expansion chamber design ensured that less exhaust residue remained to impede combustion. Thus modern water-cooled two-strokes often operate at a compression ratio no higher, and at times lower, than the old air-cooled engines. Some tuners have not been able to adjust their thinking to accept this simple fact, so consequently their engines are down on power and reliability because they have to run richer jetting and less ignition advance to keep some sort of control of the combustion flame.

Because so much confusion exists in the motorcycle industry relating to compression ratio we need to define exactly what we are talking about when we use the term. Ever since the first internal combustion engines, regardless of whether the engines were two-stroke, four-stroke, diesel, petrol, etc., compression ratio was taken

Fixed gear kart engines usually run a very wide squish band of about 73 per cent bore area. This engine, with 49.65mm bore, has a 12mm squish band cut at 15°.

The cylinder head

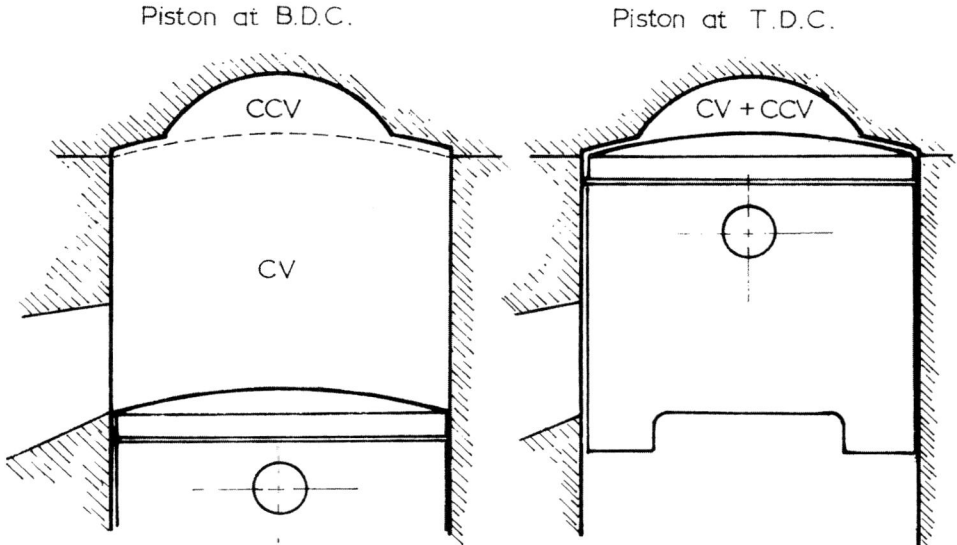

Figure 2.7 The uncorrected compression ratio compares the cylinder volume at bottom dead centre with the cylinder volume at top dead centre.

to mean the ratio of the volume of the cylinder with the piston at BDC to the volume of the cylinder with the piston at TDC (Figure 2.7). This relationship is expressed in the formula:

$$CR = \frac{CV + CCV}{CCV}$$

where CR = compression ratio, CV = cylinder volume, and CCV = combustion chamber volume.

Cylinder volume is found using the formula:

$$CV = \frac{\pi D^2 \times S}{4000}$$

where π = 3.1416, D = bore diameter in mm, and S = stroke in mm.

CCV, combustion chamber volume, is made up of the volume of the combustion chamber, plus any space existing between the piston crown and the top of the cylinder, plus the head gasket. This volume can be worked out geometrically but it is much simpler to bring the piston up to TDC. Seal around the edge of the piston with a thin layer of grease. Fit the head and head gasket and measure the volume with water or paraffin, using a burette graduated in 0.1cc.

As an example of how these formulas work we will consider the long stroke Bultaco Pursang 125. This engine has a bore of 51.5mm, a stroke of 60mm and, according to the manufacturers, a compression ratio of 14:1.

Two-stroke performance tuning

$$CV = \frac{\pi \times 51.5^2 \times 60}{4000} = 124.98 \text{cc}$$

Measured with a burette CCV is found to be 9.8cc:

$$CR = \frac{CV + CCV}{CCV} = \frac{124.98 + 9.8}{9.8} = 13.75:1$$

Therefore the engine has a compression ratio just a touch lower than specification. As this engine will be running at the speedway using 110 octane fuel (Avgas 100/130) the compression ratio will be increased to 15:1. The standard motor is designed to run on 95 octane fuel.

The formula to find the required combustion chamber volume is:

$$CCV = \frac{CV}{CR - 1}$$

$$= \frac{124.98}{15 - 1} = \frac{124.98}{14} = 8.93 \text{cc}$$

Therefore the combustion chamber volume must be reduced by 9.8–8.93 = 0.87cc.

To find how much the head must be skimmed to reduce the volume by 0.87cc we use the cylinder displacement formula transposed to read:

$$S = \frac{CV \times 4000}{\pi D^2}$$

$$= \frac{0.87 \times 4000}{\pi \times 51.5^2} = 0.42 \text{mm}$$

The above compression ratio is now referred to as the uncorrected compression ratio. The Japanese have introduced a new way of measuring the compression ratio, called in various circles effective, corrected, actual or trapped compression ratio. This can be very confusing because an 8:1 corrected compression ratio is about equivalent to a 15:1 compression ratio calculated by the old method.

The Japanese theory is that compression does not begin until the piston closes the exhaust port. Therefore the corrected compression ratio is taken to mean the ratio of the volume of the cylinder with the piston just closing the exhaust port relative to the volume of the cylinder with the piston at TDC (Figure 2.8). This is expressed in the formula:

$$CCR = \frac{ECV + CCV}{CCV}$$

where CCR = corrected compression ratio, ECV = effective cylinder volume, and CCV = combustion chamber volume.

The cylinder head

The combustion chamber volume is measured using a burette graduated to 0.1cc.

Figure 2.8 The corrected compression ratio compares the cylinder volume at the time the exhaust port closes with the cylinder volume at top dead centre.

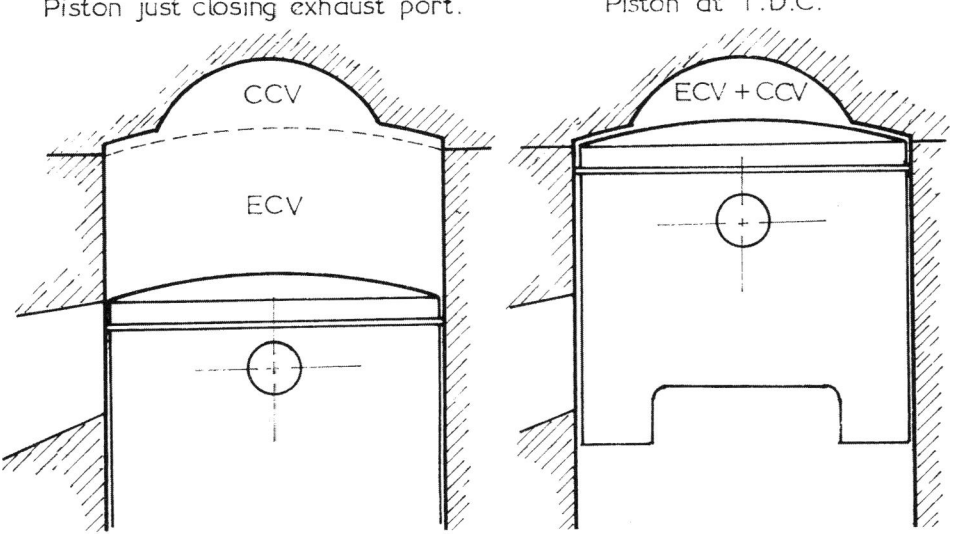

To determine the effective cylinder volume, the distance from the top of the exhaust port to the top of the piston stroke (ie the TDC point) must be known. The ECV is found using the formula:

$$ECV = \frac{\pi D^2 \times ES}{4000}$$

where $\pi = 3.1416$, D = bore diameter in mm, and ES = effective stroke in mm.

Two-stroke performance tuning

In this example we will use the Suzuki PE175C enduro bike. It has a bore of 62mm, a stroke of 57mm and a corrected compression ratio of 7.6:1, according to Suzuki. By measurement I have found that the exhaust port is 31.5mm from the top of the barrel, but at TDC the piston is 0.3mm from the top of the barrel, which means the effective stroke is 31.5–0.3 = 31.2mm.

$$ECV = \frac{\pi D^2 \times ES}{4000}$$

$$= \frac{\pi \times 62^2 \times 31.2}{4000} = 94.19\text{cc}$$

Measured with a burette, the combustion chamber volume was found to be 14.7cc. Therefore the corrected compression ratio is:

$$CCR = \frac{ECV + CCV}{CCV}$$

$$= \frac{94.19 + 14.7}{14.7} = 7.4:1$$

Instead of working with corrected, actual, true, call them what you like compression ratios, I prefer to convert back to the old uncorrected figures which make sense to me, even if they don't make sense to the Japanese. If there were any basis to the Japanese system we should be able to race a PE175 in the 100cc class as its effective displacement is only 94cc, but try getting any motorcycling or karting organization to swallow that one! What I would really like to know is why isn't the PE175 called a PE100?

As the Japanese motorcycling industry has been working with expansion chambers as long as most, I am sure they are aware of the fact that the return wave actually forces any fuel/air charge which has escaped into the exhaust port back up the port and into the cylinder. Therefore the 72cc of fuel/air mixture that appears to be lost out of the PE175's exhaust is not really lost, it is just temporarily misplaced.

Going back to the old system we find the PE175 has a compression ratio of:

$$CR = \frac{CV + CCV}{CCV}$$

$$= \frac{172 + 14.7}{14.7} = 12.7:1$$

To find how the compression ratio of 12.7:1 compares with what the manufacturer intended it to be, we have to convert the manufacturer's corrected figure of 7.6:1 back to the old system using the formula:

$$CR = 1 + \frac{(CCR - 1) \times S}{ES}$$

where CR = uncorrected compression ratio, CCR = corrected compression ratio, S = stroke in mm, and ES = effective stroke in mm.

$$CR = 1 + \frac{(CCR - 1) \times S}{ES}$$

$$= 1 + \frac{(7.6 - 1) \times 57}{31.2} = 13.06:1$$

As this bike is designed to run on 90–95 octane fuel the compression can safely be bumped up to 13.7:1 for competition use on 100 octane racing fuel.

After this confusion over all sorts of compression ratios, you are probably bewildered and wondering at what compression ratio your engine should be running for best performance and reliability. Really this cannot be predicted accurately, however the following comments will provide useful guidance.

Generally I would expect to be able to increase the manufacturer's compression ratio by 0.6 to 1.0 points for each 5 or 6 numbers the Research octane number (RON) of the fuel was increased. Thus if the standard compression ratio was 14:1 on 95 RON fuel then it should be expected the maximum safe compression ratio will rise to around 14.6 to 15:1 when using 100 RON fuel. If the manufacturer's figure was 13.5:1 on 98 RON fuel then the maximum safe compression ratio on Avgas 100/130 (110 RON) will be in the area of 14.7 to 15.5:1.

Note that I stated these figures as the 'maximum safe compression ratio' – not the compression ratio for maximum performance. The more time you spend with the throttle wide open in 5th and 6th gears the lower the compression ratio, and also the less spark advance there must be. Sure, the engine will not make so much mid-range power but the engine will sustain maximum power for 20–30 seconds right along the full length of the longest straights. With more compression it might make full power for only 5–10 seconds under load before a steep drop in power. If the straights have a rise in them, or you run into a headwind the internal overheating can be even more

The standard head gasket is 0.5mm thick, however a thicker 0.8mm gasket enables the compression ratio to be quickly lowered on high speed circuits if the engine is found to be detonating on the long straights. Gaskets of varying thickness can also be used to adjust the squish clearance.

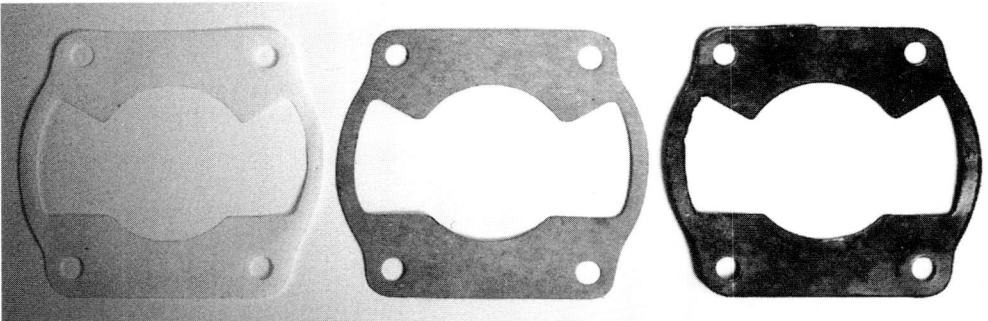

These three base gaskets are 0.125mm, 0.4mm and 0.8mm thick respectively. Different thickness base gaskets enable fine adjustments to the squish clearance and the compression ratio. They are also used to raise or lower the barrel to keep the port timing legal in classes requiring specific port heights.

rapid. Clearly the preferred option would be to run a lower compression engine, or swap to a lower compression head, on tracks where the engine is heavily loaded in 5th and 6th gears for long periods and to go for more compression on short tracks. If that is outside the budget opt for less compression to suit the high speed 5th and 6th gear circuits. Then on the short, twisty tracks dial in more spark advance to push up the mid-range.

To keep all that tightly compressed fuel/air charge in the engine there must be a perfect seal between the head and barrel. The more usual method is to use an annealed copper or aluminium head gasket inserted between the two. McCulloch go-kart engines have used a thin 0.4mm aluminium head gasket for years without too many problems, providing the engine remains stock, but I can't really recommend aluminium for any other two-stroke as there always seems to be a problem with leakage. McCulloch experienced a reverse problem in that they had used aluminium gaskets for years then, on the MC-92 motor, they switched to copper. The copper gasket always seemed to leak as you couldn't get the head tension high enough to crush the copper gasket without the head distorting. On the later MC-93 they reverted back to the old aluminium gasket.

When a copper gasket is used, there is always a temptation to re-use the old one. My advice is don't, unless the gasket has exactly the same inside and outside diameter as the top lip of the barrel. Even then the gasket should be heated with a low heat gas flame and allowed to cool and anneal.

Generally, I prefer to run air-cooled motors without a head gasket if the head is recessed to give a spigot fit with the top of the barrel. In this instance I lap the head on to the barrel, using valve grinding paste. When you are finished, be very careful to get all traces of paste out of the cylinder and then clean the head and barrel so that the gasket sealant will take. Remember that removing the head gasket will raise the compression ratio significantly.

Regardless of the type of gasket used, or even if you choose not to use a gasket, I recommend the use of either Permatex No.3 or Hylomar SQ-32M gasket sealant. Both sealants will provide a good seal at the elevated temperatures experienced in air-cooled two-stroke engines. With water-cooled engines I prefer Permatex No.3 to seal

The cylinder head

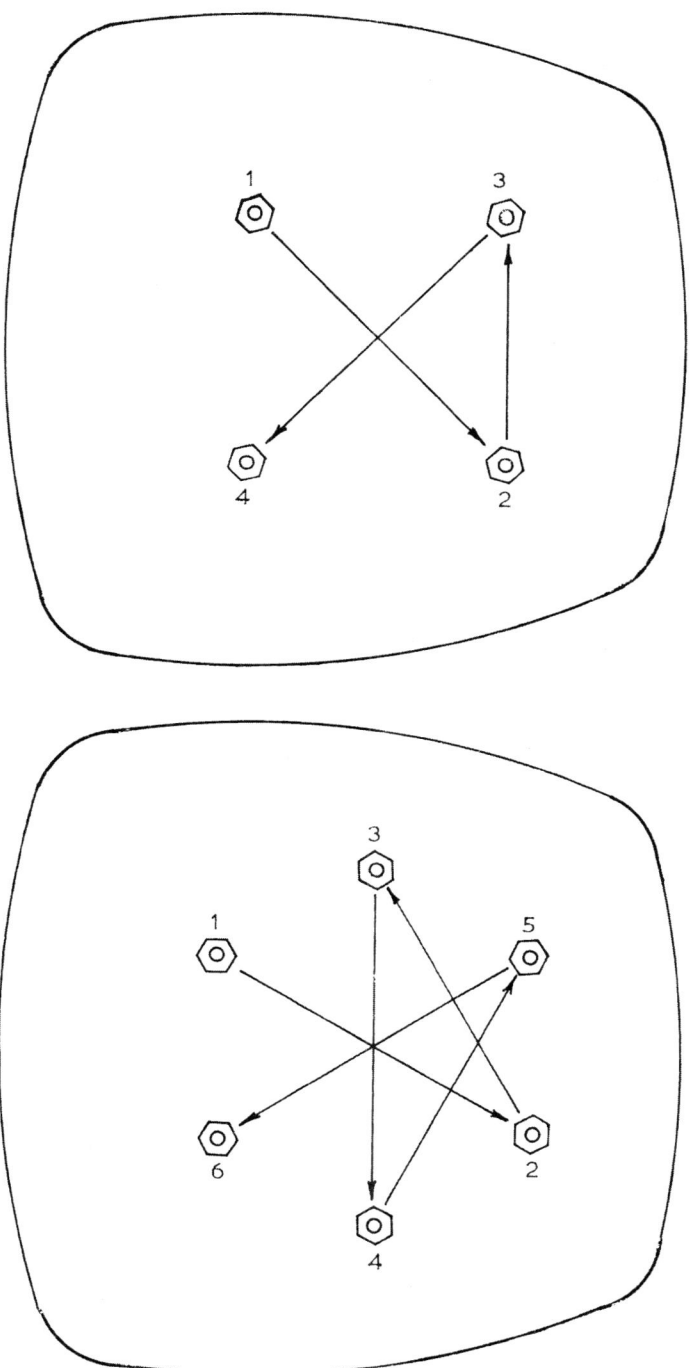

Figure 2.9 The cylinder head studs must be tensioned progressively in the correct sequence to ensure a good seal and to avoid head warpage.

the cylinder to head joint, and to stop external water leaks I use a silicone sealant such as Silastic RTV to seal the outer edge of the head to the barrel.

To ensure the head-barrel joint will not leak, the head bolts must be progressively tensioned in sequence, to the tension recommended by the manufacturer. A typical sequence is shown in Figure 2.9. This sequence must be reversed when the head is being removed.

Two-stroke cylinder heads are easily distorted, so you have to be very careful not to tension the studs more than recommended. Overtensioning will always cause the head to warp. You must be careful to tighten the studs in at least three progressive steps. If the head required 20ft/lb tension, you should take all the nuts down finger-tight and then to 10, 15 and finally 20ft/lb. After about 15 minutes go over the nuts again and then, after the motor has cooled from a run for a minimum of one hour, tension the nuts again.

Chapter 3

Porting and cylinder scavenging

Today, when we take a look down the cylinder of a two-stroke engine, we find its walls literally filled with ports to handle the induction, transfer and exhaust phases of gas flow through the engine. Those of us who have grown up in the Japanese two-stroke era take it for granted that every cylinder has a huge exhaust port flanked by anything from four to six transfer ports. However, it hasn't always been this way. As far back as 1904 Alfred Scott patented his original two-stroke vertical twin. Then in 1906 the French Garard motor appeared with a rotary disc inlet valve. Scott also developed a rotary valve engine in 1912, winning the Senior TT in that year and the following year. However in spite of some very innovative designs being incorporated in two-stroke engines they continued to be embarrassingly unreliable and this single factor stifled development right up until the time of World War II.

In the mid-1930s, the DKW company set out to make two-strokes respectable. They were in the business of manufacturing economical two-stroke motorcycles and stood to profit from changing the two-stroke's image. They engaged the services of an engineer named Zoller to build a 250 racer, which ultimately won the Isle of Man TT in 1938. This led to the development of a 125 single employing a porting arrangement originally invented for two-stroke diesels by German engineer Dr E. Schneurle. It was this concept which ultimately brought success to the two-stroke, both as an economical power source for transport and as a powerful, light-weight power source for competition. Schneurle's loop-scavenging method, patented in 1925, employed a single exhaust port flanked by two small scavenge or transfer ports, whose air streams were aimed to converge on the cylinder wall opposite the exhaust (Figure 3.1). Being aimed away from the exhaust, the transfer streams had a natural resistance to short-circuiting straight out the exhaust. Earlier designs had used deflector-dome pistons to keep the fuel/air charge away from the exhaust port. This increased the piston's heat gathering area and meant that only low power outputs could be aimed for without continually risking piston seizure.

After the war DKW moved to Ingolstadt in West Germany, while their old plant

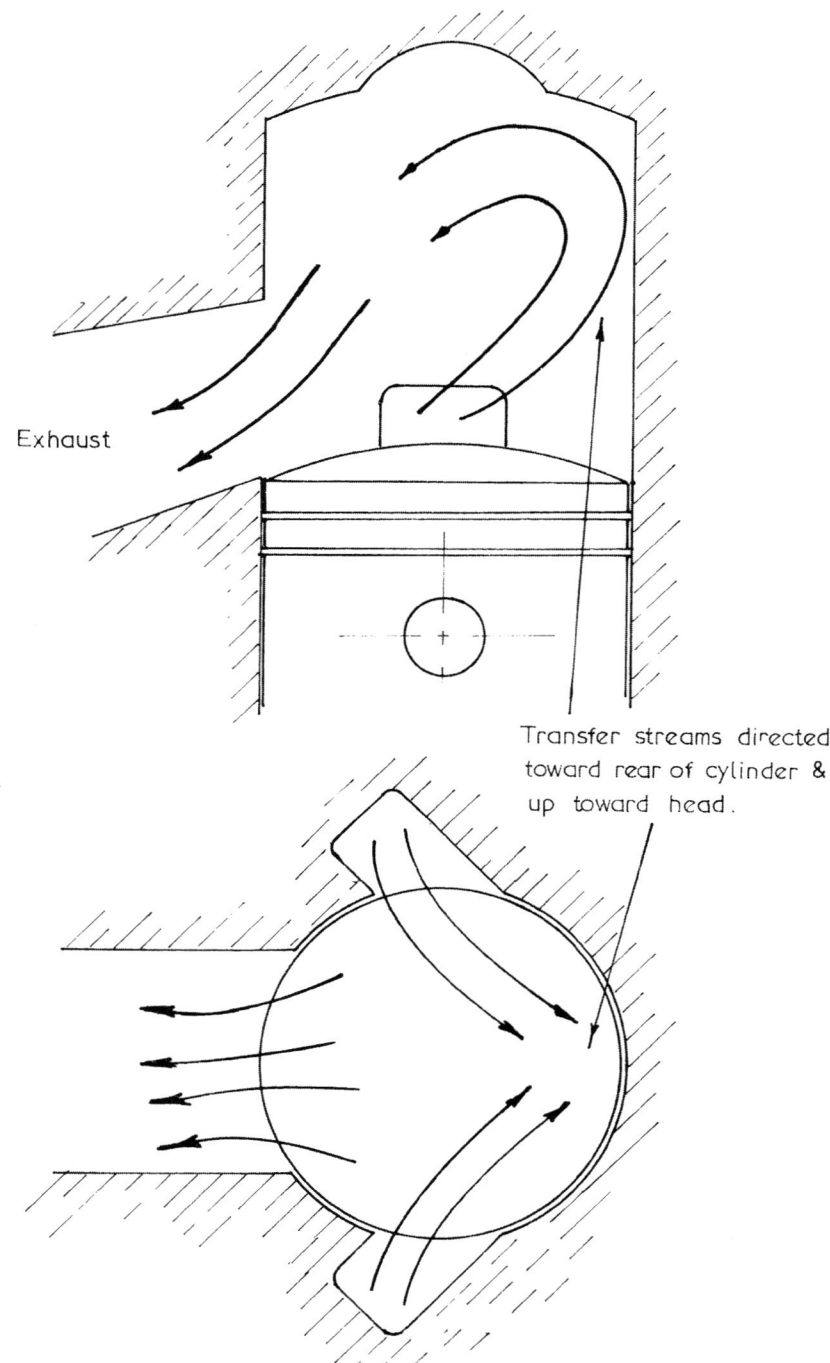

Figure 3.1 The Schneurle loop scavenge method relies on transfer port flow to rid the cylinder of exhaust gas.

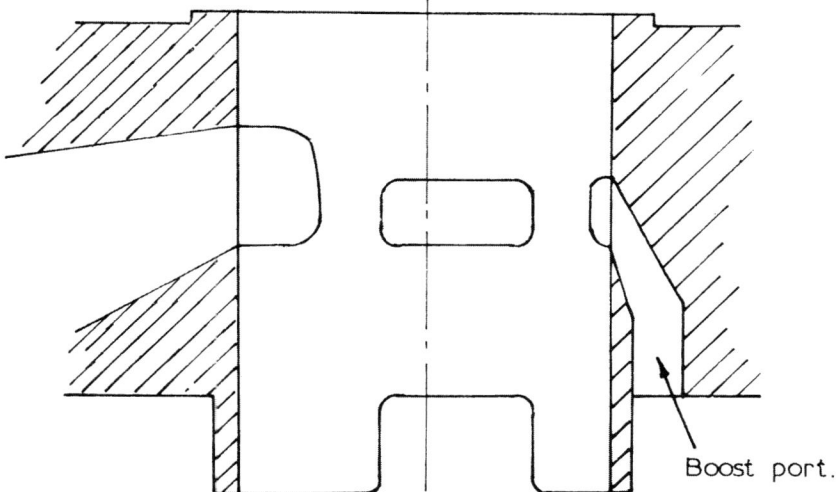

Figure 3.2 The Komet K78 TT employs a rear transfer port.

at Zschopau in East Germany was rebuilt as Motorradwerke Zschopau, or MZ. In 1952 Walter Kaaden joined MZ to take over development. His early work concentrated on exhaust development and alternate scavenge methods. After much experimentation he proved that the Schneurle loop-scavenge system yielded the best power and reliability. Then in 1957 he added a third transfer port, opposite the exhaust. Its air stream joined with the two main transfer ports, directing flow up toward the head (Figure 3.2).

Contemporary two-stroke technology was introduced initially to Suzuki, and later to Yamaha in Japan when Ernst Degner defected from East Germany to join Suzuki. By combining designs which Degner brought from MZ with Japanese technology in the field of metallurgy two-stroke power outputs and reliability took a leap forward. During the 1960s Suzuki and Yamaha both won world championships using exotic porting and rotary valve induction systems originally developed by DKW and MZ. The Yamaha engineers, however, went one step further. They added a pair of auxiliary transfer ports alongside the main transfers, which also directed mixture flow toward the rear of the cylinder and up (Figure 3.3). The Japanese engineers then realised, as did Walter Kaaden back in 1957, that there was a section of cylinder wall at the rear which could also be filled with another one or two ports. Transfer flow improved and, as the velocity of the fuel/air charge entering the cylinder was reduced, mixture loss out of the exhaust was decreased (Figure 3.4).

Back in Europe two-stroke engineers were battling excessive ring and cylinder wear, due to the exhaust port width being too great. A narrow port reduced power but improved reliability. A taller port restored lost power but made the power band unacceptably narrow. To get around the problem Rotax engineer Dr Hans Lippitsch added a pair of small auxiliary exhaust ports alongside the large oval exhaust port and above the main transfers. The two auxiliary ports connect with the main exhaust port before the exhaust flange (Figure 3.5).

Yamaha engineers tackled the problem with their power valve system (Figure

Two-stroke performance tuning

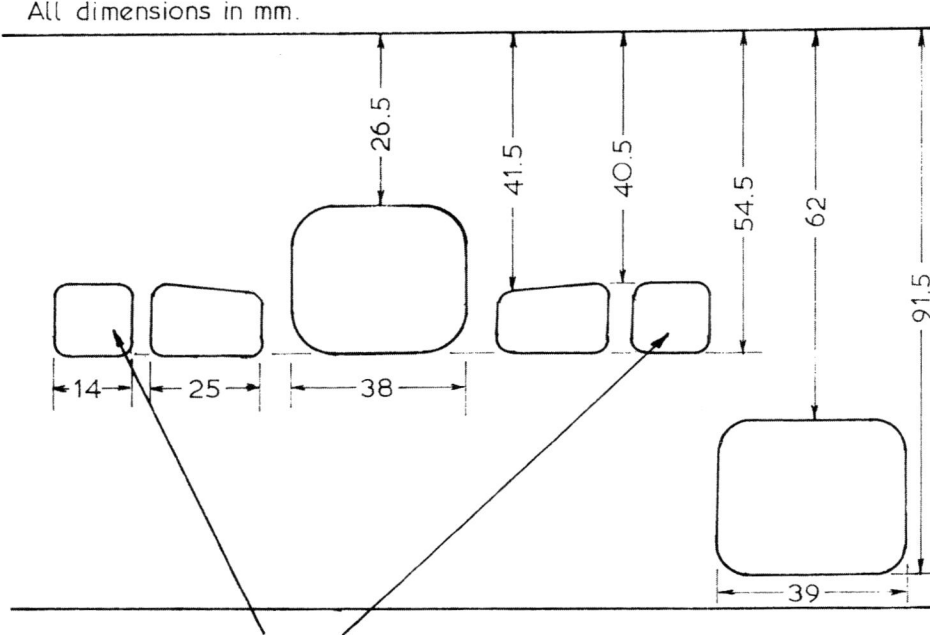

Figure 3.3 Yamaha TZ250 D/E/F cylinder with auxiliary transfer ports.

Figure 3.4 Suzuki PE175 has additional boost ports in the rear of the cylinder.

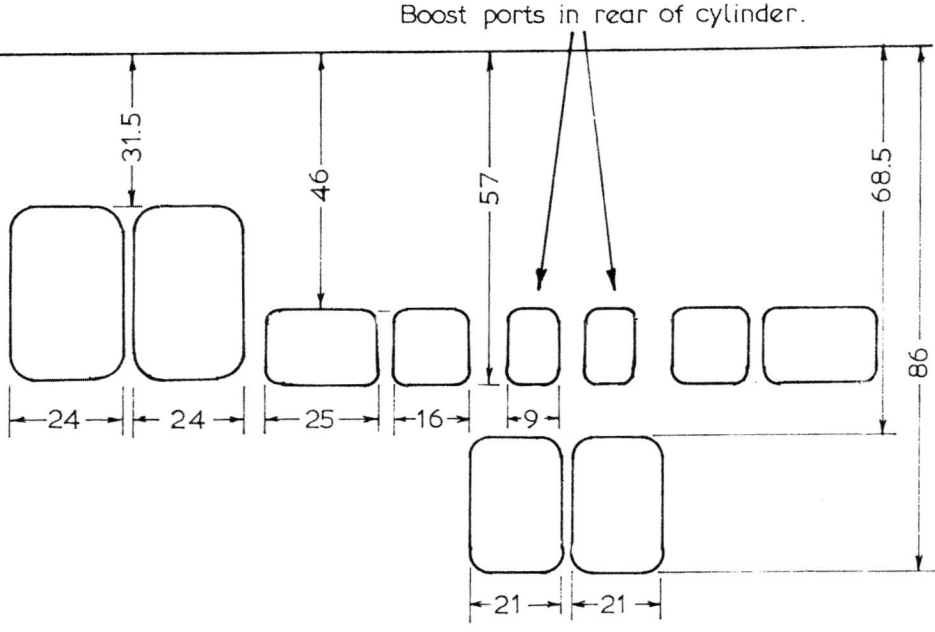

Porting and cylinder scavenging

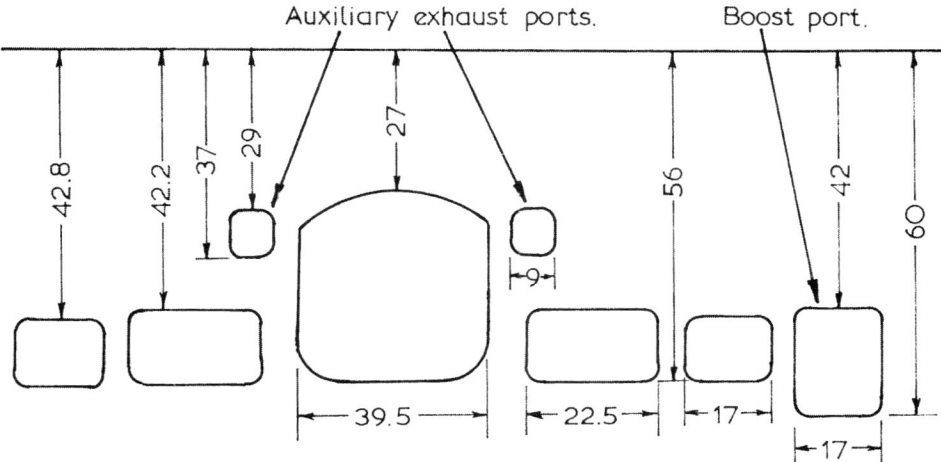

Figure 3.5 Rotax 124LC cylinder employs a pair of auxiliary exhaust ports to increase exhaust port area.

3.6), which is basically a mechanism to vary the exhaust port height without narrowing the power band. As you can see, there is a drum-like valve up against the cylinder wall. At high rpm the port is raised, increasing hp while permitting a relatively narrow port width for good ring life. At lower speeds the port is lowered, which improves mid-range power and widens the power band. Yamaha have used two methods of power valve control. The works racers, some production racers and street bikes utilise electronic control and a battery-powered motor to rotate the power valve into the required position. In other applications a purely mechanical system is employed with a centrifugal governor, driven from the crankshaft, raising and lowering the exhaust port in harmony with engine rpm. On the road racers exhaust

Figure 3.6 The Yamaha barrel-type power valve changes the exhaust port area and the exhaust duration.

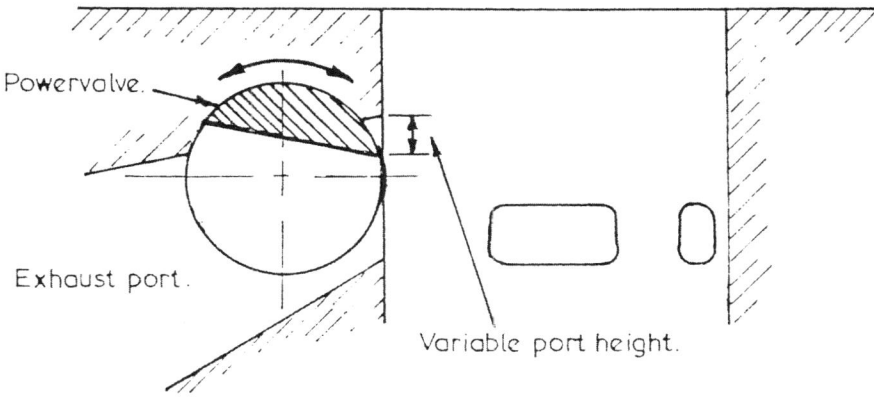

duration at higher speeds (ie above 10,500–11,000rpm) is 202°. Low rpm duration is about 180°, or similar to that of a 400 motocross engine. These figures represent about 6mm change in port height, while on some enduro and motocross engines it can be up around 8mm.

Soon after this Kawasaki introduced a different type of power valve system which they called KIPS (Kawasaki Integrated Powervalve System). The Kawasaki system incorporated two auxiliary exhaust ports, both higher than the main exhaust port, with each small port controlled by its own power valve. At higher rpm the power valves opened the auxiliary ports, effectively increasing exhaust duration and the exhaust port area. Then at lower speeds both power valves closed, however in the closed position one power valve opened into a small resonant chamber. This changed the frequency of pressure waves in the expansion chamber, which along with reduced exhaust open duration, served to improve mid-range power.

In 1985, Rotax unveiled their new 250 Grand Prix engines with a simplified type of Yamaha power valve. Rather than use a rotating valve their racing development head, Hans Holzleitner, developed a flat guillotine slide pneumatic valve which would rise and fall in the top of the exhaust port, increasing exhaust duration at high rpm and decreasing it at speeds below about 11,000rpm. The spring-loaded guillotine is controlled by exhaust pressure. At lower engine speed and load, exhaust pressure is reduced so the flat guillotine slide drops to the fully lowered position, pushed down by spring pre-load. As engine speed and load increases, which increases exhaust pressure, the guillotine rises in response to that pressure rise. This is accomplished very simply by two passages in the roof of the exhaust port directing exhaust pressure to the underside of a diaphragm which is bolted to the side of the barrel. Pressure causes the diaphragm to move and it pulls against the spring-loaded guillotine, thus raising the exhaust port. The full extent of slide lift on the 250 tandem twin was about 6mm.

During the 1985 GP season the Yamaha OW81 also ran with guillotine type exhaust power valves. However, on this 500 racer Yamaha relied on a computer to control the power valves via a servo motor and cable system. This allowed very precise programming of the valves to suit various riding styles, circuit layouts and track surface conditions. Additionally, unlike the Rotax system, exhaust development is not hampered by the effects that different expansion chamber pressure characteristics are having on power valve operation.

Just how effective the power valve was in allowing for much higher power outputs through the use of more radical porting and/or expansion chambers, without sacrificing mid-range power, can be seen by comparing the Yamaha RD350 LC and the RZ350. The engines were nearly identical except that one had power valves and the other did not. Looking at Figure 3.7 it is not difficult to see why many two-stroke manufacturers adopted a Yamaha-type power valve arrangement. Some of course weren't looking for more hp at the top of the power curve – they wanted more mid-range which was easily won by adopting more conservative porting and a matching expansion chamber.

When it comes to modifying a cylinder, the most logical place to start is the exhaust port. A little grinding (or filing) at the sides and top of the port will yield large power increases if approached correctly. Exhaust ports come in all shapes and sizes; each type has its advantages and disadvantages. The port in Figure 3.8 is really

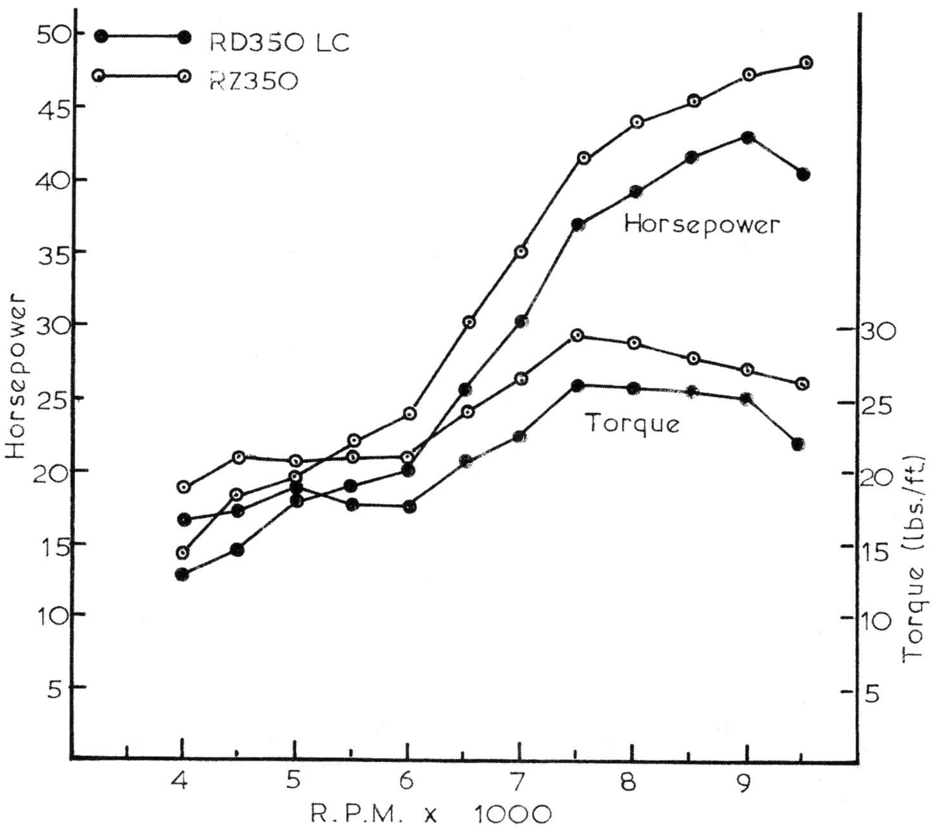

Porting and cylinder scavenging

Figure 3.7 Yamaha RD350 LC and RZ350 power curves illustrate how the exhaust power valve aids performance.

rectangular but it is usually referred to as a square port. This is the type that you will find in many low performance engines. The size of it has to be small so that the rings won't catch on the top of the port and break. There are two ways this port can be modified: either it can be widened at the top or it can be ovalised. We have to be careful that the exhaust port doesn't get too close to the transfers, otherwise there will be excessive loss of fuel/air mixture out of the exhaust. I like to see 8mm separation

Figure 3.8 A square exhaust port must be correctly reshaped to avoid snagging or damaging the piston rings.

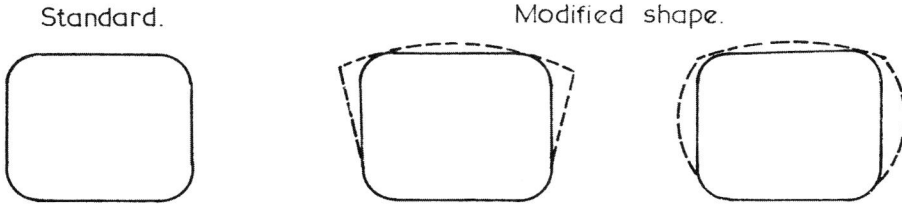

between these ports, but at times it is possible to go down to as little as 5mm without ill effect.

If port spacing is a problem, you will have no alternative other than to widen the exhaust port at the top. This type of port will give the engine good power from the upper mid-range to maximum hp. When you grind this type of port, the centre of the port should be 4° to 5° higher than the ends. The reason for this is that when the engine is on the compression stroke the ring bulges out into the port to its greatest extent just as the port is being closed. However, by raising the centre of the port, the ring has less chance of hanging up on the edge of the port and breaking because the ends of the port actually begin pushing the ring back into the piston groove before the port closes.

The elliptical or oval port is the one which I prefer if the port spacing is suitable. It is the type which you will find in most competition two-strokes. The shape of the port is fairly gentle on rings providing it isn't made excessively wide. What is an excessive width? Well, I'm not sure; but I have found that a port 0.71 of the bore diameter is a good compromise for most road race and motocross engines using ductile iron rings (the maximum safe port size is about 0.65 with brittle cast iron rings). Some tuners take the port size up to 0.75 but ring, piston and port damage is unacceptable. I have been able to take some ports out to 0.73 of the bore size, but this is the exception rather than the rule.

The square bridged port is fairly common in large displacement motocross and enduro engines (Figure 3.9). It has a very large port area, but then it has to have a large area as it flows only about 85 per cent as well as an unbridged port of equivalent area. In past years this type of port gave a lot of trouble as the bridge would overheat and bulge, pressing hard against the piston and causing a seize-up. However, the bridge gives little trouble now, providing it is not narrowed down. If heavy bridge-to-piston contact does occur, the piston should be relieved where it scuffs against the bridge. As bridged ports are usually quite close to the transfers, there is only one way to increase the port area, and that is by making the top of the port wider. Modify the port as shown, and don't copy the 'eyebrows' type exhaust port discussed next.

The 'T' or eyebrows port is seldom seen these days although it was used by Suzuki, Kawasaki and Honda in the past (Figure 3.10). This type of port has very little going for it as the sudden change in shape above the main transfers is very harsh on both piston and rings. Usually there is very little that can be done to improve this type of port.

Bridged exhaust ports can be made very wide, but there is a limit to how far you

Figure 3.9 A bridged exhaust port may be widened as illustrated.

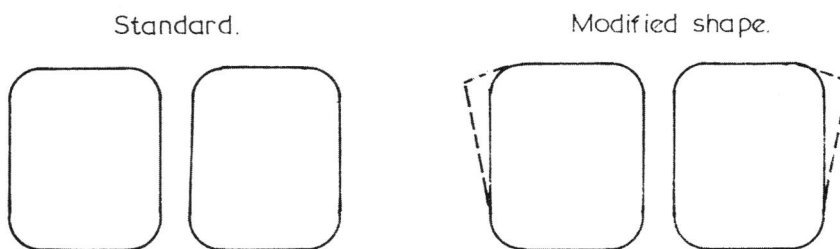

Porting and cylinder scavenging

All dimensions in mm.

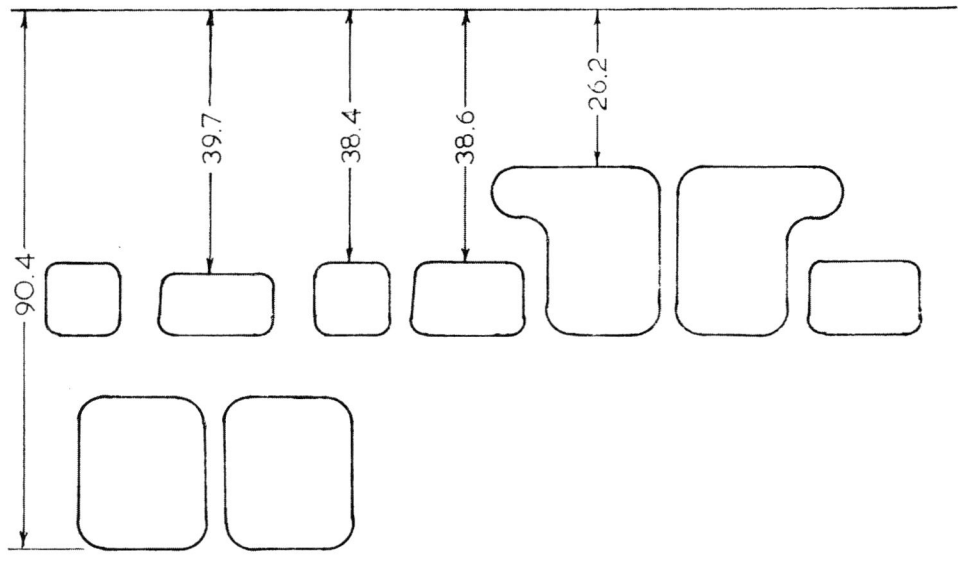

Figure 3.10 The Honda MT125 RIII employs an eyebrows type exhaust port.

Figure 3.11 The Suzuki RM125C uses a piston design which limits exhaust port widening to the dimensions shown.

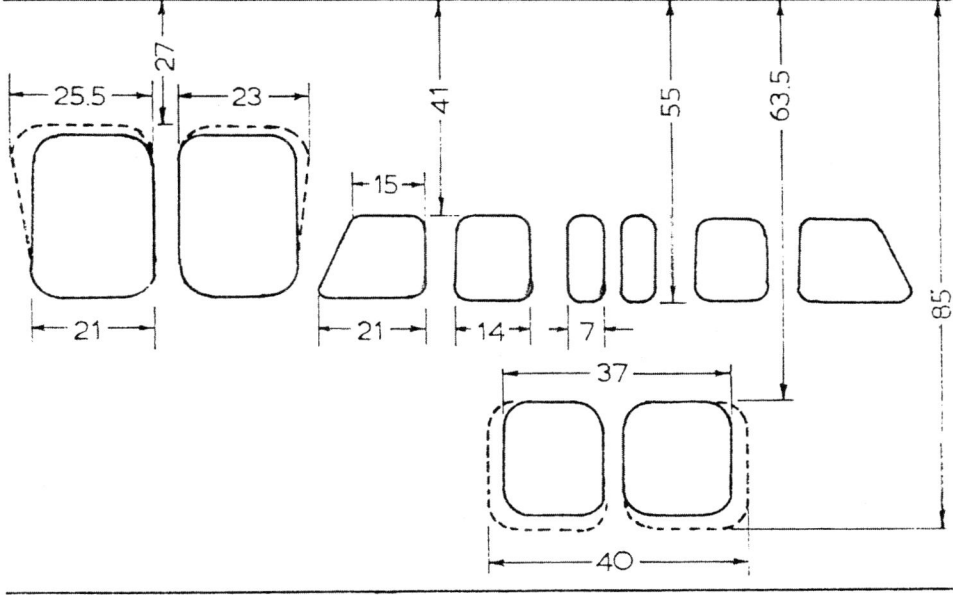

These reliefs around the pin bosses limit how far the exhaust port can be widened. There should always be at least 2mm width of cylinder wall on both sides of the exhaust port for the piston skirt to seal against.

can go. With the Suzuki RM125 engines (all models A to T), the maximum width is 23mm for the left port window (viewed from the front of the bike) and 25.5mm for the right half of the port. If you go any wider than this, the piston will not be able to seal the crankcase from the exhaust port because the skirt is relieved around the pin bosses. There should always be sufficient cylinder wall on the sides of both exhaust and inlet ports for a 2mm width of piston skirt to bear against and effect a seal (Figure 3.11).

To ensure that you don't go too far in widening the exhaust port, you will have to carefully scribe the outline of the port windows on the piston skirt with the crankshaft rotated to TDC. Then remove the barrel and measure the distance from the scribed lines to the relieved area around the piston pin bosses. Subtract 2mm from the measurement and this is the amount the port can be increased in width. The amount which can be removed from bridged inlet ports can be ascertained in a similar way, but with the piston at BDC.

Thus far we have talked about changing the shape and width of the exhaust port but not the height. Increasing the width of an exhaust port will always result in a power increase from the upper mid-range to peak rpm. Usually there will be little or no loss in mid-range power. Raising the port, on the other hand, will always knock bottom end power. Increasing the duration, the port open period, by just a couple of degrees can make a bike unrideable in some instances. Just how far you can raise the exhaust port is the million dollar question everyone would like to know. Some tuners work to a time-area/angle-area formula devised some time back. Frankly, I have found this method of calculating port timing completely useless. The geometry and mathematics involved is very tedious and, when you have finished the entire routine, you find that the answer bears little relationship with present-day two-stroke technology.

I have certain ideas on exhaust port timing, but blindly following my suggestions could get you into a lot of trouble. My theory is that an engine requires a certain exhaust duration to attain a specific engine speed. Therefore, if an engine is required to make maximum hp at, say, 12,000rpm, the exhaust duration required will be the same (±1°) regardless of whether the engine is an 80cc motocross engine or a twin cylinder 250 road racer. From experience I have a fair idea of just how much duration specific engines need (see Table 3.1). However, if the cylinder has a shorter transfer open period than I like, the exhaust duration will have to be reduced, otherwise the bike will be too 'pipey' to ride. On the other hand, I may choose to raise the transfer ports and use the suggested exhaust timing.

Table 3.1 Exhaust port duration

Engine size (cc)	Application	Engine speed (rpm)	Exhaust duration (°)
2 x 62	Road race	14,000	206–208
1 x 80	Motocross	12,500	199–201
1 x 80	Road race	13,500	202–204
1 x 100	Motocross	11,800	196–198
1 x 100	Kart*	10,800	178–180
1 x 125	Motocross	11,500	196–198
1 x 125	Road race	12,000	200–202
2 x 125	Road race	12,500	202–204
2 x 125	Road race**	10,250	193–195
4 x 125	Road race	12,250	201–203
1 x 175	Enduro	9,500	186–188
2 x 175	Road race**	10,750	194–196
1 x 200	Enduro	9,000	184–186
1 x 250	Enduro	7,500	180–182
1 x 250	Motocross	8,000	182–184
1 x 250	Road race	10,000	192–194
2 x 250	Road race	10,500	194–196
1 x 400	Enduro	6,000	174–176
1 x 400	Motocross	6,500	176–178

 * Fixed gear kart, hence short exhaust open period.
 ** Street sports modified for road racing.

You can easily tie yourself in knots when you tackle cylinder porting. I've known tuners who have moved exhaust ports up and down and all over the place, searching for more power or a better spread of power. After months of hard work they have achieved nothing, basically because the transfer duration was too short and/or the expansion chamber was all wrong. While it may appear rather arbitrary to select an exhaust timing figure and stick to that, I feel that this is currently the best way to go about two-stroke tuning. Then, if the engine does exhibit some undesirable trait, like a narrow power range, I change the expansion chamber design to produce the required power characteristics. What I'm saying is that expansion chamber design is far more critical than exhaust port duration. The exhaust open period determines to some extent what the maximum hp will be and at what engine speed it will be produced. The expansion chamber, on the other hand, 'adjusts' the power characteristics of the engine at speeds above and below maximum hp revs.

The formula which I use to calculate exhaust open duration (and transfer duration) is as follows:

Two-stroke performance tuning

$$D = \left(180 - \text{Cos}\, \frac{T^2 + R^2 - L^2}{2 \times R \times T}\right) \times 2$$

where $T = R + L + C - E$, R = stroke divided by 2 in mm, L = con rod length in mm centre to centre (usually the stroke multiplied by 2), C = deck clearance in mm (ie the distance the piston is below the top of the barrel at TDC), and E = distance from the top of exhaust port to top of barrel

For example, the exhaust duration of the Morbidelli 125 twin production racer (Figure 3.12) is as follows:

$R = 20.5$mm, $L = 87$mm, $C = 0$mm

$$T = R + L + C - E$$
$$= 20.5 + 87 + 0 - 18.2$$
$$= 89.3$$

$$D = \left(180 - \text{Cos}\, \frac{T^2 + R^2 - L^2}{2 \times R \times T}\right) \times 2$$

$$= \left(180 - \text{Cos}\, \frac{89.3^2 + 20.5^2 - 87^2}{2 \times 20.5 \times 89.3}\right) \times 2$$

$$= (180 - \text{Cos}.22553) \times 2$$
$$= (180 - 77) \times 2$$
$$= 206°$$

Figure 3.12 Morbidelli 125 road racer porting.

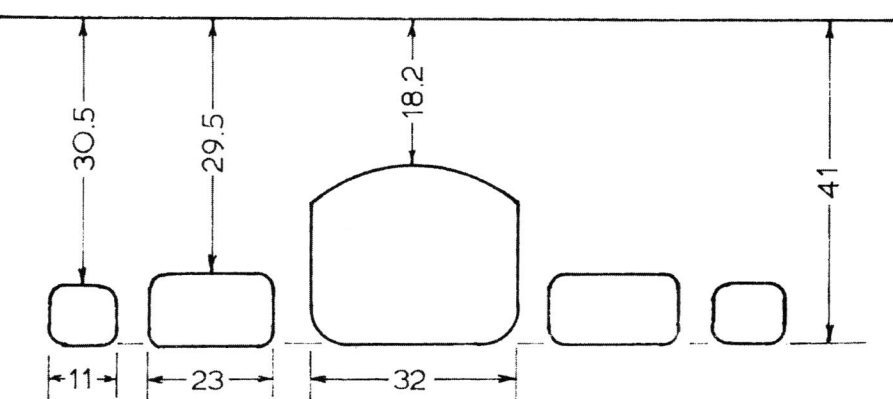

Porting and cylinder scavenging

Looking at Table 3.1 you can see that the exhaust duration is right where we want it for peak hp at 13,500–13,700rpm. However, if we were going to modify this engine extensively by boring the Mikunis 1mm to 29mm and fabricating a new set of expansion chambers, we would want the power peak at a little over 14,000rpm, which would mean that the duration would have to be increased to 208° to take advantage of the engine's improved breathing. Therefore we would raise the exhaust port 0.35mm. E will now equal 17.85mm and T will equal 20.5 + 87 + 0 – 17.85 = 89.65.

$$D = \left(180 - \mathrm{Cos}\ \frac{89.65^2 + 20.5^2 - 87^2}{2 \times 20.5 \times 89.65}\right) \times 2$$

$$= (180 - \mathrm{Cos}.24169) \times 2$$

$$= (180 - 76) \times 2$$

$$= 208°$$

On some engines fitted with Dykes rings, the top piston ring and not the piston crown controls the opening and closing of the exhaust and transfer ports. With these engines, the exhaust duration is calculated using the same formula, however dimension C (the deck clearance in mm) must be very carefully measured using a depth gauge otherwise your calculations will be several degrees out. In engines where the Dykes ring actually determines the port opening and closing, dimension C is the distance the ring is below the top of the barrel at TDC. Referring back to Figure 3.5 you will note that the Rotax kart engine appears to have mild porting for a road racer. This engine, in fact, has a single Dykes ring located very close to the top of the piston. Dimension C is 1.8mm, so what looks like motocross porting is truly road race porting. In this case the exhaust duration is 201°.

If you have not had any previous experience tuning two-strokes it is a lot safer to modify the piston crown to increase exhaust duration rather than raise the port. Once you have taken the metal away you can't put it back, but fortunately pistons are a good deal less expensive than barrels so all you have to do is keep accurate notes and then retrogress one step when you have gone too far (Figure 3.13). The idea is to progressively file 0.5mm off the exhaust side of the piston crown until you reach a point where you are happy with the power output. If you accidentally go one step too far, it is easy to back-track. All you need is a new piston and then, when you modify the exhaust port proper, raise it 0.5mm less than the amount you filed off the piston. This type of tuning is back to front to the way in which I prefer to do things, but if you don't want to get involved in expensive and time-consuming expansion chamber fabrication, it is the safest way out. You will never get the best possible power out of the motor by shifting the exhaust port around to work within the limitations imposed by the expansion chamber fitted to your bike. However, this is one of the safer places to begin modifying two-strokes, and even within the boundary set by the stock expansion chamber you should end up with an engine which works better than the stock item.

When working on the exhaust port, there are two checks which should be made. Firstly, with the piston at BDC, the bottom of the port window should be level with, or lower than, the piston crown, otherwise high speed gas flow will be disrupted (Figure

Two-stroke performance tuning

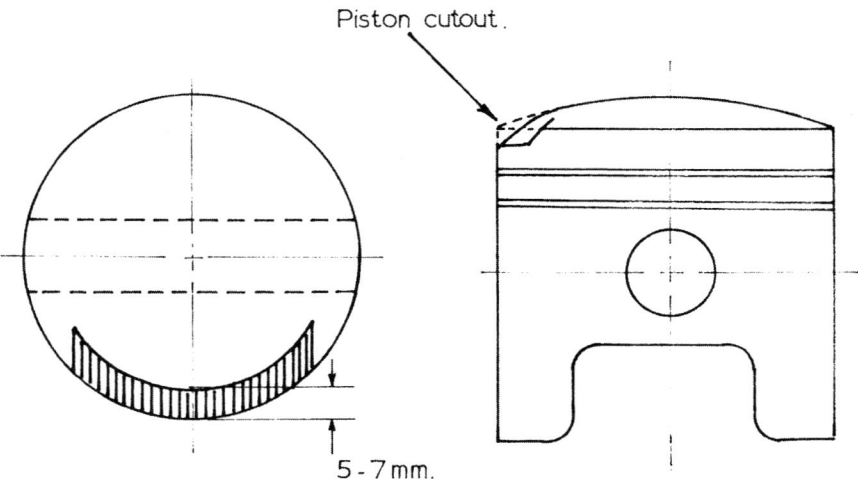

Figure 3.13 Piston crown may be filed as shown during exhaust port timing experiments.

3.14). Secondly, in the case of bridged ports, ensure that both halves of the port open simultaneously. If one side opens a little before the other, gas flow is disrupted to some extent, but worse the pressure waves transmitted to the expansion chamber are of a lower amplitude. This reduces the effectiveness of the exhaust pulses in evacuating and recharging the cylinder with fresh mixture (Figure 3.15).

If you own a power valve type engine there is an additional inspection which must be made. Regardless of whether the exhaust port is standard or has been raised check that the power valve opens fully to align with the exhaust port roof. Manually

Figure 3.14 The exhaust port must be lower than the piston at bottom dead centre to ensure best exhaust flow.

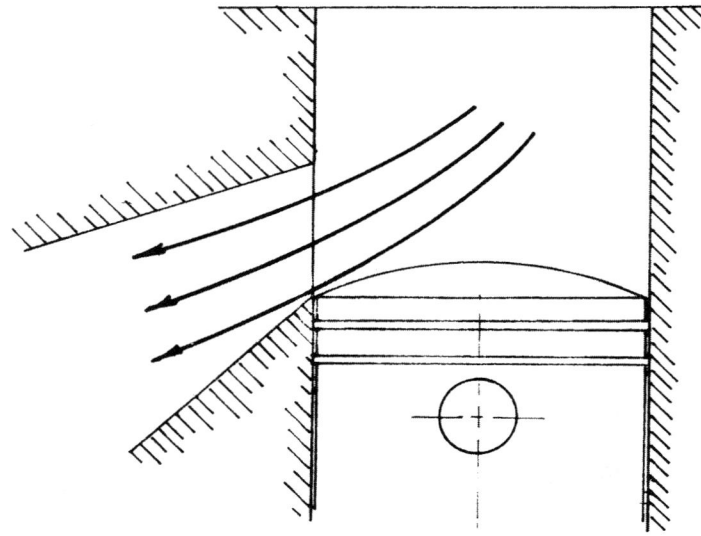

Porting and cylinder scavenging

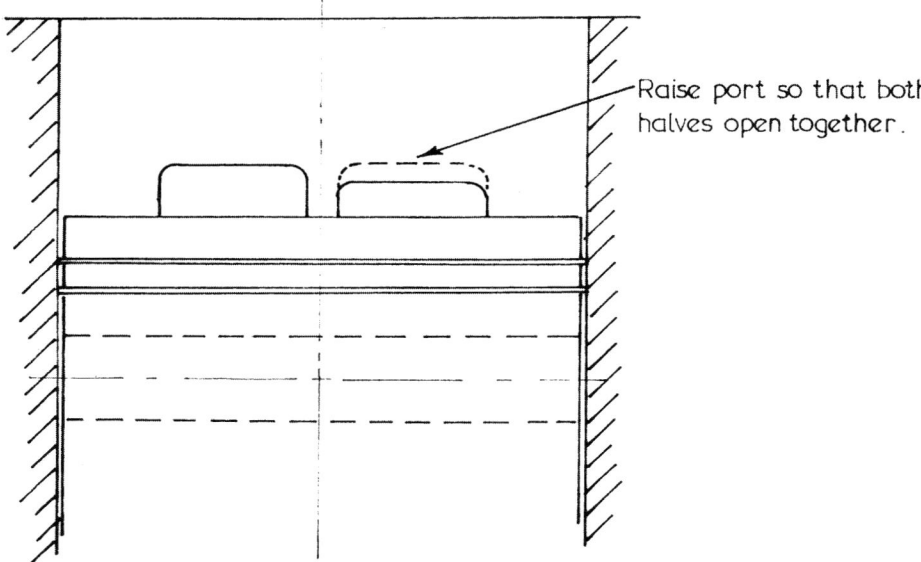

Figure 3.15 Both halves of bridged exhaust ports must open simultaneously to ensure maximum pulse activity in the expansion chamber.

push the actuator arm as far as it will go to see if the valve and port align. Usually some adjustment is required. After loosening the adjusting nut and moving the valve to the correct position be sure to Loctite the nut so that it does not vibrate loose. Finally verify that the valve timing is correct with the engine running. This is accomplished by marking the full extent of the actuator arm travel on the cylinder and revving the engine in short bursts to see if the valve actually opens that far. If it does not you will have to adjust the valve to a position slightly higher than the exhaust port roof with the actuator arm pushed to the full open position. Then recheck for full open with the engine running.

The only other power valve adjustment which is permissible alters the governor spring preload and changes the mid-range rpm valve timing. When spring preload is increased, lower rpm exhaust duration is increased. As this has the effect of raising top end power and narrowing the power range, it is a modification recommended only for expert riders on fast circuits. Begin testing with an additional 0.020in shim fitted behind the governor spring. If the power comes on too quickly or the power range is too narrow, try a 0.012in shim. Alternatively, fitting a softer spring will delay the rpm at which the power valve commences rising, and the valve will open more slowly. If power delivery is still too brutal for your riding ability or track conditions are very demanding, the actuator can be adjusted so that the power valve does not go fully open. When the power valve is computer controlled the ECU can also be reprogrammed both to change the revs at which opening commences and how quickly the full-open position is reached, but this is more expensive than experimenting with governor spring pre-load.

In recent years, the physical size and shape of the exhaust port between the port window and the flange where the expansion chamber connects, is under close scrutiny.

Attempts are now being made to keep the diameter of the port as small as possible, without impeding the flow of gas out of the cylinder. Whereas the port diameter of a typical 125cc cylinder was 40 to 42mm several years ago, most exhaust ports for a 125 are now about 37 or 38mm diameter. This is being done to keep the exhaust pulse wave at a high amplitude so that the cylinder is scavenged and recharged more completely. It has been found that allowing the exhaust gases to expand and cool too quickly, as occurs when the exhaust port is large, actually diminishes the strength of the exhaust pulse.

Naturally the tuner's desire to keep the exhaust gas confined so that a strong pulse wave is transmitted through the expansion chamber, has to be balanced against the need for a free-flowing exhaust passage, which allows the burnt gases to stream unimpeded out of the cylinder. To this end, the exhaust port must be relatively straight and without abrupt directional changes to eliminate eddying, and the exhaust flange must match the port perfectly and not change the direction of exhaust flow. When an exhaust port meets these requirements, gas flow out of the cylinder will be good, even though the port diameter is relatively small to keep pulse intensity at a high value.

A quick look through the exhaust flange and port will indicate how straight is the exhaust passage. However, unless you are very experienced in the science of gas flow, you will not know if the exhaust gases are eddying or not. If you are using castor oil or some other oil which produced a fair build-up of carbon, you will be able to see where the exhaust port is 'dead'. Any place where there is a layer of carbon in a port which is basically carbon-free is a place of little flow activity. In such an area you can be fairly certain that the gases are eddying and disrupting flow out of the cylinder.

At times, the low pressure area can be eliminated by grinding metal out of the port, but more often than not the port will require welding up. The exhaust port illustrated in Figure 3.16 is a particularly nasty one. The flange changes the direction of flow very abruptly, which produces an eddy current in the top of the flange. Also the floor of the port drops away too quickly, causing eddying in this area.

Carbon deposits in the exhaust port (left) disrupt gas flow out of the cylinder, but they also show where flow activity in the port is poor and can provide clues as to what port modifications may increase power output. This engine (right) has covered many miles on Castrol R30 castor oil, however the exhaust port is still relatively clean indicating good exhaust port flow.

 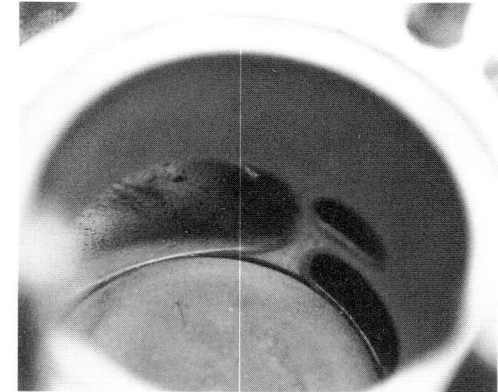

Porting and cylinder scavenging

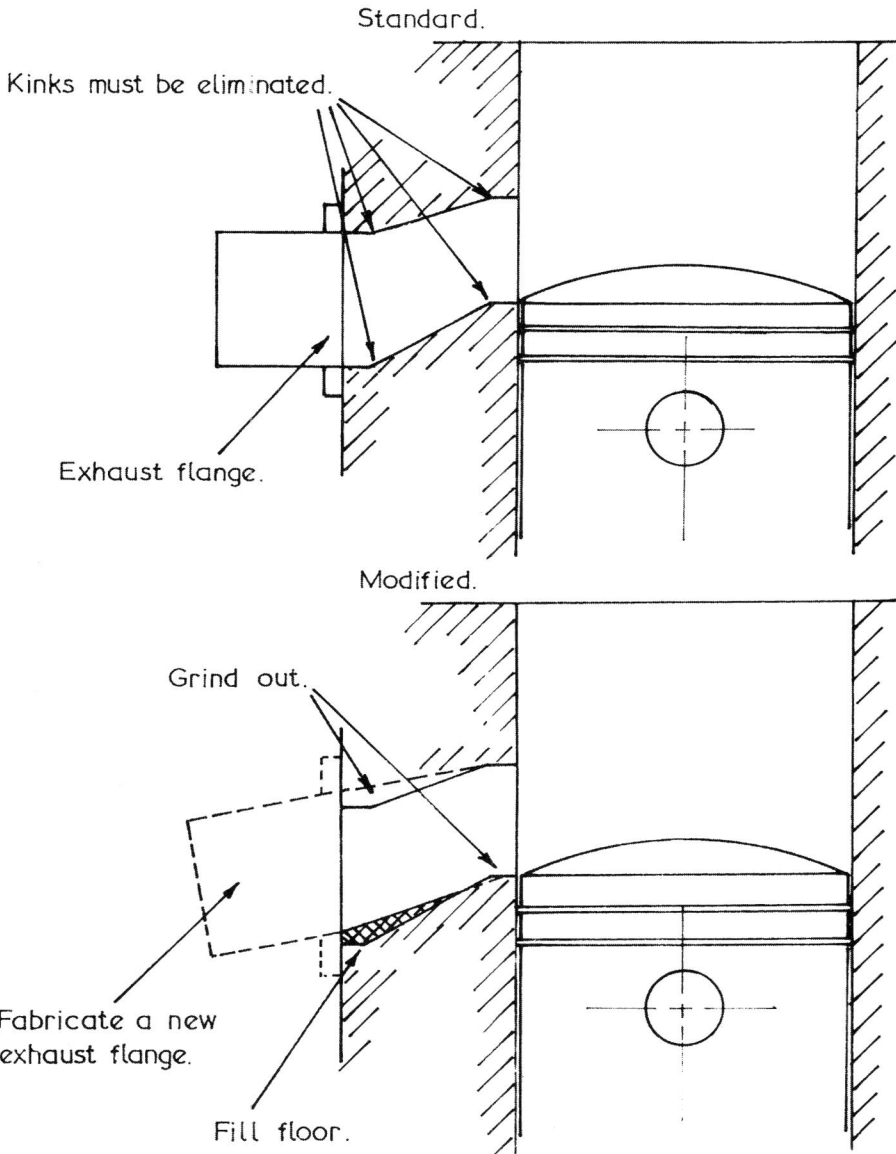

Figure 3.16 The exhaust port must be correctly modified to assist gas flow out of the cylinder.

There are two ways to tackle the problem with the flange. The roof of the port may be ground higher and the flange raised to reduce the kink in the port's roof. On the other hand a new flange can be fabricated with the roof in line with the roof of the exhaust port. Either way, the floor of the port, and perhaps the floor of the flange too, will have to be welded up to improve the profile. The aluminium floor naturally will have to be argon-arc welded. Fill in only a little at a time and allow the cylinder plenty of time to cool between each run, otherwise it will distort.

As shown in Figure 3.17, the exhaust flange may be out of line when viewed from above. Again this must be corrected by fabricating a new flange which aligns with the exhaust port.

From the aspect of two-stroke engine design, I feel that the transfer ports are the most important. Unfortunately, from the average tuner's viewpoint, the transfers are the most difficult to modify and the least understood. By definition, the transfer ports have the job of transferring the fuel/air mixture from the crankcase into the cylinder. That sounds simple enough but, after we consider all of the factors involved, you will better appreciate what a mammoth task this really is.

In an average racing engine the induction cycle will take place during around 190° of crankshaft rotation. The exhaust cycle will occur over a period of 200°. The transfer phase, however, has to be completed through 130° of crankshaft movement. Not only do the transfers have an extremely short time in which to recharge the cylinder with fuel/air mixture, they must also control the flow pattern of the charge to prevent mixture loss out of the exhaust, and drive exhaust gases from the rear of the cylinder towards the exhaust port.

During the 1960s, when Suzuki and Yamaha dominated Grand Prix racing, their engineers revived a myth which surfaced from the development of BSA Bantam and Villiers engines for racing just after the Second World War. These engines had massive spaces in the crankcase and tuners reasoned, rightly enough, that filling the crankcase with a variety of 'stuffers' would reduce crankcase volume and hence increase crankcase compression when the piston descended to BDC. Increasing crankcase compression naturally enough results in higher crankcase pressure which, all else being equal, raises transfer flow and improves maximum hp output. Tuners cited the

Figure 3.17 The exhaust flange should be in line with the exhaust port to stop eddying.

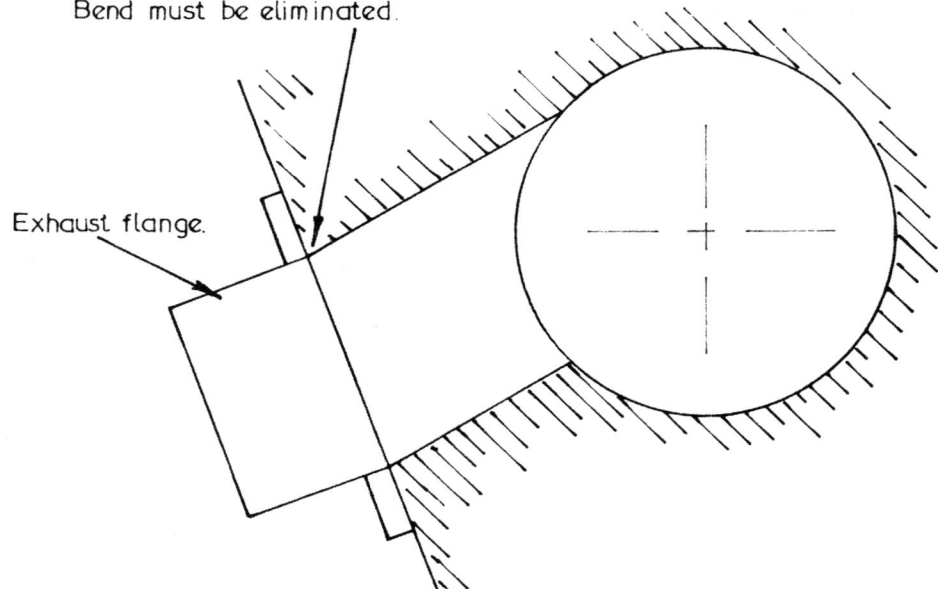

reason for this as being due to the transfer streams erupting under considerable pressure into the cylinder. Because of this the fuel/air charge tended to behave like a wedge on entering the cylinder. It didn't break up and mingle with the exhaust gases, but pushed them out of the cylinder with considerable force.

So effective was this method of cylinder scavenging that the fuel/air 'wedge' was actually being partly lost out of the exhaust before the port closed. Two-stroke tuners overcame this problem by opening the transfer ports later and closing them earlier, reducing traditional transfer duration from 130° down to 120°. Because of more fuel charge being contained within the cylinder, power increased. This encouraged engineers to further increase crankcase compression and reduce the transfer open period to less than 110°. Horsepower again rose, instilling in Japanese engineers the idea that dominance in Grand Prix racing would depend on them reducing transfer duration to contain charge loss out of the exhaust and increasing crankcase compression to ensure efficient pumping of the fuel/air mixture from the crankcase into the cylinder.

The theory sounds good, but in practice there were problems. True, power outputs rose to levels previously unknown from two-strokes, but the power bands became razor thin and engine speeds rose to incredible levels. Not to be deterred, the Japanese engineers embarked on a scheme of cylinder size reduction to enable very high rpm to be attained reliably. Again power levels increased, providing a further stimulus to reduce cylinder displacement. This led to the development of such machines as the three-cylinder 50cc Suzuki and the four-cylinder Yamaha 125 which produced 40hp at 18,000rpm. At this time road racers had from 10 to 18 gears, such were the power characteristics of these engines.

The problem was that in spite of the very limited transfer open periods employed, at lower engine speeds too much charge was being lost out of the exhaust. This occurred because the transfer charge entered the cylinder under so much pressure that it had time to spurt right out of the exhaust at low rpm. Hence little power was produced at speeds below maximum hp revs. At higher rpm, power was again restricted, due to the transfer ports being too small to flow a larger volume of fuel/air mixture in the available time.

Today, the very same problem occurs when very short transfer periods are employed. Generally, you will find that bikes which are 'pipey', coming on to the power too quickly or exhibiting a narrow power range, are that way because the transfer ports are too low (ie short duration) or because the ports are incorrectly aimed.

Fortunately, manufacturers have mostly got away from the idea of using high crankcase compression to push the fuel charge through the transfers into the cylinder, so we can forget about crankcase compression and concentrate on the transfer ports. However, for those who are interested, primary compression or crankcase compression is calculated using this formula:

$$PC = \frac{CCV}{CCV - CV}$$

where CCV = crankcase volume at TDC, and CV = cylinder volume.

To measure the crankcase volume (CCV), first turn the engine on to its side, with the inlet port facing up, and rotate the crank to bring the piston up to TDC. Then, using a burette filled with liquid paraffin (kerosene) and engine oil, mixed 50–50, fill the crankcase up to the cylinder wall face of the inlet port. If this equals, say, 425cc, and the engine has a 125cc cylinder, the primary compression ratio will be 1.42:1.

At this time, instead of relying solely on crankcase pressure to push the fuel/air mix into the cylinder, we also use the suction wave produced in the expansion chamber to pull the intake charge up through the transfers. If we use an expansion chamber with shallow tapers, maximum power will be suppressed, but the suction wave will be active in drawing mixture into the cylinder over a wide rpm range. On the other hand a chamber with steeper cones will produce a stronger suction wave, raising peak hp, but it will be effective over a much narrower rev range.

Obviously the longer we leave the transfer ports open, the larger the rpm range will be over which the exhaust pulses effectively pull up fresh mixture from the crankcase. Conversely, if transfer duration is kept short, we have to rely more on crankcase compression to shift the fuel/air charge, as the suction pulse in the exhaust will only arrive at the right time to draw up fuel over a limited rpm range. It stands to reason, if the transfer port is closed when the pulse wave arrives, it will not do any good. On the other hand, if we keep the port open for as long as possible we have a better chance of having pulse waves arrive at the right time, over a wider range of engine speeds.

With this idea in mind, we should realise that the transfer duration will vary for high and low speed engines. A high speed engine (ie 13,500rpm) will want the transfer ports open for 140–142° while an engine running at 6,500rpm will be happy with a duration of 120–124° when exhaust port open periods are close to those in Table 3.1. At higher engine speeds there is less time for cylinder filling so we need a longer transfer period, but at lower speeds a long transfer period will allow too much charge to escape out of the exhaust so a shorter duration is in order for low speed engines. Table 3.2 sets out the transfer durations which I have found to allow good engine breathing at the speeds indicated. To pick up mid-range power the shorter duration should be chosen. The engine won't rev far past maximum hp revs but the power output below maximum will be superior. For good power past maximum rpm the longer transfer period is desirable. If exhaust port durations longer than those indicated in Table 3.1 are used, then more transfer timing may be necessary otherwise the engine could become too 'pipey'.

One ploy which is very effective in giving the engine good power over a wide range is to use staggered transfer durations. The old MZ 125 racer had the two main transfer ports open for 136°, while the third transfer port in the rear of the cylinder had a much shorter duration of 128°. Many of the Italian go-kart engines also used this type of porting in past years. When Honda introduced the MT-125RII production racer in 1977, they took this principle one step further. The main transfers opened 39.2mm from the top of the cylinder (126° duration), the secondary transfers opened a little earlier at 38.5mm (130° duration) and the boost port in the back of the cylinder opened the last, 39.7mm down (123° duration).

Tuners reasoned that as the back port aimed its flow towards the exhaust port there would be some loss of charge, unless steps were taken to prevent this occurring. Therefore the back port was opened around 1mm after the main transfers, so that flow from the main transfer ports, being aimed towards the rear of the cylinder, would

Porting and cylinder scavenging

Table 3.2 Transfer port duration

rpm	Transfer duration (°)
6,500	120–124
8,000	124–128
9,000	126–130
10,000	128–132
11,000	130–134
12,000	132–136
13,000	134–140
14,000	136–142

Note – The transfer duration refers to the open period of the main transfer ports in particular. The secondary transfers and the boost port may beneficially use durations longer than shown.

actually form a wall of mixture in front of the boost port and thus prevent a loss of charge out of the exhaust. Furthermore, it was felt that delaying the opening of the rear port would allow crankcase pressure to 'blow down' through the main transfers. Hence a high pressure stream would not erupt from the back port and head right out of the exhaust.

Today those theories have been forgotten. The majority of engines come from the manufacturers with all the transfer ports at the same height. However, this does not mean that staggered porting does not work. Most tuners recognise that it does; but the transfers are staggered in reverse to the old school of thought. At this time, when a cylinder is modified, the back port is often opened 1.0 to 1.5mm earlier than the other

This cylinder employs mildly staggered transfer porting. The boost port (on left) opens 0.2mm before the main transfers, while the secondary transfers (centre) open 0.6mm later.

transfers. Also I have found that opening the secondary transfers up to 0.8mm before the main transfers benefits the power curve as well.

There are several reasons why staggered-type porting works so well at this time. For one thing the manufacturers have forgotten their preoccupation with high crankcase pressure. Therefore, the transfer charge enters the cylinder in a more orderly and controlled manner. Additionally, the transfer ports have been re-aimed. Whereas the ports were tilted upwards so that the mixture streams from opposite sides of the cylinder gently met at a point in the cylinder just slightly higher than mid-stroke, today's ports are tilted very little or not at all (Figure 3.18). This means that the flow streams hug the piston crown, rather than shooting up towards the head to mingle with exhaust gases. Instead, the streams crash into each other, dissipating much of their energy. The mixture then rises relatively slowly in the cylinder, where it is trapped as the exhaust port closes. For these reasons, we can open the boost port and the secondary transfer ports a little earlier, as there is less risk of mixture escaping out of the exhaust, even at lower speeds when there is more time for this to occur. If the main transfers were opened earlier, exhaust flow would tend to turn the transfer flow around and direct it out of the exhaust port, but flow through ports further away from the exhaust port are not influenced to such an extent by the direction of exhaust flow.

When staggered porting is employed, it is usual for mid-range and maximum power to increase, due to the longer transfer periods improving cylinder filling, particularly at high rpm. Much of the mid-range power gain, I feel, is due to the cylinder being scavenged better. With the new type of transfer porting, a pocket of exhaust gas can be left unscavenged high up in the cylinder at lower engine speeds. Opening the boost port early would tend to get this pocket of stagnant gas moving, because its flow stream is still directed upwards at 45° to 60°. Some fuel charge is possibly lost out of the exhaust but, because this pocket of exhaust gas is purged out of the cylinder, there is less dilution of the remaining fuel/air mixture. Consequently combustion will be faster and more complete, raising the hp output.

Figure 3.18 Old style transfers direct flow upward while new design directs flow across the piston crown.

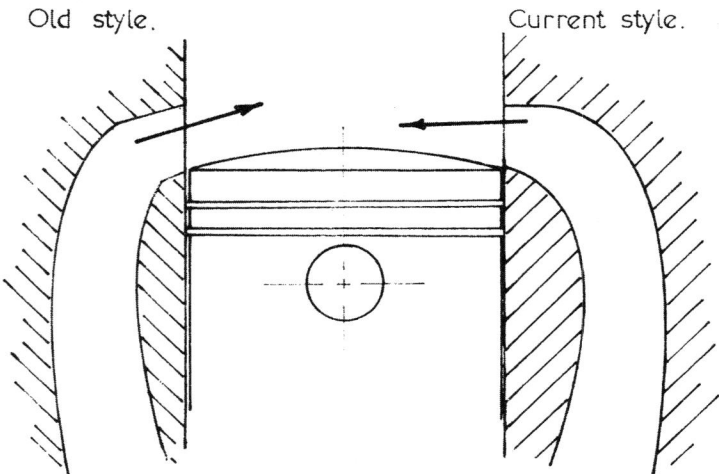

Naturally, in our quest to improve the engine's breathing potential, we can go too far in extending the transfer open period. This is where an adequate exhaust 'blow down' period must be allowed for and a correctly designed expansion chamber fitted otherwise, except at certain, usually very high rpm, the exhaust gas will reverse flow back into the transfer ports. This not only contaminates the inlet charge with exhaust gas but also disrupts transfer flow by blasting a positive pressure wave down the transfer passages. Again, the value of a good expansion chamber can be clearly appreciated. Such a chamber will assist rapid exhaust flow out of the cylinder over a wide rpm range such that when the transfers open, cylinder pressure will be lower than that in the transfer ports. A properly designed expansion chamber cannot accomplish this task all by itself; it needs assistance in the form of an adequate time period to allow cylinder pressure to 'blow down'. This usually means that in a road racer the transfers will be timed to open around 32–35° after the exhaust port opens.

Because the direction of transfer flow is so very important in obtaining a high power output and a good power range, only very experienced tuners should attempt to modify the top section of transfer ports. If you don't know what you are doing, you could easily render the cylinder useless. When the transfer duration is too short, raise the barrel using an aluminium spacer of the required thickness, and fit a base gasket on each side to ensure a good seal. Naturally the compression will have to be restored by turning an amount equal to the thickness of the spacer, plus the thickness of one base gasket, from the barrel or cylinder head. Keep in mind, when the cylinder is raised, that the piston rings may become exposed in the inlet port. This is of no consequence providing the top of the port is correctly shaped and providing the ring ends are not exposed. If just the bottom ring is opening into the inlet port, it can be removed if the engine is usually operated at 8,000rpm plus. In piston-ported engines, raising the barrel will shorten the inlet open period so the inlet port will have to be lowered to compensate.

Cylinders employing the type of boost port usually found in reed valve engines are quite easy to modify. This type of back port can be raised or increased in width, using hand files. Take care that you don't nick the bore wall with the file and do not make the port so wide that it opens out to the piston ring pegs. A width equal to that of the main transfer port is close to what is required, but always check to be sure. (Figure 3.19).

The secondary transfers should be raised by a professional tuner with good knowledge of the subject and good equipment to do the job. The alternative, which works very well, is to file metal off the piston crown (see Figure 3.13) in the manner described for increasing exhaust port duration. If the piston is fitted with a Dykes ring high up this method will not work, as the piston ring and not the piston crown actually controls the exhaust and transfer opening.

The safest part of the transfer port for you to modify is the bottom of the port where it joins the crankcase. Cut the base gasket to match the crankcase cut-outs and then match the transfers to the base gasket. This will ensure that there is no step in the port to disrupt flow. Then carefully smooth the transfers, removing all casting imperfections. The piston cut-out below the gudgeon pin is also a part of the transfer port, so dress it up too.

When the piston cut-outs are very narrow they can restrict flow into one pair of transfer ports as the piston approaches and passes BDC. In high rpm engines the

Two-stroke performance tuning

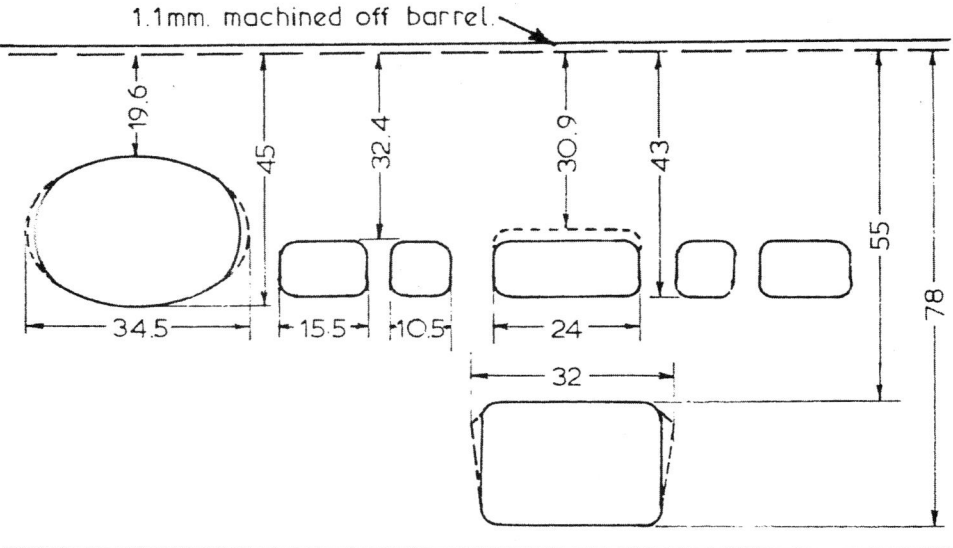

Figure 3.19 Yamaha cylinder and boost port have been raised to increase transfer open period.

situation can be improved marginally by modifying the piston as shown in Figure 3.20. Note that care is necessary so as to not weaken the piston. Therefore it is usually inadvisable to modify the inlet port side of the cut-out as the piston skirt is very heavily loaded as it passes down over the floor of the inlet port. On the exhaust side

Figure 3.20 When piston skirt obstructs flow into the transfer ports the cut-away may be enlarged as shown. However take care that enlarging the cut-away will not connect the exhaust port to the crankcase when the piston is at TDC.

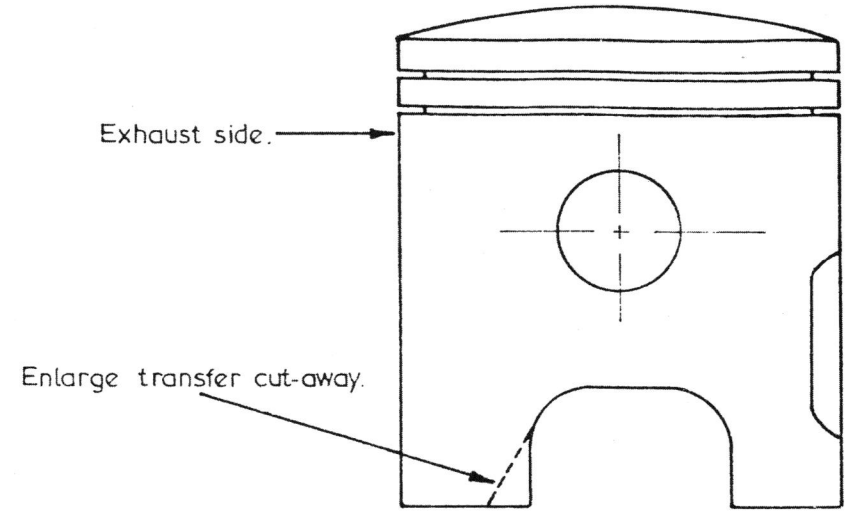

Porting and cylinder scavenging

also, care is needed otherwise, if the cut-out is extended excessively, the crankcase could become connected to the exhaust port with the piston at TDC.

In a number of Rotax rotary valve engines the fifth transfer port, the boost port, is almost closed off from the crankcase when the piston descends to BDC. To increase mixture flow through this port at high rpm, the piston skirt on the 124LC kart engine for example, should be modified as shown in Figure 3.21. This modification weakens the piston but, on the positive side, little end lubrication and cooling is very much improved.

Thus far we have only discussed working with the ports provided by the manufacturer, but on older engines extra transfer ports can often be added. Here there are two approaches which we can take, depending on whether we want a small increase in performance and good piston cooling, or a larger power rise without the benefit of improved piston cooling.

We will deal with the cool piston approach first, which can be applied to some engines regardless of the type of induction system employed. I first saw porting like that illustrated in Figure 3.22 on the old 250 Bultaco Pursang and Matador. As you can see, two boost ports are machined the depth of the cylinder liner about 7–9mm wide, on either side of the inlet port. These ports are fed through two holes in the piston. The flow of mixture past the little end and under the piston crown does much to reduce their temperatures. Desert racing engines in particular benefit from this type of porting. There isn't a huge increase in power, but usually a couple of horsepower will be picked up at the top end of the power band.

The next type of boost porting also improves little end lubrication and piston cooling (Figure 3.23). It is intended for piston-ported engines which have a lot of cylinder wall height between the top of the inlet port and the piston crown at BDC. Two boost ports are machined into the cylinder, generally with a 13mm cutter tilted at

Figure 3.21 Rotax piston modification to enhance transfer flow in rear boost port also improves little end lubrication and cooling.

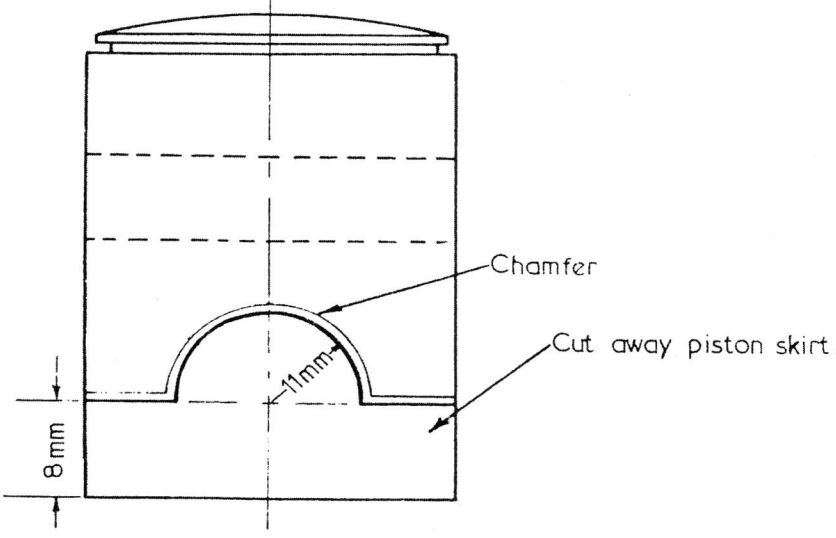

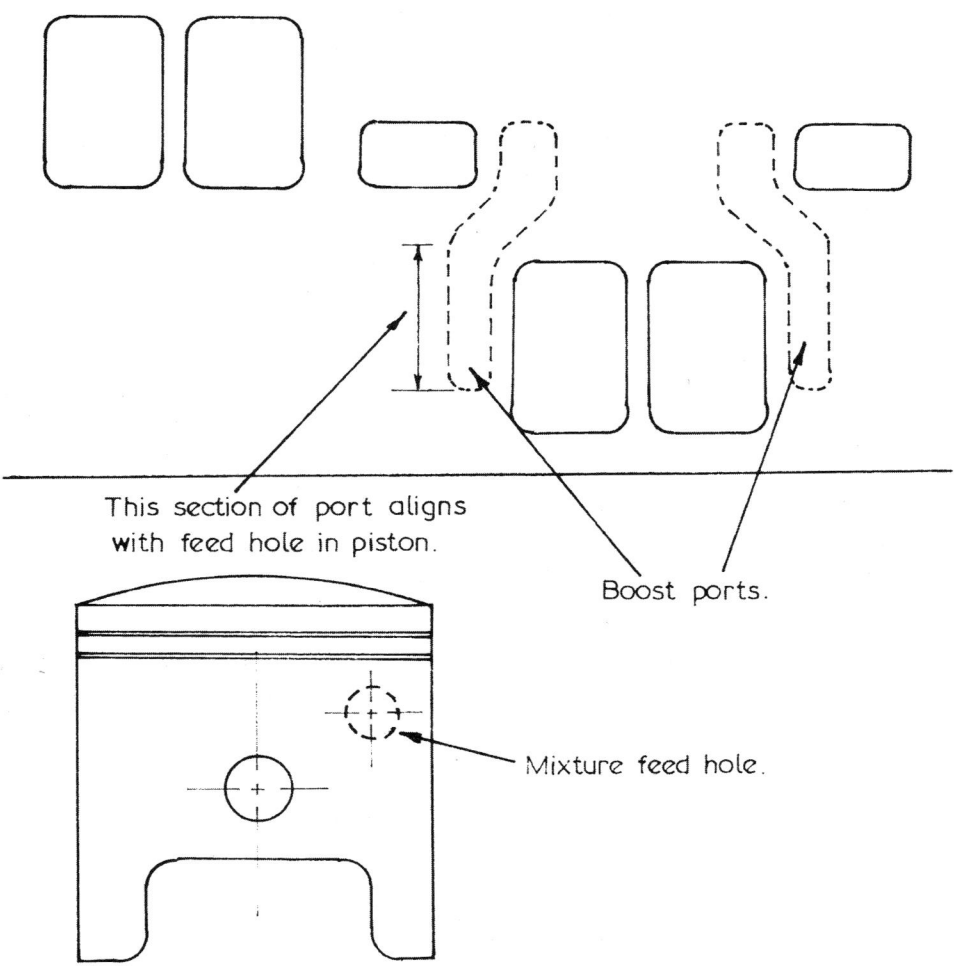

Figure 3.22 The style of boost porting found in the old Bultaco Pursang 250 also aids piston cooling.

25°. Ensure that the boost ports are at least 1.5mm above the inlet port, to ensure an effective seal.

The third type of boost porting shouldn't really be called boost porting (Figure 3.24). It doesn't do anything to increase hp output, but it will extend piston and little end life in desert bikes. I call it 'last resort' porting. Two 9mm wide slots are machined the depth of the cylinder liner to join with the main transfers. Holes in the piston feed these ports as in the first example.

The final type of boost porting can only be used with port type reed valve induction. (Figure 3.25). When the inlet port is bridged, two ports are milled with a 13mm cutter tilted at 25 to 35°. If the cylinder had a single inlet port, overlapping cuts would be made to form a single port of about 18 to 20mm width.

Porting and cylinder scavenging

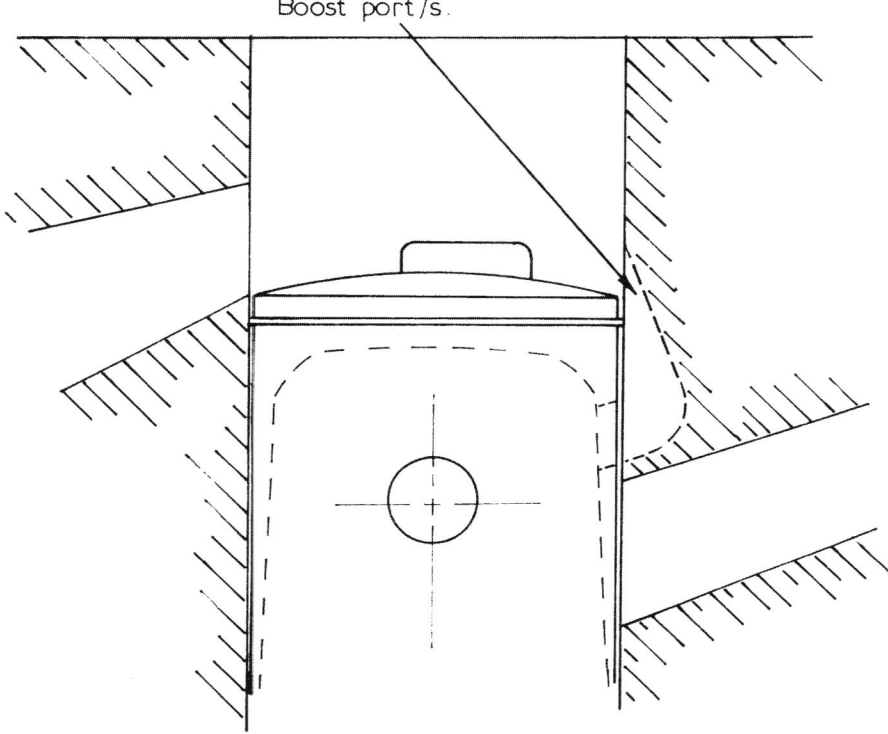

Figure 3.23 Some piston ported engines may utilise a pair of boost ports above the inlet port.

Figure 3.24 This type of boost porting aids piston cooling.

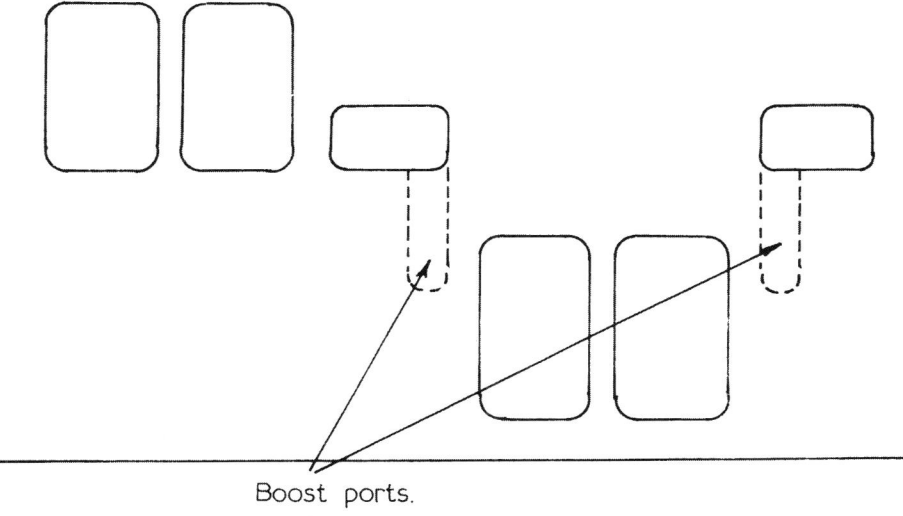

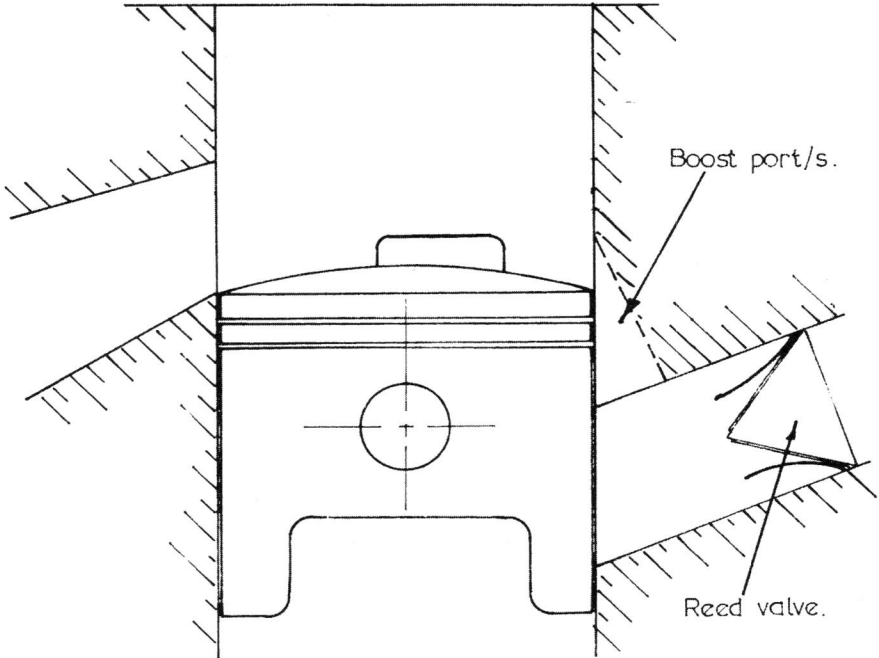

Figure 3.25 Boost porting for port type reed valve engines.

When there is sufficient cylinder wall space available, two types of boost porting may be employed together. The porting shown in Figure 3.22 can often be combined with the arrangements shown in either Figure 3.23 or Figure 3.25. The resulting increase in transfer area improves transfer flow and it reduces the velocity at which the fuel charge enters the cylinder. This minimises charge loss out of the exhaust and improves cylinder scavenging.

Today, even in road racing, piston controlled induction systems have fallen from favour; but, as it is the most basic two-stroke inlet arrangement, we will consider it before reed valve and rotary disc valve systems. In this way you will better appreciate why these other designs have been developed and what their respective advantages and disadvantages are.

Piston controlled inlet ports have the advantage of simplicity, but they are handicapped to some extent due to the port opening and closing points being symmetrically disposed before and after TDC. As the piston rises in the cylinder, the inlet port opens, usually at around 70° before TDC in low speed engines, and 100° before TDC in high speed engines. The rising piston creates a depression in the crankcase, thus air rushes down the inlet tract to fill the crankcase. However, at TDC the port is still open so, as the piston descends, fuel/air mixture will be pushed out of the crankcase through the open inlet port. Fortunately, reverse flow occurs only after the piston has travelled about 50° past TDC at engine speeds around 4,000rpm. Therefore, if the inlet port closes at 70° after TDC, only a small amount of fuel charge will be lost. At higher engine speeds there won't be any loss of mixture, as the combined force of pulse waves and the inertia of the high velocity mixture is stronger

Porting and cylinder scavenging

than the pressure created in the crankcase by the descending piston. For this reason we can employ longer inlet durations in high speed engines, but at lower rpm they suffer from such a bad dose of the blubbers that they will hardly run.

The poor low speed running is partly due to not enough fuel/air mixture being available in the crankcase to adequately fill the cylinder, but there is another reason. The low rpm blubbers and stumbles are basically due to flooding. When the mixture is pushed out of the crankcase and up the inlet tract, it eventually passes through the carburettor. On its way through it picks up another load of fuel, then when the inlet port again opens the fuel/air mixture reverses and travels back through the carburettor, collecting yet another load of fuel. The rich mixture which results burns slowly and wets the spark plug.

The inlet durations set out in Table 3.3 will give good power at the speeds indicated. The shorter duration will improve mid-range pulling power and the longer duration for each speed will enable the engine to produce more power at rpm in excess of maximum hp revs. Motocross and enduro engines such as the RM and PE series Suzukis, with crankcase power reed valves, would normally want inlet durations 15–20° and 25–30° shorter respectively. When power reed engines are used for flat track and road racing, the inlet open period is reduced 10–15°, as mid-range power is not so important.

Table 3.3 **Inlet port duration**

rpm	Inlet duration (°)
7,000	150–155
8,000	155–160
9,500	165–170
11,000	185–190
12,000	195–200

The inlet duration is calculated using the formula:

$$D = \left(\cos \frac{P^2 + R^2 - L^2}{2 \times P \times R}\right) \times 2$$

where $P = R + L + H + C - F$, R = stroke divided by 2 in mm, L = con rod length centre to centre in mm (usually the stroke multiplied by 2), C = deck clearance in mm (ie the distance the piston is below the top of the barrel at TDC), H = piston height in mm (ie the length of the piston on the inlet side), and F = inlet floor depth (ie the distance from the top of the barrel to the bottom of the inlet port).

For example, the inlet open period of the Yamaha KT-100S kart engine (Figure 3.26) is as follows:

$R = 23$mm, $L = 100$mm, $C = 0.2$mm, $H = 56$mm, and $F = 77$mm.

$$P = R + L + H + C - F$$
$$= 102.2$$

$$D = \left(\cos \frac{P^2 + R^2 - L^2}{2 \times P \times R}\right) \times 2$$

$$= \left(\cos \frac{102.2^2 + 23^2 - 100^2}{2 \times 102.2 \times 23}\right) \times 2$$

$$= \left(\cos \frac{973.84}{4701.2}\right) \times 2$$

$$= \cos .20715 \times 2$$

$$= 78 \times 2$$

$$= 156°$$

Because of the bad effect long inlet periods have on mid-range power, it is always preferable to first enlarge the inlet port and see if that change gives the required improvement in high rpm power. It is impossible to say how wide an inlet port can be, as cylinder designs vary so much. However, I will say that if the port has a nice concave floor like that shown in Figure 3.26, even cylinders with very weak lower cylinder walls will be reliable with a port 0.65 the bore size, whilst cylinders with the lower wall well supported will accept port widths up to 0.75 the bore diameter. If the inlet port is bridged, the port width can be up to 0.85 the bore size.

The piston bears quite heavily against the inlet side of the cylinder, so always increase the width by no more than 2mm initially and progress slowly from there. Before you widen the port, check to see that the piston skirt is wide enough to cover and seal the port window. There must be 2mm down each side of the inlet port against which the piston will effect a seal. If the rings run into the port at BDC, you will have to ensure that you do not increase the width so much that the ends of the ring become exposed. However, if you decide to run just the top ring, and it is the second ring which is running into the inlet tract, you won't have to worry about this.

Beside reducing frictional losses and bore wear, discarding the second ring can also have another benefit. With the second ring out of the way it is possible, in many instances, to increase the inlet port height. At times this won't work without also increasing the port timing, as the piston skirt will block the top of the port at TDC, unless it is shortened. Actually, the first check that you should make before lowering the inlet port to increase the port open period is to see that the lower edge of the piston skirt does not protrude into the top of the port with the crank rotated to TDC. When the skirt is shortened, cut off just the inlet side and be sure to put a good chamfer on the skirt so that it encourages lubricant to stay on the cylinder wall.

A lot of tuners lengthen the inlet timing just by shortening the piston. Sometimes there is no alternative, as the cylinder may be too weak to stand having metal removed, but, generally, skirt cutting is the easy way out. Even though cutting 3mm off the skirt will increase the inlet duration to the same figure as lowering the inlet floor by 3mm, you will find that maximum hp will not be as high and the engine will not rev as far past maximum hp revs. The simple truth is that the port area, as well as the duration, must be increased to flow the amount of air necessary to improve the

Porting and cylinder scavenging

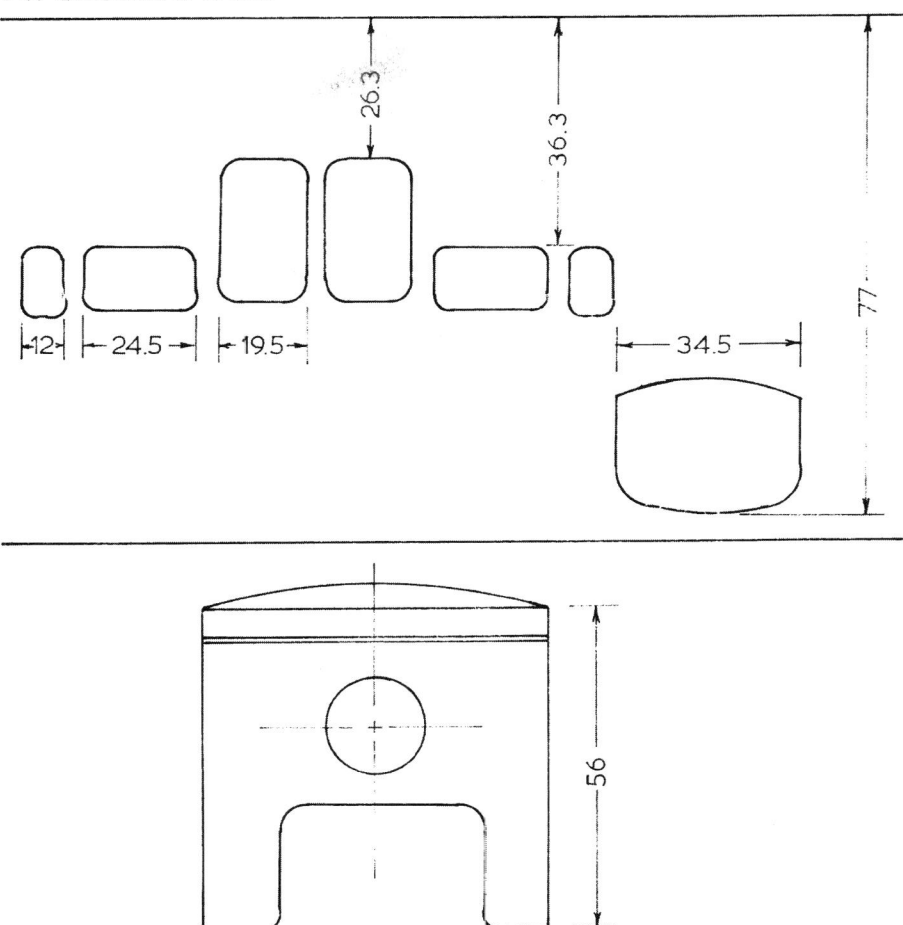

Figure 3.26 Yamaha KT-100S porting and piston dimensions.

power output. I have found, as a general rule, that the piston skirt will have to be shortened by 4mm to give the same high speed power characteristics as obtained by lowering the port 3mm. However, mid-range power is not as good, due to increased blow-back caused by the longer duration. For maximum power, the inlet port area should be about 10 to 15 per cent larger than the area of the carburettor bore.

When the inlet floor is lowered, the full length of the port floor right back to the carburettor will have to be reworked. (Figure 3.27). If this isn't done, the port will not flow any better than the standard port. In fact, an abrupt increase in area close to the port window will cause eddying and actually decrease flow.

To encourage air flow in the inlet tract, it must be smooth and free of obstructions. Casting flaws, bumps and hollows must be eliminated by grinding or filing. The carburettor should be matched to the inlet manifold and in turn the port must be matched with the manifold. In the case of road race engines, high speed air flow is very important, so investigate if it is possible to raise the carburettor to

Two-stroke performance tuning

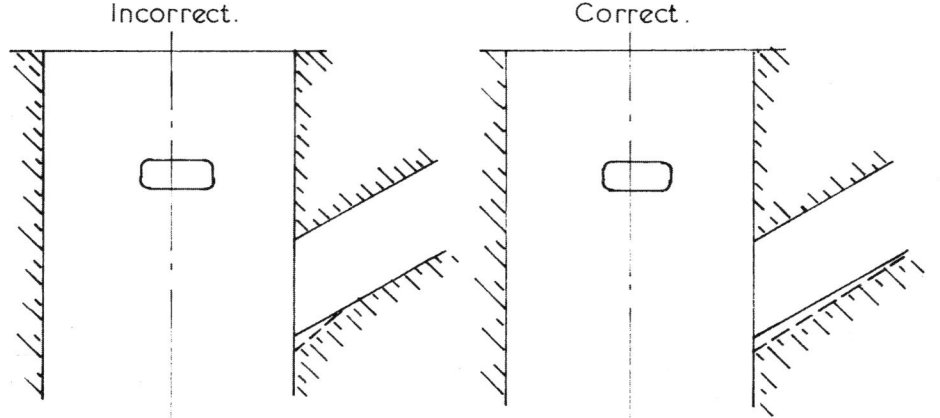

Figure 3.27 The inlet port floor must be correctly lowered to improve inlet flow.

eliminate the kink which usually exists in the inlet tract (Figure 3.28). If there is room to raise the carburettor, this can be done either by inserting a wedge-shape spacer between the inlet manifold and barrel, or by remachining the face of the inlet port at a suitable angle.

Rotary disc valve induction, normally called rotary valve induction, neatly solves the problem of low speed blow-back experienced with piston-port induction, but in doing so introduces a couple of other problems. Because of additional mechanical complications, rotary valve engines cost a good deal more to produce than piston-port or reed valve engines and, unless special measures are taken, they are

Figure 3.28 The kink in the inlet tract can be improved by raising the carburettor.

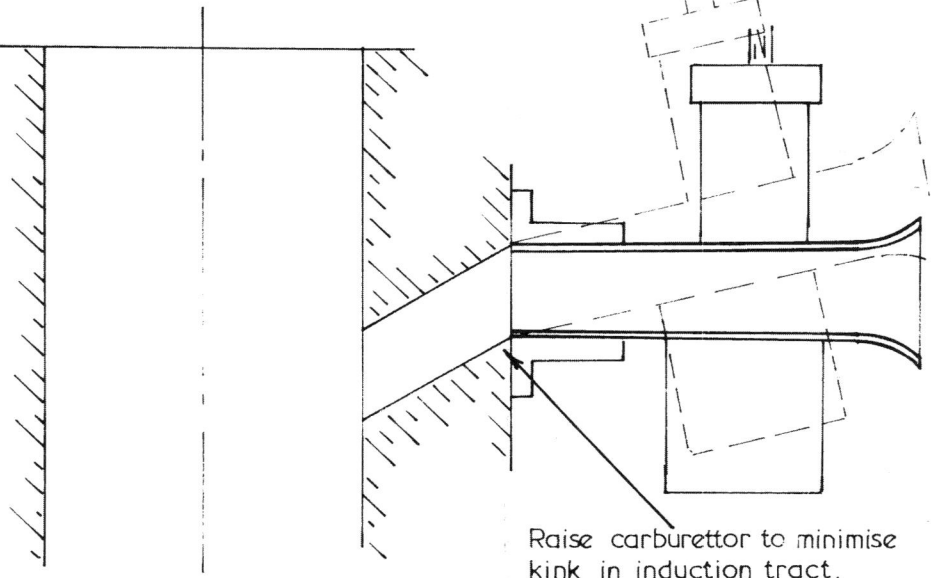

Porting and cylinder scavenging

much wider, physically, than other types of two-strokes. These considerations aside, rotary valve induction is still marginally the best type of inlet system currently available for two-stroke engines. A rotary valve engine will produce more power and a better spread of power. This is principally because it doesn't present the high speed flow resistance which stifles maximum hp output in reed motors, but as reed designs keep on improving the performance gap between rotary and reed is now extremely narrow. Additionally, because rotary valve induction is free of low rpm blowback problems, the induction period can be longer than for engines with piston-controlled inlet ports, so high rpm crankcase filling is superior.

As shown in Figure 3.29, the rotary valve is most commonly driven off the end of the crankshaft, with the induction tract entering the side, rather than the rear of the engine.

However at times, some multi-cylinder road race engines have utilised just one rotary valve disc shared between two cylinders. In a parallel twin for example, with the cylinders firing alternately 180° apart, one disc located transversely at the rear of the two cylinders can control the two inlet ports which are located 180° apart. Obviously with this arrangement the carbs and inlet ports are at the rear of the engine. With either design the cutout in the disc valve and the width of the inlet port actually determine when the induction period will commence and finish. Because of this, we

Figure 3.29 Rotary valve induction arrangement utilises a thin metal disc driven by the crank to open and close the inlet port.

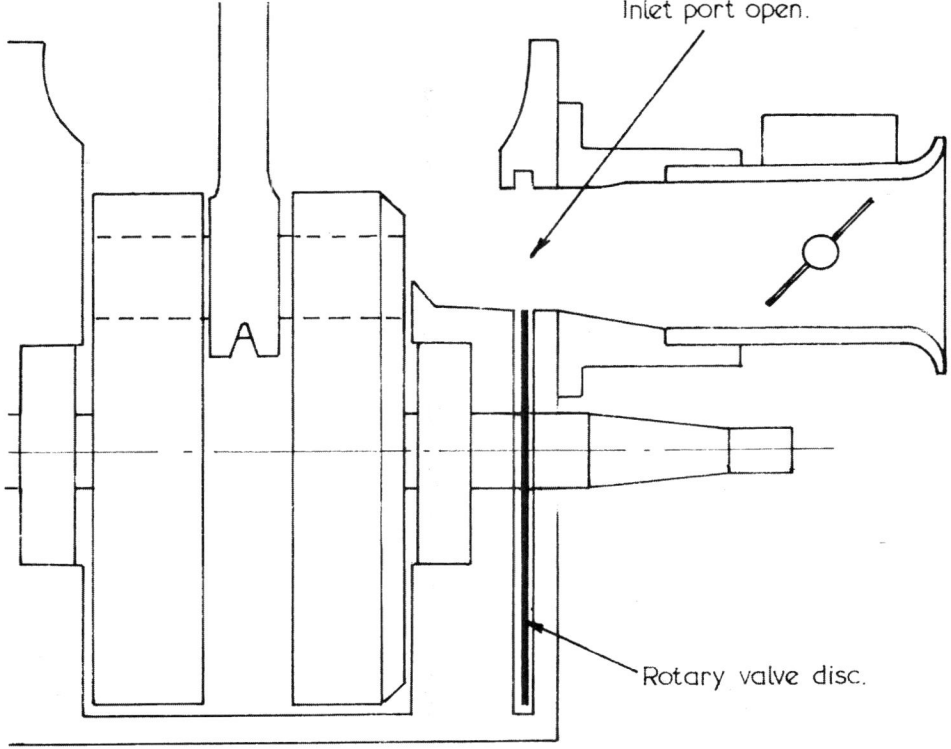

Two-stroke performance tuning

can start the inlet cycle much earlier than in a piston-ported engine and we can also close the inlet port much sooner after TDC.

It is usual, when good low speed power is required, to open the inlet port about 5 to 10° before the transfer port closes (ie 120 to 130° before TDC) and to close the inlet port at about 55 to 60° after TDC. This results in an inlet duration of around 180 to 190°. For more power at the top end of the power curve, the duration is increased to something like 200 to 210°. There will, however, be some loss of low speed power and the engine won't take a fistful of throttle at low revs without stumbling. The increase in duration can be obtained in two ways. Either we can have the rotary valve open a little earlier at 135 to 140° before TDC and close a little later at 65 to 70° after TDC, or we can leave the valve opening point alone and pick up the extra duration by closing the port at 70 to 80° after TDC. The effect on the power curve will be quite different, even though the inlet open period is the same. Opening the valve at, say, 140° before TDC and closing it at 65° after TDC (205° duration) will tend to lift maximum power a little, but the main effect will be to considerably increase power in the upper mid-range. Leaving the opening point at 125° before TDC and shifting the moment of closing to 80° after TDC (205° duration) will reduce mid-range power due to increased blow-back, but there will be a good power rise right at the top end of the power curve (Figure 3.30).

In high rpm road racing engines, where mid-range power is of only minor

Transverse rotary valve on an Ital Sistem SV21 kart engine improves flow into the crankcase and provides equalised flow into the right and left side transfer ports.

Porting and cylinder scavenging

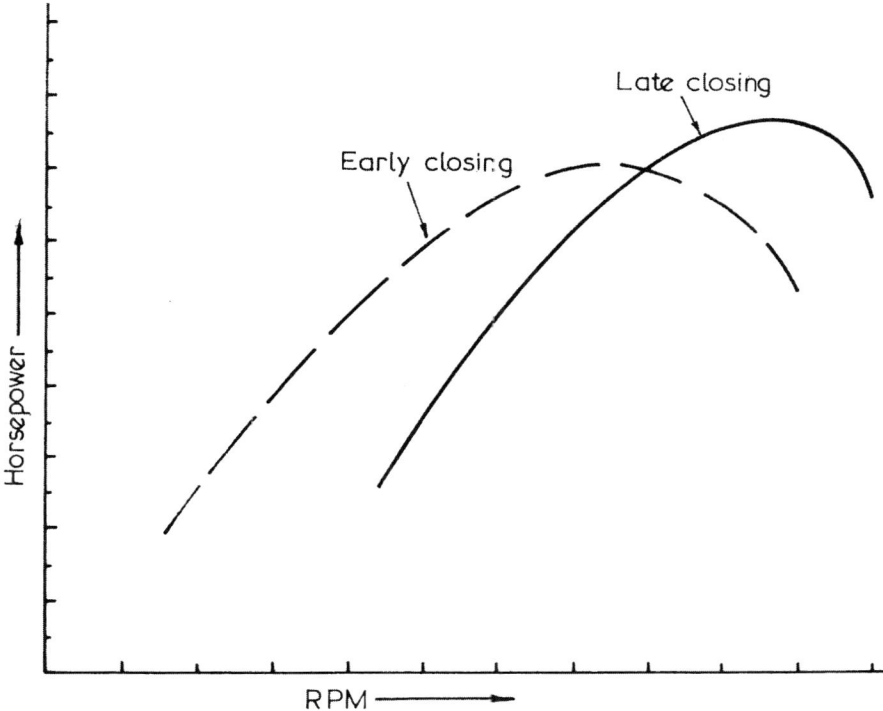

Figure 3.30 Effect on power curve of changing rotary valve closing angle.

concern, the inlet duration is increased to about 215 to 230°. The rotary valve will open at 135 to 150° before TDC and close at 80 to 90° after TDC. The main concern here is that the inlet duration is of sufficient length to ensure complete crankcase filling at the rpm where maximum horsepower is desired. If we want peak power at 14,000rpm then the duration will be around 230°, but if we want peak power at 12,500rpm the duration will be close to 220°.

Table 3.4 sets out the rotary valve timing for a number of go-kart and bike engines. All of the 100cc kart engines have fixed gearing.

Before you set about altering the valve timing, check to see that the inlet port is of the correct shape and that the valve cover perfectly matches the inlet port in the crankcase. Any obstruction here will disrupt air flow. You will find in many engines that the port in the valve cover does not align with the crankcase port. Grinding the port in the valve cover or the crankcase will affect the inlet timing. In some engines the inlet port opens and closes slowly because the sides of the port are the wrong shape. The port illustrated in Figure 3.31 should be reshaped as shown. The port area is increased and it will open and close more abruptly, generating beneficial pulse waves in the inlet tract.

The actual side profile of the inlet port is very poor in many rotary valve engines. In Figure 3.29 you can see a common mistake made by manufacturers which is very disruptive to air flow. The mixture rushes straight down the inlet port and proceeds to bang right into the crankwheel, losing a good deal of inertia. Some of the mixture will

Table 3.4 Rotary valve timing

Engine Type	Capacity (cc)	Valve timing	Transfer closing
Arisco C-75 kart	100	155/43	124
BM K96-3 kart	100	115/60	123
BM FC-52 kart	100	115/60	120
Can-Am MX-6 bike	125	140/85	113
Can-Am MX-3 bike	250	140/85	125
Can-Am MX-6 bike	250	140/85	113
Can-Am Qualifier bike	175	137/75	113
Can-Am Qualifier bike	250	137/75	116
Can-Am Qualifier bike	350	137/75	116
DAP T81 kart	100	132/58	117
DAP-JM T71 kart	100	120/55	113.5
Komet K78 kart	100	132/60	118
Komet K78 TT kart	100	132/60	117
Morbidelli 125 bike	2 x 62	150/79	109
MZ 125 bike	125	135/70	112
Rotax 124 LC kart	125	120/87	113
Rotax 250 LC bike/kart	2 x 125	136/80	112.5
Sirio ST50 kart	100	134/75	116.5
Sirio ST504 kart	100	135/65	120
Sirio ST52 kart	100	134/75	117.3
Zip ZED1 kart	100	140/66	121.5

Note – The first valve timing figure refers to the opening point in degrees before TDC and the second figure is the closing point after TDC. The transfer closing figure refers to the closing point in degrees before TDC.

slowly rise up and around the crankwheel into the crankcase and a little of the air will form into a turbulent eddy current. When this kind of situation exists, air flow into the engine is severely restricted at high rpm. To increase air flow, and consequently high speed hp, there are two options open. Either the inlet port open period can be increased, which will reduce mid-range power, or we can reprofile the inlet port and increase air flow in this way. Top end power will improve and often mid-range power rises too.

What we must do is change the shape of the inlet port, so as to encourage the mixture to turn up and over the crankwheel. In effect, the edge of the crankwheel has to become a part of the inlet tract floor, instead of a barrier at the end of a hole. In Figure 3.32 you can see the shape we have to aim for. The floor of the port is built up to blend into the crankwheel, and the lip formed by the port roof and the transfer cutout is radiused. The port can be built up using Devcon F aluminium epoxy. It contains 80 per cent aluminium, is heat resistant to 250°F and is not attacked by petrol, methanol, oil or toluol.

Ideally, manufacturers should turn to the use of larger disc valves so that the inlet port floor could be in line with the top of the crankwheel. In this situation the fuel/air charge would flow straight into the crankcase unimpeded. In addition to this, there is another advantage in the use of large diameter discs, which is the primary reason for

Porting and cylinder scavenging

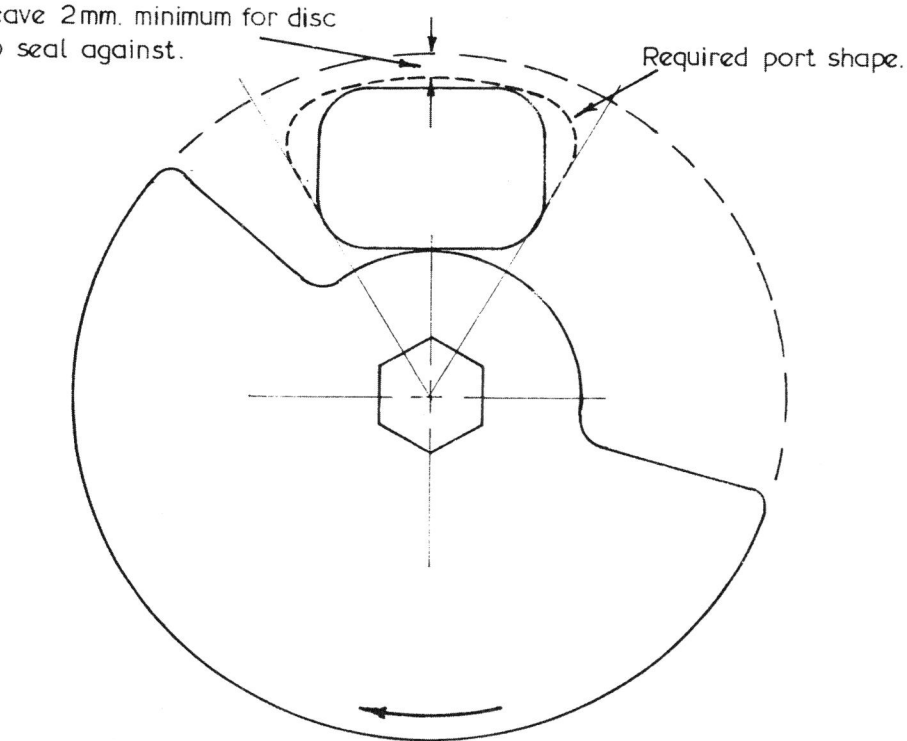

Figure 3.31 Changing the inlet port shape as shown improves flow without affecting rotary valve timing.

their existence on the works Minarelli and Morbidelli Grand Prix racers. When the rotary disc diameter is increased, there is a corresponding decrease in the duration angle actually taken up by the inlet port itself, assuming the inlet port width is not altered. This allows for a longer duration angle with the large disc without increasing the actual inlet port open period. The power output then goes up, because the inlet port is fully open for a greater number of degrees, without being partially closed by the disc, or as the engine sees it for more time, so more air flows into the crankcase. Conversely, if the engine is already producing ample power at the top end of the rpm range, then the inlet open period can be reduced with the large diameter disc. In this way peak power will remain the same, but the mid-range to upper mid-range will rise appreciably.

Trying to get the sense of this is quite hard just using words, so I will help you to reason it out with an example and an illustration (Figure 3.33). As you can see, both engines have an inlet port 34mm wide and an inlet duration of 200°. The engine with the small 100mm diameter disc (engine A) has an inlet port and rotary disc which takes up 40° and 160° respectively of the 200° inlet cycle. On the other hand the inlet port and disc occupy 27° and 173° respectively when a 150mm disc (engine B) is used. This means that the inlet port is not obstructed in any way by the rotary valve for 120° (200 − [2 x 40] = 120°) in the case of engine A and for 146° for engine B

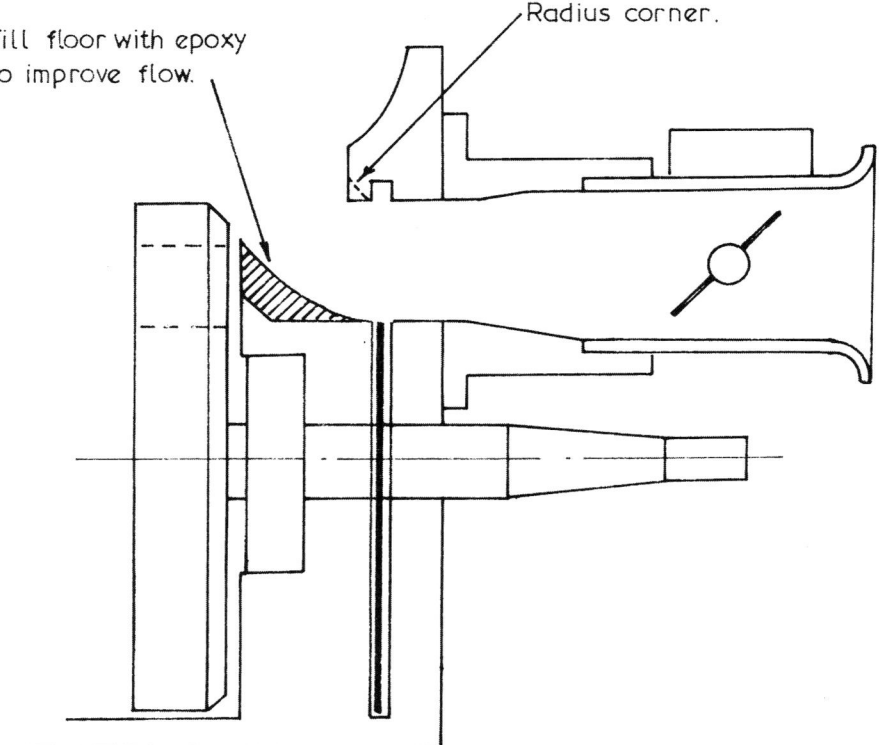

Figure 3.32 The rotary valve inlet port can be modified to enhance inlet flow.

(200 − [2 x 27] = 146°). In other words the inlet port will be fully open for 26° or 22 per cent longer.

Before you modify the rotary valve to change either the inlet opening or closing points, it is a good idea to find out exactly what the standard timing is for your engine and then compare that with the manufacturer's specifications. At times there can be variations, because a keyway or master spline is cut slightly out or, in some engines, it is possible that the disc valve has been fitted one tooth out on the drive gear either during manufacture or when the engine has been repaired.

To check the valve timing you will require a 360° timing disc or, if you can't obtain one of these in your area, buy a large 200mm diameter protractor and drill a suitable size hole exactly in the centre so that it fits the end of the crankshaft. You will also need a good solid pointer which can be fixed under a stud in the crankcase.

Using a dial gauge find TDC and rotate the timing disc to align the zero mark with the pointer. Lock the disc in place on the crank and again check that the pointer points to zero when the dial gauge indicates TDC. Then simply rotate the crankshaft in the normal direction of rotation, noting at what angle the inlet port opens and closes. When making this check it is necessary to shine a light down the inlet port so it can be clearly seen when the valve opens and closes.

Instead of using a degree wheel to physically determine the valve timing, it may be calculated mathematically using this formula:

$$A = \text{Cos}\left(\frac{T^2 + R^2 - L^2}{2 \times R \times T}\right)$$

where T = R + L + C – E, R = stroke divided by 2 in mm, L = con rod length in mm centre to centre (usually the stroke multiplied by 2), C = deck clearance in mm (ie the distance the piston is below the top of the barrel at TDC), and E = distance from the top of the barrel to the piston at the instant of inlet opening or closing.

For example the inlet timing of the Rotax 124LC is as follows:

R = 27mm, L = 110mm, C = 1.8mm, E = 44.7mm (valve opening) and E = 31.9mm (valve closing), T = 27 + 110 + 1.8 – 44.7 (valve opening) = 94.1, and T = 27 + 110 + 1.8 – 31.9 (valve closing) = 106.9.

Figure 3.33 Comparison of rotary valve discs illustrating how a larger disc diameter increases inlet port flow duration.

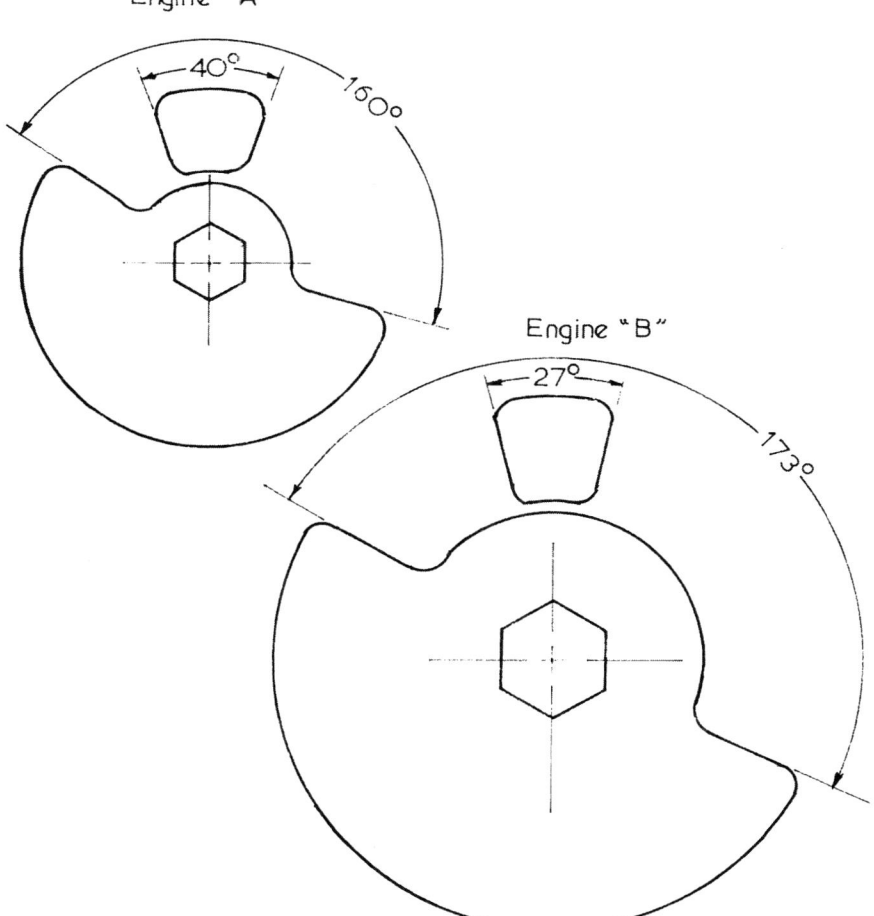

$$\text{valve opening A} = \text{Cos}\left(\frac{T^2 + R^2 - L^2}{2 \times R \times T}\right)$$

$$= \text{Cos}\left(\frac{94.1^2 + 27^2 - 110^2}{2 \times 27 \times 94.1}\right)$$

$$= \text{Cos } .49518$$

$$= 119.7° \text{ before TDC}$$

$$\text{valve opening A} = \text{Cos}\left(\frac{106.9^2 + 27^2 - 110^2}{2 \times 27 \times 106.9}\right)$$

$$= \text{Cos } .00981$$

$$= 89.4° \text{ after TDC}$$

When you have the timing figures for your engine, check them against the manufacturer's figures. If the makers state that the valve opens 130° before TDC and closes 65° after TDC and yours opens 132° before TDC and closes 63° after TDC, then you know that the timing has been advanced 2° due to manufacturing errors. This will have the effect of slightly increasing mid-range power at the expense of a reduction at the top end. Obviously, if you are after more top end power, the first move should be to machine the disc to move the closing angle to 65° after TDC. If, after this, you want still more power at high engine speeds, move the closing point 2° at a time, but stop once you reach about 76°. Then go back and add 4° to the opening angle to bring it up to 136° and see how the engine responds. If the engine reacts favourably, but you are after still more power, move the opening angle another 4° to 140°. After this, you can go back to delaying the angle of closing in increments of 2° at a time.

If, after checking your timing against the maker's figures, you find that the disc valve has been retarded by, say, 6° and the bike is very 'pipey', coming on to the power with a sudden rush, then it is probable that the power curve can be improved by getting the inlet timing back to what the manufacturer originally intended. (A disc retarded by 6° would be indicated by manufacturer's figures of, say, 130°/65° and your figures being 124°/71°.) One way to cure a problem such as this, which fortunately occurs infrequently, is to relocate the rotary valve cover, moving it around 6° in the opposite direction to crankshaft rotation. To calculate how far the cover has to be rotated, first measure across the valve cover from the centre of one retaining screw to the screw opposite it. Say this dimension is 145mm. In this instance the cover will have to be rotated by 7.6mm which is calculated using this formula:

$$X = \frac{D \times \pi \times A}{360}$$

where D = diameter across cover retaining screws, A = angle of timing error.

In this example D = 145mm and A = 6°, therefore

Porting and cylinder scavenging

$$X = \frac{145 \times \pi \times 6}{360}$$

$$= 7.6mm$$

The idea is to then drill a new set of fixing holes in the valve cover 7.6mm from the original hole centres. Having done this, refit the cover and check when the rotary valve opens and closes the port in the valve cover. Actually, the port timing figures should always be taken off the valve cover, never the crankcase port. If the timing is correct, it is advantageous to take the machine for a test run before you spend a lot of time matching the ports. Naturally, the engine will be down on top end power, but it is its 'pipeyness' which you are checking, not top end power. If the results are satisfactory then match the ports, filling one side with Devcon F as shown in Figure 3.34 and grinding the other side out.

It is not always necessary to relocate the rotary valve cover to correct valve timing errors. Some engines have the rotary valve driven by a hub which is located on the crankshaft by a key. In the case of Rotax motors the hub is cut with 22 external gear teeth so moving the rotary valve one tooth on the hub will alter the timing by 16.4° (360° ÷ 22 = 16.4°) which isn't of much use to us. However, by machining a new keyway in the hub and by moving the valve around the appropriate number of teeth, timing errors can be corrected.

For example, machining a new keyway 90° around from the original and moving the valve around by five or six teeth (depending on whether the timing is advanced or retarded) will correct an 8° timing error. That is quite an easy one to work out, but what if the timing is retarded by 6°? To calculate how many degrees around the new

Figure 3.34 Match the inlet port after relocating the rotary valve cover.

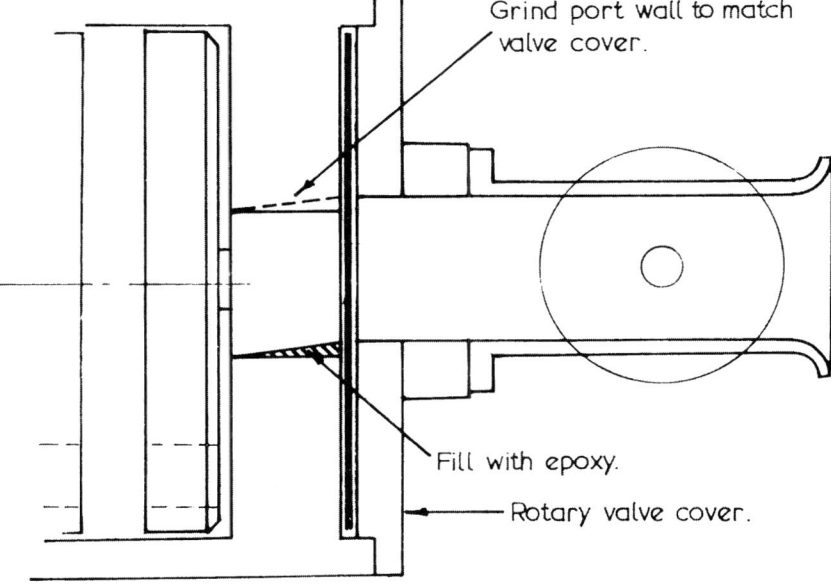

keyway has to be cut, add the angle of error to 90° and subtract 8°, which equals 88° (90° + 6° − 8° = 88°). The new keyway will therefore have to be cut 88° around from the original, but will it be to the left (counter-clockwise) or right (clockwise) of the original? Since the valve is presently retarded it will have to be advanced to correct the timing error. The engine turns counter-clockwise so the new keyway will have to be machined 88° to the left of the original keyway, which will advance the hub by 88°. Retarding the rotary valve (ie moving it in the same direction as crank rotation) by five teeth will retard the timing by 82° (16.36° x 5 = 82°), consequently the timing will finish up being advanced by 6° (88° − 82°= 6°) from what the original timing figure was, which should be the figure specified by the manufacturer. You must be very careful that this type of work is carried out by only a top class machinist as it is exceedingly difficult to work to such fine tolerances in a bore as small as that found in the hub of a rotary valve.

To change the opening and closing points of the rotary valve disc proper is a little difficult, unless you make a special setting-up template of white cardboard or cartridge paper. In the centre of the template draw a cross (+) with lines about 150mm long intersecting at exactly 90°. Using the cross as the centre, draw a circle exactly the same diameter as the rotary valve disc. Carefully lay the rotary valve on the template within the confines of the circle which you just drew and using a sharp pencil draw the disc cut-out. (Be sure that the outside face of the disc is facing up.) Now draw another circle, using the cross as the centre, about 50mm larger in diameter than the rotary valve. After this, set a large (100mm or larger) protractor exactly on centre and note the angle of the disc's opening and closing points. Now carefully mark in the new opening or closing angle which you want. Draw a line from the centre through this point right out to the edge of the large circle. Lay the disc back on to the template, the correct way up, being careful to line it up within the boundary formed by the small circle and the original opening and closing lines. Now scribe a line across the disc exactly in line with the line you drew to show the new opening or closing angle. With that done, the disc can be modified to change the inlet timing.

The clearance between the valve cover and the rotary disc is very important. If the clearance is too tight, power is lost due to friction and, if the clearance is excessive, low speed and mid-range power is lost due to fuel/air charge leakage out past the valve. Unless otherwise specified, the clearance should normally be between 0.25mm and 0.35mm. If it is less than 0.25mm the face of the valve cover will have to be machined the appropriate amount. On the other hand, if the clearance is greater than 0.35mm the mating surface of the cover will require machining.

Reed valve induction was first introduced to the motorcycle world in 1972 when Yamaha released their range of 'Torque Induction' bikes (Figure 3.25). The reed valve functions as a simple check valve and prevents blow-back in the inlet tract. Therefore, a reed engine can be lugged down to very low rpm (depending on the exhaust timing), as the air flowing down the inlet tract is trapped once it passes the reed valve. Low speed cylinder filling improves and, because the air passes through the carburettor just once, the fuel/air ratio remains correct. This results in good low speed combustion.

Reed valve induction, however, is not entirely free of problems. Until very recently, the stiffness of the reed petals was severely compromised. To ensure good low speed crankcase filling, the reed petals must be thin and flexible so that they open easily and do not unduly restrict air flow. On the other hand the petals must be thick and stiff, otherwise crankcase filling at high speeds is not good. At high speeds thin, flexible petals flutter, allowing reverse flow out of the crankcase. They tend to close and then rebound from their seats due to inertia and/or resonance in the induction tract.

A dual reed assembly patented by Eyvind Boyesen reduces this compromise a considerable amount. The presence of a reed cage and petals in the induction tract still reduces high speed air flow below that possible with rotary valve or piston ported induction, but the difference is not so great as before. The Boyesen assembly comprises a thin 0.25mm reed, riding on top of a thicker 0.7mm reed. The thin reed opens easily under a low pressure drop and the thicker one takes over at higher rpm. This gives the benefits of good low speed air flow, as well as an absence of high speed petal flutter.

The next development from Boyesen was the introduction of the Pro Series petals. These petals combine carbon fibre and fibreglass to give a light, but relatively stiff petal. The light weight allows the petal to open and close more rapidly, while the stiffness imparted by the carbon fibre and fibreglass ensures that the petals do not bounce when they close. Additionally this stiffness means that on closing the petals do not bend and collapse into the reed block openings.

Over the years a lot has been said about the benefits of reed valve induction, but it seems that very few people realise the very high power outputs now being produced by motocross and enduro engines are not a direct result of reed valve induction. A lot seem to think that, because the piston is cut away, or has windows in it, allowing in some engines up to 360° inlet open period, this automatically results in a high power output. I can assure you that this is not so. In itself a reed valve improves low speed and mid-range power only, by preventing blow-back.

When Yamaha were developing their early 125 motocross reed valve engine they tested a whole range of inlet open periods and found that once past about a 230° inlet duration, there was no change in hp anywhere in the power range. With the piston skirt allowing an inlet duration of 280° the reed petals commenced lifting at around 120° before TDC and closed at 40° after TDC, a duration of 160° at lower engine speeds. Then, at close to maximum engine rpm, the reeds stayed open for 200°, with opening commencing at 80° before TDC and continuing to 120° after TDC (Figure 3.35). When the piston skirt was opened right up to allow a continuous 360° inlet open period the reed petal opening and closing points stayed pretty much the same but interestingly maximum reed lift was reduced to a little under 8.5mm; about 0.5mm less than with a 280° inlet period.

To give you some additional proof of why I say that in itself a reed valve primarily only improves low speed and mid-range power, consider the effect of adding a reed valve to an old 250 Bultaco Matador enduro bike. In standard tune the engine had exhaust, transfer and inlet open periods of 170°, 126° and 150° respectively. As shown in Table 3.5, the engine has a gentle power curve. It pulls very well at low speed and produces a maximum of 25.8hp at 7,000rpm. In Test 2 a reed valve was added and four 16mm holes were drilled in the piston skirt, increasing the inlet timing to 360°. As you can see, there has been very little increase in power at the top end in spite of the fact that a 34mm Bing carburettor was fitted to replace the stock 32mm Amal. Note, too, that there has been only a marginal decrease in low speed power, due to the reed valve offsetting to some extent the bad effect the larger carburettor would have had at low rpm.

Two-stroke performance tuning

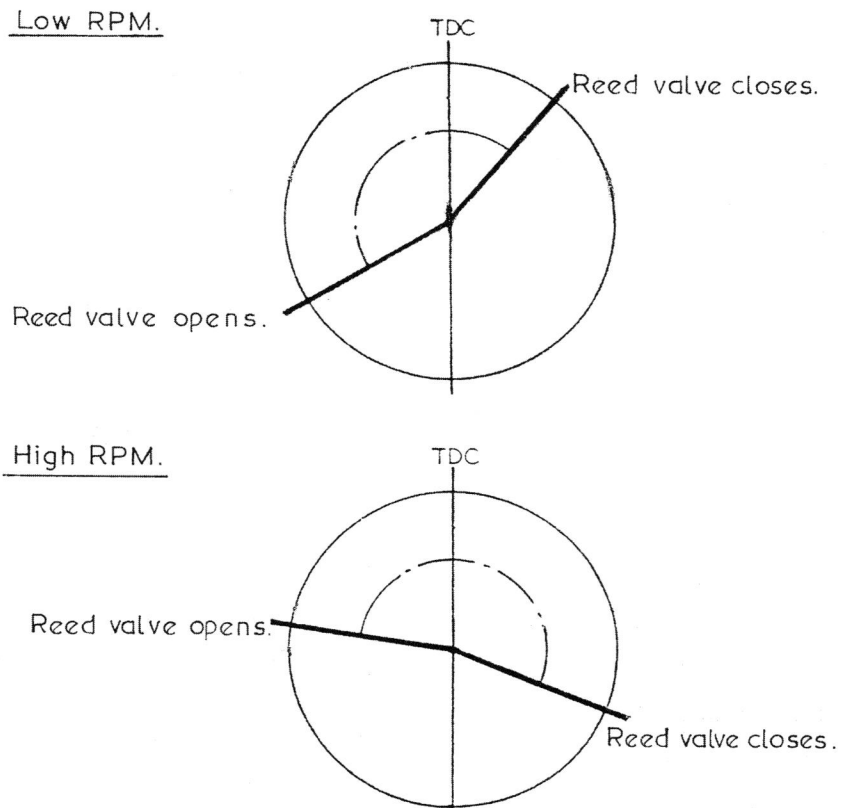

Figure 3.35 Reed valve opening and closing angles vary with changes in engine speed.

In Test 3, however, you can see that power right through the range has risen by an average of 1.5hp below 5,500rpm, and by up to 3.1hp between 6,000 and 7,000rpm. What brought about such a sudden power increase? In this test, two boost ports were added in the rear of the cylinder. The ports were cut with a 13mm cutter tilted at 30°. So it was the increase in transfer port area which picked up power significantly, not the addition of a reed valve.

In Test 4, there was an increase in power above 6,500rpm, but a decrease at lower speeds. For this test, a new piston was fitted which had 13mm removed from the bottom of the inlet skirt to give an inlet open period of 200°. This means that the piston exercises control over which direction transfer flow will take. In Test 3, the boost ports are always connected with the crankcase (ie for 360°) but in Test 4 the boost ports are isolated from the crankcase (see Figure 3.25) once the piston skirt drops below the level of the inlet port floor. Thus, any flow through the reed valve will be diverted up through the boost ports once the piston closes the inlet tract off from the crankcase. With this arrangement, low speed power falls away, because the boost ports flow only if the exhaust pulses create a depression low enough to open the reed valve and pull fuel/air mixture up through the boost ports. However, at higher speeds, peak power is increased with this system, because the piston closes off the crankcase

Table 3.5 Effect of reed valve induction

rpm	Test 1 (hp)	Test 2 (hp)	Test 3 (hp)	Test 4 (hp)
3,000	6.8	6.4	8.3	7.9
3,500	7.9	8.1	10.9	10.4
4,000	11.9	11.3	12.1	11.8
4,500	14.2	13.6	14.8	14.6
5,000	16.0	15.6	17.0	16.6
5,500	18.1	18.0	19.7	19.3
6,000	22.6	22.9	26.0	25.7
6,500	23.3	24.9	27.2	27.1
7,000	25.8	26.7	27.8	28.4
7,500	25.6	25.1	26.3	27.6
8,000	23.7	24.8	25.5	26.2
8,500	18.1	20.6	22.1	22.8

Test 1 – standard Bultaco Matador 250cc

Test 2 – reed valve assembly and 34mm Bing carburettor added; piston modified to give 360° inlet open period.

Test 3 – as above, with the addition of two boost ports in rear of cylinder.

Test 4 – as above but with piston modified to give 200° inlet open period, ie 'power ported'.

from the inlet tract, preventing back flow out of the crankcase as the piston descends to BDC. Without the effects of reverse flow to fight against, mixture will continue on flowing through the reed valve and up through the boost ports until cylinder pressure equals pressure in the inlet tract, causing the reed valve to close.

When this latter type of 'power porting' (ie Test 4) is applied to more modern two-stroke engines, there is often little or no loss of low speed power because of larger transfer port areas being employed today. However, on some bikes the power curve can become very peaky, making the bike difficult to ride. This is why you will seldom see this arrangement employed on anything but small displacement motocross bikes and reed valve road race engines. Of course, with many engines there isn't much you can do to convert from the conventional type of boost porting to power porting, unless you can find a suitable piston from another engine which doesn't have windows in the skirt. However some engines have two small passages, instead of piston skirt windows, which connect the inlet tract with the crankcase. If these boost passages are filled with an epoxy such as Devcon F, the inlet tract will be isolated from the crankcase when the piston skirt closes the inlet port allowing the engine to operate as a power ported engine.

In Table 3.6 you can see the effect which such a modification had on an old Honda CR125R equipped with a Mugen air-cooled hot-up kit. As you can see, low speed power has not been affected by blocking up the two small crankcase feed passages. From 7,500rpm up to maximum rpm, there is a steady power increase. Peak power is up 0.9hp, but at higher speeds the power rise is more dramatic. It is up by 2.2hp and 5.1hp at 10,500rpm and 11,000rpm respectively, and at 11,500rpm the engine is still making 15.8hp. I must point out that some of this high speed power increase also comes about due to changes in the transfer open period. When the cylinder was converted to power porting the auxiliary transfers were raised 0.8mm and the boost port was raised 1.2mm. These modifications probably accounted for about 50 per cent of the power increase from 10,500rpm up. In both tests the engine was equipped with a 34mm Mikuni carburettor bored out to 35.3mm and a special expansion chamber was used. Without these additions, power above 10,500rpm would have been suppressed in both tests.

Table 3.6 Effect of power porting

rpm	Standard boost porting		Power porting	
	hp	Torque (lb/ft)	hp	Torque (lb/ft)
3,000	2.5	4.4	2.6	4.5
3,500	3.0	4.5	3.9	5.8
4,000	4.0	5.3	4.5	5.9
4,500	4.8	5.6	5.0	5.8
5,000	5.9	6.2	6.4	6.7
5,500	6.7	6.4	7.1	6.8
6,000	7.7	6.7	7.8	6.8
6,500	8.9	7.2	8.5	6.9
7,000	9.3	7.0	8.9	6.7
7,500	10.4	7.3	11.3	7.9
8,000	14.5	9.5	16.1	10.6
8,500	18.0	11.1	18.3	11.3
9,000	19.0	11.1	20.0	11.7
9,500	21.3	11.8	22.1	12.2
10,000	22.1	11.6	23.0	12.1
10,500	20.8	10.4	23.0	11.5
11,000	16.5	7.9	21.6	10.3
11,500			15.8	7.2

Note – Inlet port open period was 230° with power porting.

In 1976, Suzuki introduced us to a new type of reed valve system with the release of their 'A' series RM motocross bikes. The 'Power Reed Intake System', as it is called by Suzuki was an attempt to combine good features of both reed induction and piston-ported induction (Figure 3.36). With the power reed system, both the reed valve and the action of the piston opening and closing the inlet port controls mixture flow into the crankcase. Even a very potent engine like the little RM 125 had an inlet open period of only about 150°, which is very short when compared with the average 125 piston port motocrosser from that era employing 170° inlet duration. When a short inlet duration is used, there is very little blow-back at low speeds, so throttle response and low speed running is good. As shown in Figure 3.37, the reed valve operates even at low rpm, ensuring good crankcase filling. Then at higher speeds the reed valve stays open until after the piston timed inlet tract has closed, ensuring good high speed hp.

What is the difference in performance between power reed induction and conventional port reed induction? When I first saw the Suzuki system, I was convinced that it would enable much higher power outputs than a conventional reed

Porting and cylinder scavenging

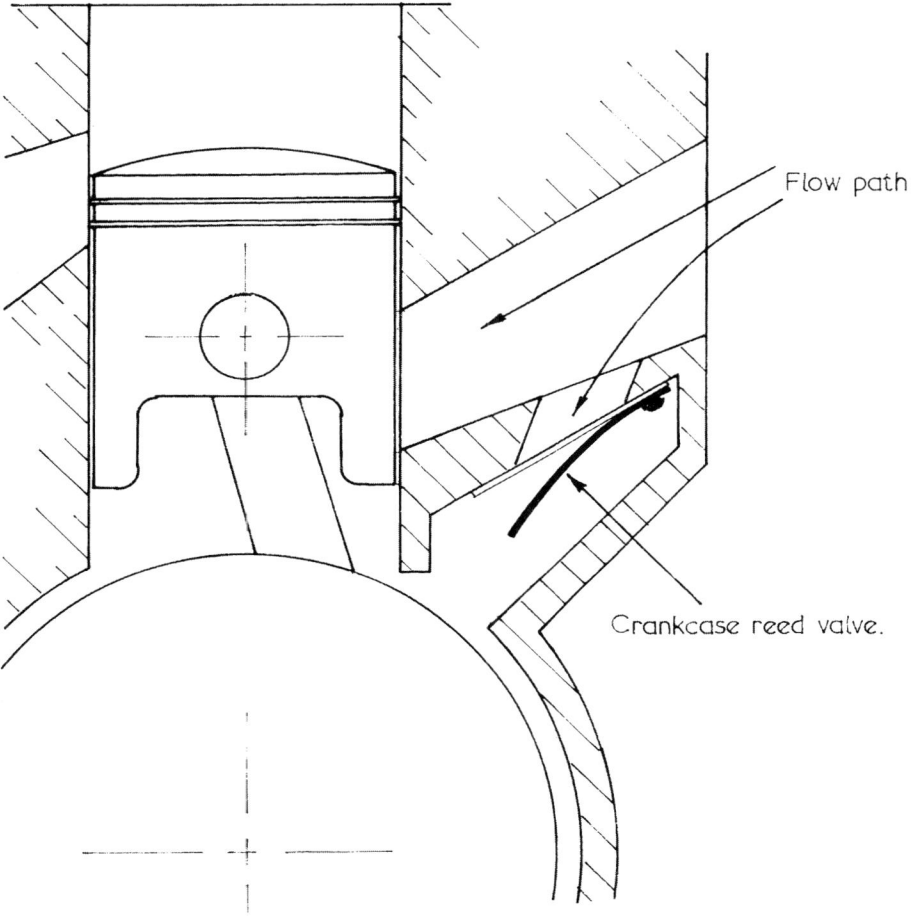

Figure 3.36 Suzuki power reed induction arrangement combines piston ported inlet and crankcase reed valve.

engine, but I was soon proved wrong. Even in road race go-kart applications, where the unobstructed inlet tract of the Suzuki should assist air flow, an engine like the YZ Yamaha with port reed induction would pick up one or two lengths on a Suzuki down the longer straights, or up hills.

When you make a more careful examination of a power reed engine like the Suzuki (KTM and Rotax also use a power reed on some of their engines), you can see why it would not make as much power as a Yamaha or Honda in all-out applications like flat track racing or go-kart road racing. It highlights once again the importance of getting all of the elements correct and not overemphasising a clear improvement in just one area. All I saw when I first compared the two systems was that in the case power reed arrangement the inlet charge had a clear unobstructed shot into the crankcase for a major part of the inlet period. What I failed to see was the clearly inadequate flow area of the small crankcase reed block and petals, and the very poor back transfer ports located above the inlet port. With conventional port reed engines

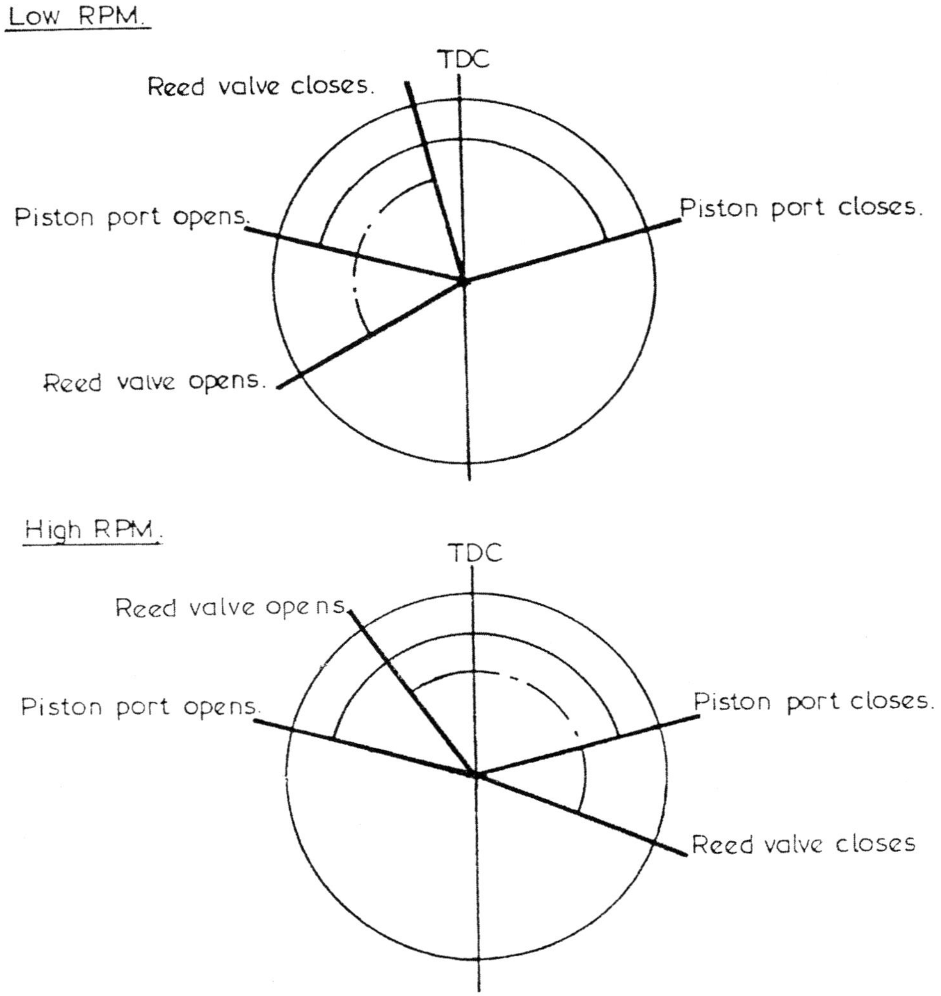

Two-stroke performance tuning

Low RPM.

- Reed valve closes.
- TDC
- Piston port opens.
- Piston port closes.
- Reed valve opens.

High RPM.

- Reed valve opens.
- TDC
- Piston port opens.
- Piston port closes.
- Reed valve closes

Figure 3.37 Suzuki power reed valve opening and closing angles.

the back transfer port, or boost port, can be made very wide and, as it enters right into the roof of the inlet port and is free of obstruction, it flows very well. The power reed engines by comparison had two tiny boost ports fed from the crankcase. Because the transfer passages which fed these back ports were compromised, being very small and entering at a strange angle, transfer flow was poor and directional control of the emerging streams was not good. Of course I knew these deficiencies existed but I underestimated their effect while at the same time overestimating the importance of the 'clean' inlet port.

By the mid-1980s a different type of crankcase reed system appeared. To differentiate it from the early Yamaha port reed system and also the Suzuki case power reed, this simplified system illustrated in Figure 3.38 was referred to as a case reed. In reality it is little different from the original Yamaha port reed. However, instead of the

Porting and cylinder scavenging

reed valve controlling flow in an inlet port which enters the side of the cylinder, in the case reed system the inlet port enters directly into the crankcase. This has two advantages over the port reed: first the piston is stronger because it no longer has large sections cut out of the skirt to permit inlet flow and secondly, with no inlet port entering the side of the barrel, the rear side of the barrel can now accommodate anywhere from one to three carefully aimed rear transfer ports with a high flow potential. In effect this permitted case reed engines to run the very same advanced boost port/s only previously possible in rotary valve engines. The net result was that case reed engines could now compete on equal terms with rotary valve units.

Regardless of the type of reed employed, the same basic principles apply for the modification of reed valve systems. Naturally, the entrance of the reed cage must be perfectly matched to the inlet manifold to ensure minimal air flow disruption. If the reed fixing screws protrude into the air stream, grind them off flush with the reed cage. However, if the screws are well below the surface, fill the holes with Devcon F. Be sure to put a drop of Loctite (blue grade) on the fixing screws each time that they are refitted.

Within the reed cage you will often find little steps and ridges which can be filed out. When you do this, be sure to leave a 1mm wide seat for the reed to seal against and, when you refit the reed petals, carefully line them up with the cage openings. Unfortunately, some manufacturers make the petal fixing holes so much larger than the diameter of the fixing screws that the petals can be fitted seating over an area 2mm wide on one side and barely covering the cage opening on the other side.

Figure 3.38 With case reed induction the inlet port enters into the crankcase, rather than the cylinder.

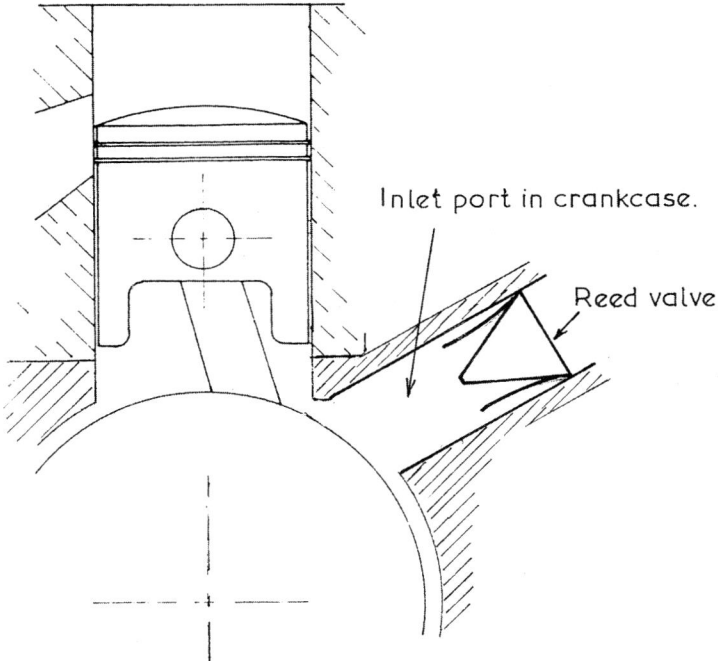

Two-stroke performance tuning

On tight sprint kart circuits case reed engines now frequently out-perform rotary valve engines. This has come about as a result of improvements in reed petal performance giving better high speed flow. Additionally because the inlet tract now enters the crankcase the rear of the cylinder can accommodate a generously sized boost port identical to that previously restricted to rotary valve engines.

Porting and cylinder scavenging

If you are not after a huge increase in high speed air flow, then I would recommend that you retain the standard reed cage and fit phenolic reed petals or, better still, a set of dual Boyesen petals or the later Boyesen carbon fibre/fibreglass Pro Series petals. Phenolic petals wear out fairly quickly, which is why some manufacturers prefer stainless steel petals. However, phenolic petals respond to the air requirement of the engine more quickly and they do not flutter and rebound off the cage to the same extent as stainless steel petals. To increase the life of phenolic petals they should be carefully sanded around the edges, using 600 grit wet and dry carborundum paper, before being fitted. Some petals are smooth on one side only, so be sure to fit these with the smooth side sealing against the reed cage.

Figure 3.39 Yamaha YZ80 reed block modification prior to fitting Boyesen reeds.

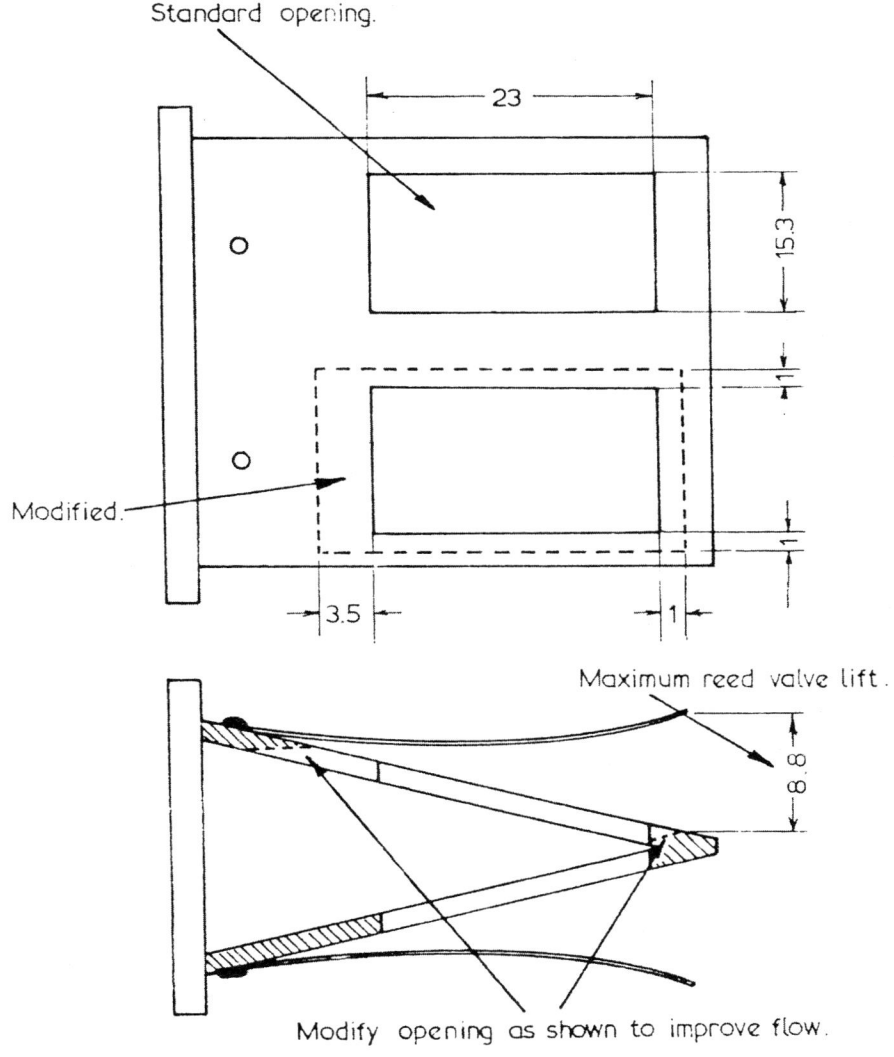

Two-stroke performance tuning

Looking at a standard reed block it is obvious that much can be done to open up its flow area. The Yamaha block shown in Figure 3.39 illustrates what can be done to greatly improve the situation. The standard block at a maximum reed lift of 7.3mm has four 'windows' just 15.3mm wide by 7.3mm high, a total of 447 square millimetres through which to admit the fuel/air mixture. Clearly this is not a very large flow window; it is about equilavent to the hole in a 24mm carburettor. When modified with streamlined and enlarged openings, and the reed stops bent to increase maximum reed lift to 8.8mm, the flow window with the reeds at full lift becomes 609 square millimetres; a 36 per cent increase or an area now about equal to a 28mm carburettor.

In this application maximum reed lift was increased progressively during dyno testing after dual Boyesen petals were fitted. Because of their design characteristics these reeds will often maintain good control at increased lift. In fact, in some engines the stop can be removed completely without adverse effects. However, if the standard factory petals are being retained the reed stops should never be bent to increase reed lift and a spacer must never be fitted under the stop to increase reed lift. Either practice will cause petal flutter at higher rpm because the reed becomes unstable or out of control, and unable to seal the fuel mixture in the crankcase. On the average 125 motocrosser, bending the stops to increase lift by just 0.7mm will knock around 2hp off the top end.

Another 'no no' is the cutting out of the 'ribs' in the reed block. To the eye the ribs appear to obstruct air flow but dyno results prove otherwise. For example when

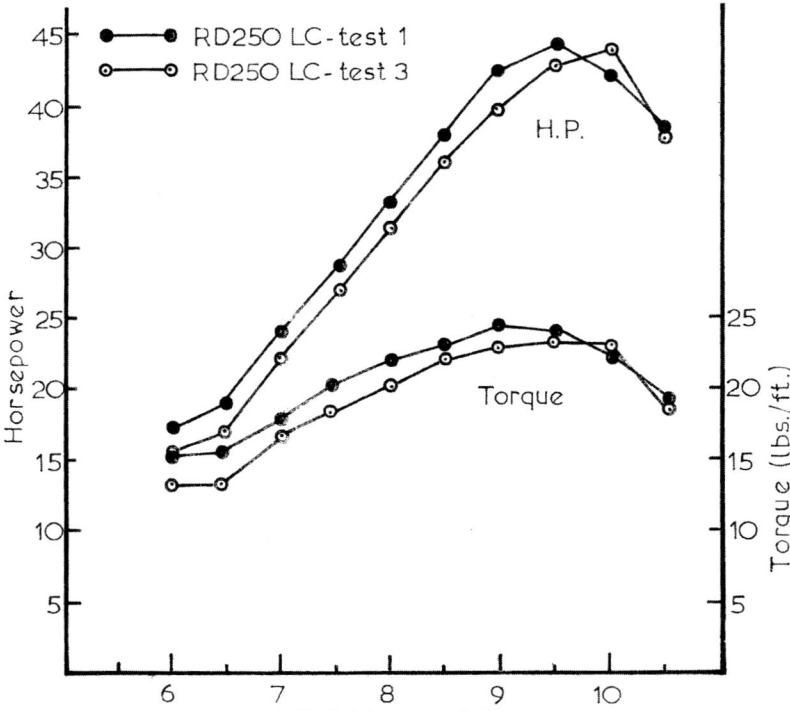

Figure 3.40 Yamaha RD250 LC power curves.

Porting and cylinder scavenging

dual Boyesen reeds are fitted it is quite permissible to cut all the ribs out of the reed cage, but on the dyno you generally won't find any hp change and usually the reeds will chip and require replacement sooner.

Table 3.7 shows how a Yamaha RD250 LC street bike, modified for production racing, responded to different types of reed blocks and reed petals. Summarising the results it is clear that in this class of competition, where stock exhausts and carburettors are mandatory, that the Boyesen dual petals with stops bent to allow 11.5mm lift are superior, allowing the engine to make in excess of 40hp over a 1,700rpm band. When the reed block ribs were cut out there was virtually no change in power at any point, but when the stops were removed the engine lost close to 2hp in the upper part of the power band. This appears to be a big loss but in fact it is only

Table 3.7 Yamaha RD250 LC dyno test

	Test 1		Test 2	
rpm	hp	Torque (lb/ft)	hp	Torque (lb/ft)
6,000	17.4	15.2	16.8	14.7
6,500	18.9	15.3	19.4	15.7
7,000	24.1	18.1	25.2	18.9
7,500	28.8	20.2	28.3	19.8
8,000	33.5	22.0	33.0	21.7
8,500	38.2	23.6	36.9	22.8
9,000	42.3	24.7	40.6	23.7
9,500	44.3	24.5	42.5	23.5
10,000	42.1	22.1	40.3	21.2
10,500	38.8	19.4	36.2	18.1

	Test 3		Test 4		Test 5	
rpm	hp	Torque (lb/ft)	hp	Torque (lb ft)	hp	Torque (lb ft)
6,000	15.6	13.7	15.9	13.9	17.5	15.3
6,500	16.9	13.7	17.5	14.2	18.9	15.3
7,000	22.5	16.9	22.0	16.5	24.3	18.2
7,500	26.8	18.8	27.1	18.9	28.7	20.1
8,000	31.4	20.6	32.1	21.1	33.8	22.2
8,500	36.1	22.3	35.5	21.9	37.9	23.4
9,000	39.6	23.1	39.2	22.9	42.5	24.8
9,500	43.2	23.9	42.7	23.6	44.5	24.6
10,000	44.2	23.2	44.7	23.5	42.5	22.3
10,500	38.2	19.1	37.2	18.6	39.9	19.9

Test 1 – standard reed block; enlarged windows, stops bent to allow 11.5mm lift, Boyesen dual reeds.
Test 2 – as above but reed stops removed.
Test 3 – standard reed block with all ribs removed, Nogucchi petals, stops bent to allow 11.5mm lift.
Test 4 – as above but with Harpowa petals.
Test 5 – as above but with Boyesen petals.

Two-stroke performance tuning

about half the loss that would have been experienced if standard Yamaha reeds were in place. The big single petal Nogucchi and Harpowa reeds produced the same peak power as the Boyesens, but 500rpm further up the range, and at the rev limit they were a couple of hp down. Figure 3.40 shows the power curves from Test 1 and Test 3.

One area where I see modern high flow reeds causing big tuning problems is as a result of the massive increase which they produce in inlet tract volume immediately after the carburettor. On passing from the carburettor the inlet tract area typically increases in a short length to almost double that of the carburettor bore. This reduces air speed, and as a consequence it also reduces the pressure differential existing between the carburettor side of the reed valve and that in the crankcase. Remembering that it is the pressure differential which determines how soon and how rapidly the reed petals lift you can begin to appreciate why this is most undesirable. That big volume delays the reed opening by several degrees and it slows the rate of opening. In practice it is akin to tying a 6ft length of elastic cord to a door and tugging the cord to open the door. The time delay from the initial tug to when the door opens is quite considerable, the result being that the initial flow into the crankcase is reduced.

Manufacturers have attempted to get around this to some degree by inserting rubber or plastic 'stuffers' in the reed block entry area. These inserts help in reducing pre-reed intake volume but I believe that the factories have not gone far enough. My conviction is that the cross-sectional area of the inlet tract at any point between the carburettor and the crankcase end of the reed block with the reeds at full lift, should ideally not be more than 10–15 per cent greater than that of the carburettor bore. Therefore if we were running a 38mm carburettor with a cross-sectional area of 1,134 square millimetres, then the flow area at any point between the carb and the tips of the reeds should not exceed 1,250–1,300 square millimetres.

The opening in this reed block is 65 x 31mm, an area of 2,015 square millimetres which is 78 per cent larger than the area of the engine's 38mm carburettor.

Porting and cylinder scavenging

These stuffers are inadequate as they only reduce the reed block entry area from 65 x 31mm down to 53 x 31mm, which is still 45 per cent larger than the carburettor bore area. The opening should be narrowed down from 31 to 25mm using stuffers in the top and bottom of the reed block.

When dyno testing an engine with a fully functional 'stuffer' you generally won't find any significant change in power readings. However it is not unusual to find that you can lean off the carburation, and riders report that the engine responds much more rapidly to the throttle so actual track performance is better.

When a large increase in high rpm air flow is required, a larger reed valve assembly will be necessary. There are special assemblies available for some engines but often it will be necessary to adapt a reed valve from a larger, or more potent, engine. This can be quite frustrating, as you won't find many dealers with reed valves in their parts bins for you to inspect for size. The best way is to go to a motorcycle wrecker, cylinder in hand, and look through his range of reed valves. Don't look for a reed valve which will drop straight into the inlet tract of your cylinder, as it probably won't be much bigger than the standard assembly. Instead, look for a valve which is a little wider and perhaps a little higher than standard. The fixing holes probably will not line up with the holes in your cylinder, but that is not a great problem providing the wrecker has the inlet manifold which matches the reed valve. Check that the inlet manifold has a hole of the correct size to suit your carburettor; if it does, you are in business.

The next problem is enlarging the inlet cavity to suit the bigger reed cage. To accomplish this, you will have to use your judgement. Start by measuring the reed assembly and comparing its size with the reed cavity in the inlet port. If it's 4mm wider, then grind 2mm off each side of the cavity and so on.

When the reed valve fits the cavity, you can then decide what has to be done to fix the reed valve and manifold to the cylinder. If the fixing holes are close, then it may be possible to elongate the holes in the reed cage and manifold to align with the

Two-stroke performance tuning

cylinder. In some cases, it will be a matter of filling the stud holes in the cylinder and then drilling and tapping new fixing holes. However a more expensive solution may be necessary such as when the Yamaha RD350 or 400 is fitted with TZ 750 reeds. In this instance, an aluminium plate with fixing holes to suit the TZ reeds is welded to the RD cylinder face.

Before some special replacement reed assemblies are fitted, they have to be modified in ways different to that outlined on preceding pages. One such reed which comes to mind is the R & R Hi-Volume reed for RM and PE model Suzukis. This reed flows very well but it falls short in two areas, which could easily catch out the unsuspecting tuner. The first problem is that the screw heads on the lower side of the reed prevent the cage from seating properly against the base of the cylinder. Thus an air leak can develop and spoil engine performance. What must be done is file the edges of the screw heads flush with the cage mounting face, so that the cage can seal against the cylinder base. The other problem involves the reed stop for the main (ie bigger) reed petal. The stop is too flexible and actually rebounds the reed petal when it comes against the stop. This sends the petal into flutter and reduces high speed power. To cure this, the R & R stop should be removed and the standard Suzuki reed stop fitted. You will note that the Suzuki stop is much thicker, so longer fixing screws are required. If these are unobtainable, the holes in the stop can be countersunk to give the screws more bite.

One of the great port reed valve engine myths is that there are big power gains in hacking the back out of the piston to open up the inlet flow path (Figure 3.41). My advice is don't waste your time cutting the back out of the piston or enlarging the skirt windows. This weakens the piston and there is little or no gain in hp anywhere in the power range. The only exceptions to this rule would be in the case of desert racers, or bikes which are very pipey, if they don't have any holes high up on the piston skirt.

Figure 3.41 In port reed engines cutting the back out of the piston does not increase horsepower – rather it reduces piston life.

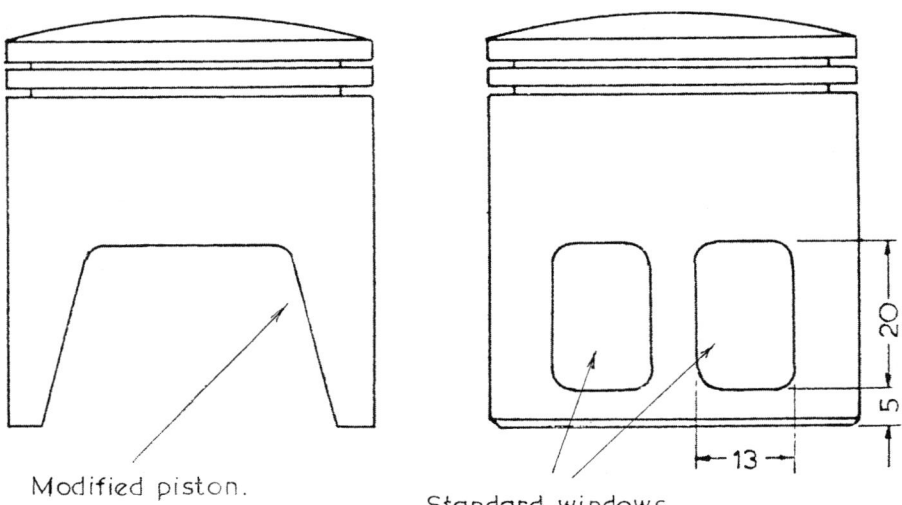

Drilling a pair of round holes just below the ring land will help cool the piston crown and little end, or if the bike is very peaky and nothing else has tamed the rush of power, maybe a pair of holes will help. The holes shouldn't be too large: 10 to 13mm is plenty big enough for a 125 or 175, and larger engines could use holes about 14 to 16mm in diameter. After the holes are drilled, carefully chamfer them on the inside and outside of the piston, then dress the holes with 180 grit wet and dry. These precautions will help prevent premature cracking of the piston skirt.

Chapter 4

The exhaust system

It is true to say that the largest contribution towards achieving the current high power levels from the two-stroke engine has come from increasing knowledge in the area of exhaust system design. Originally, the exhaust pipe was designed to get burnt gases out of the engine as quickly as possible. Then, as designers learned more about pressure waves, they attempted to make use of them to scavenge the cylinder of exhaust gases.

The basic theory of pressure waves is quite easy to understand, but the practical application of pressure wave phenomenon is very difficult to formulate. Fortunately, the experimental work done by two-stroke engineers during the past 20 years has made the task of building an effective exhaust well within the grasp of any two-stroke engine tuner. This is not to imply that the first exhaust you fabricate will be perfect. The Japanese manufacturers know as much as anyone about exhaust tuning, but you will find they are constantly updating the expansion chambers on both their works and production racers, proving they are still looking for the best design.

Understanding exhaust design begins with an appreciation of the behaviour of sonic waves travelling through a pipe. These waves travel at a speed determined by the temperature and pressure of the outflowing exhaust gas. This speed always equals the speed of sound, which averages around 1,675ft per second in hot exhaust gas.

Sonic waves have the strange property of being reflected back along the pipe they are travelling through, regardless of whether the pipe has an open or closed end. Even more peculiar is another fundamental law of acoustics which causes a pressure wave to invert its sign on reaching the open end of a pipe. A positive pressure wave, on reaching the pipe's open end, goes back up the pipe as a negative wave, and vice versa. Reflection from a pipe's closed end does not change sign, a positive wave stays positive.

The earliest exhausts were a piece of straight pipe, but these were not able to take full advantage of pulse waves to 'suck' exhaust gases out of the cylinder. In this type of system a positive pressure wave charged down the pipe immediately the

The exhaust system

exhaust port opened. On reaching the end of the pipe it was reflected as a negative wave, but with reduced intensity because much of its wave energy was lost to the surrounding atmosphere. However, some energy did remain and if, when the negative wave reached the exhaust port, the port was still open, it would assist in a small way in evacuating the cylinder. This being the case, the wave would turn round and travel back down the exhaust still negative, then, on reaching the open end of the pipe, be reflected up again as a positive pressure wave. If the exhaust was of the correct length, the positive wave should have arrived back at the exhaust port just before it closed, forcing any fuel/air mixture that had spilled into the exhaust back into the cylinder to be burned.

In theory it sounds good, but in practice the straight pipe exhaust never worked too well, primarily because so much kinetic energy was lost each time the sonic wave reached the open end of the exhaust pipe. A two-stroke engine requires strong pressure pulses to work efficiently, so engineers added a megaphone to the end of the straight pipe.

A megaphone, more correctly called a diffuser, is in effect a relatively efficient energy inverter. In a diffuser the walls diverge causing the sonic wave to react just as though it had reached the open end of the exhaust. However, the reflected wave retains most of its energy and can create a vacuum as low as 6psi. Obviously a pulse wave of this magnitude can be very effective in drawing exhaust gas out of the cylinder, and in pulling the fresh charge from the crankcase up through the transfer ports.

The problem with this system is that much of the time the strong negative pulse wave will arrive at the wrong moment, and draw a considerable amount of fuel/air mixture into the exhaust. The exhaust port will close before the reflected positive wave arrives, to force the mixture back into the cylinder.

The next step was to add a reverse cone with a small outlet to the diffuser, and this proved to be the real break-through in two-stroke exhaust design. This type of exhaust is referred to as an expansion chamber. The addition of the reverse cone with a small bleed-off hole acts as a closed pipe, giving the exhaust a double pulse action. When the positive wave reaches the diffuser, part of the wave is inverted and reflected as a negative wave to evacuate the cylinder. However, part of this wave continues on to be reflected by the reverse cone. Because of the pressure build-up caused by the small bleed hole, the reverse cone acts like a closed pipe, reflecting the wave with the same positive sign. This strong positive pulse arrives just before the exhaust port closes, forcing any escaped mixture back into the cylinder, increasing power output and reducing fuel consumption.

In 1983, Honda took resonant wave theory a step further when they introduced the ATAC (Automatic Torque Amplification Chamber) system on the works NS500 GP racer. As the name suggests, the aim of this system is to bolster engine torque so that the engine can be tuned with the appropriate porting and expansion chamber to give high maximum hp and then with ATAC switching in, pump up the mid-range.

The system is very simple with a 170cc canister attached to the header. At lower rpm a butterfly valve, 30mm diameter in the CR125 and 27mm in the CR250, opens to increase the volume of the expansion chamber. This increase in volume changes the wave action in the expansion chamber at lower rpm to either increase the volume of fuel/air mixture drawn up from the crankcase and/or to reduce the volume of mixture lost into the exhaust before the exhaust port is closed on the compression stroke. At higher engine speed the butterfly closes and expansion chamber pulse action returns to normal to allow the engine to produce maximum power and torque.

On Honda motocross bikes a centrifugal governor driven off the crank controls butterfly opening and closing. At higher rpm centrifugal force actuates the linkage to close the butterfly valve. However at low and mid-range rpm centrifugal force is insufficient to overcome pre-load spring tension so the valve is pushed open. The GP racers and street bikes such as the NS400 utilise electronic control of the butterfly valve. On the NS400 the ATAC butterflies close at 7,200rpm. The engine produced maximum torque at 8000rpm with maximum power coming in at 9,500rpm.

The Kawasaki KIPS system also features a small chamber which is opened to the exhaust at lower rpm. Rather than use a butterfly valve the Kawasaki arrangement very cleverly utilises one of its exhaust port power valves to open and close this small chamber. As explained in Chapter 3, at higher rpm both power valves open a pair of auxiliary exhaust ports to increase both exhaust open duration and exhaust port area. Then at lower rpm a centrifugal governor closes both power valves, closing off both auxiliary exhaust ports. However, when one valve moves into the closed position it opens the port to the small KIPS chamber. Like the Honda ATAC system this increases expansion chamber volume which in turn changes exhaust pulse activity to push up low and mid-range power. In effect the KIPS system attempts to combine the benefits of both the Yamaha power valve arrangement and the Honda ATAC system. After some time Honda abandoned their ATAC system as it was clearly demonstrated to be inferior to a combined exhaust power valve/sub-chamber system.

On paper expansion chamber theory sounds very simple, but there is much more involved when we actually set about designing a system. Obviously the expansion chamber must be of the correct length to ensure the pulse waves are reflected to arrive at the exhaust port at the proper time.

The formula we use to determine the tuned length of the exhaust is:

$$L = \frac{ED \times 42545}{rpm}$$

where L = tuned length in mm, ED = exhaust duration in degrees, and rpm = engine speed exhaust is tuned to work best at.

Assuming we had an engine with an exhaust duration of 196° and producing maximum power at 11,000rpm, the tuned length would be:

$$L = \frac{196 \times 42545}{11,000}$$

$$= 758mm$$

The length of 758mm is measured from the piston face to the assumed reflection point of the reverse cone. This point is half way along the cone but, because the top is cut from the cone, the point must be calculated mathematically. (Figure 4.1).

The first part of the expansion chamber, the header pipe, may be either a straight pipe with parallel walls, or a tapered pipe with diverging walls. A tapered header pipe

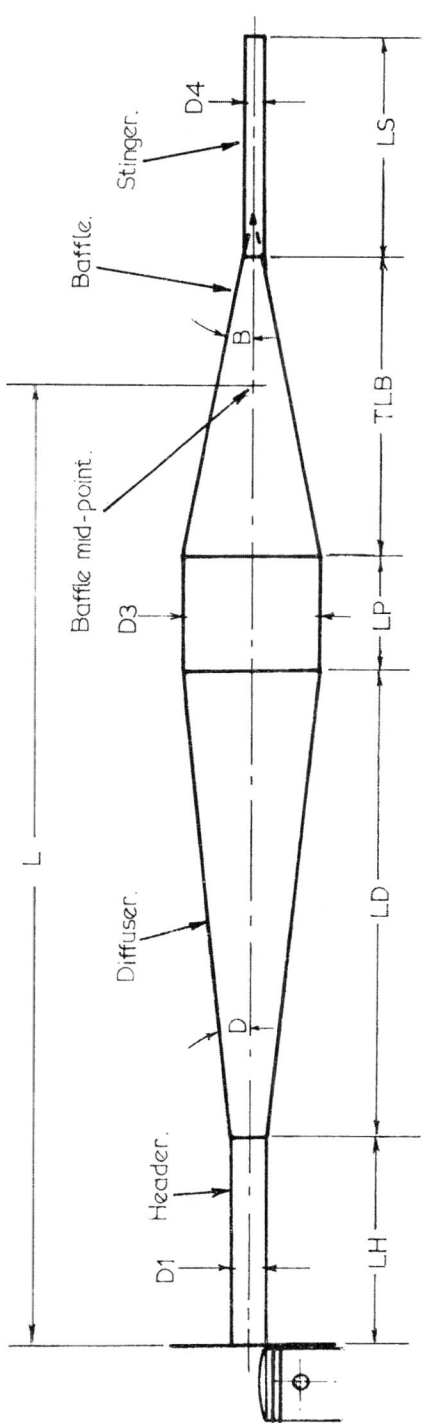

Figure 4.1 The basic two-stroke expansion chamber.

Two-stroke performance tuning

is to be preferred, as it will improve the power and power range; however, it is much more difficult to fabricate than the straight pipe (Figure 4.2).

The actual length required for the header pipe can only be determined accurately by testing. Over the years, I have devised and tried all kinds of formulae to calculate the length of the header, but I've never found one that works too well. It has been my experience that it is much quicker to make an educated guess and work from there.

In Table 4.1, I have set out what I consider to be a good starting point in working out the header pipe length. For example, if you are building a chamber for a road race 125 with an exhaust port inside diameter of 38mm and you intended to fabricate a multi-stage diffuser, then the header would be between 247mm and 285mm long. However, if you were to use a single stage diffuser, the header would be a little longer at 296 to 323mm. Usually, you will find the shorter length in both instances to be very close to what is required for best performance. Lengthening the header has the effect of increasing mid-range power at the expense of a drop in maximum hp.

Table 4.1 Calculating header pipe length

Cylinder size (cc)	Multiplying Factor			
	Road race		Motocross & Enduro	
	single stage	multi-stage	single stage	multi-stage
50–80	8.5–9.5	8–9	10–11	8.5–9.5
100–125	7.8–8.5	6.5–7.5	7.8–8.5	6.5–7.5
175–250	7.3–8.3	6.5–7.5	9–10	8.2–9.2
350–500			8.5–9.5	7.5–8.5

Note – To calculate header pipe length multiply exhaust port diameter by the appropriate multiplying factor.

Single stage refers to a single stage diffuser.

Multi-stage refers to a multi-stage diffuser.

These expansion chambers allow the engine power characteristics to be changed to suit particular circuits. The lower pipe provides a wide power range to suit tight or wet tracks, while the other has been built to give maximum top end power on high speed circuits.

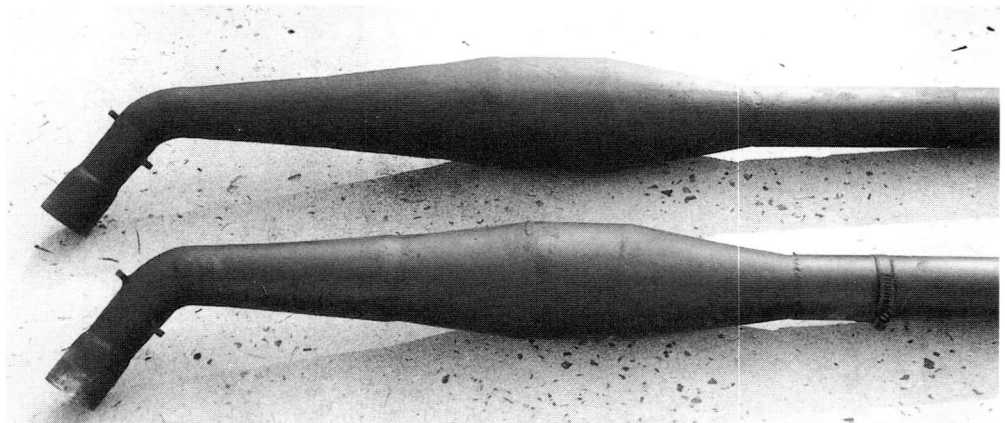

The exhaust system

Hartman go-kart

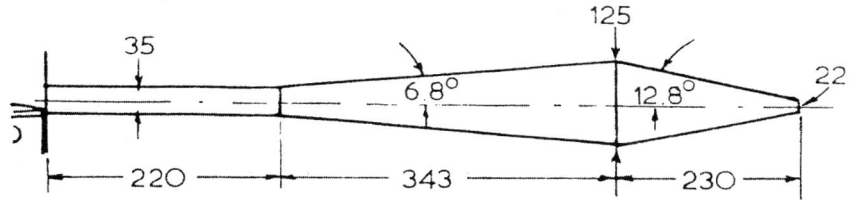

Yamaha MX360

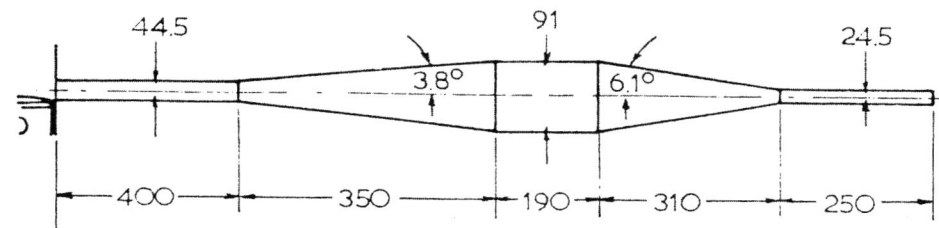

Vevey kart

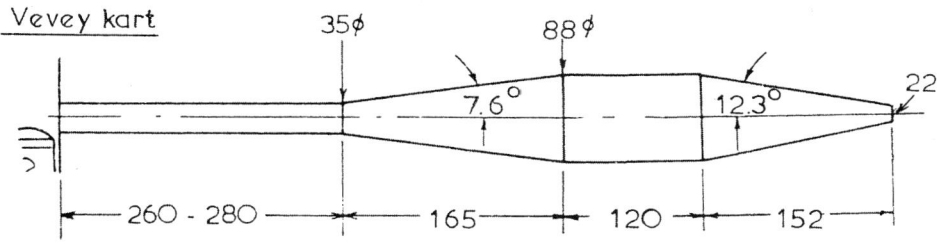

Powermac kart

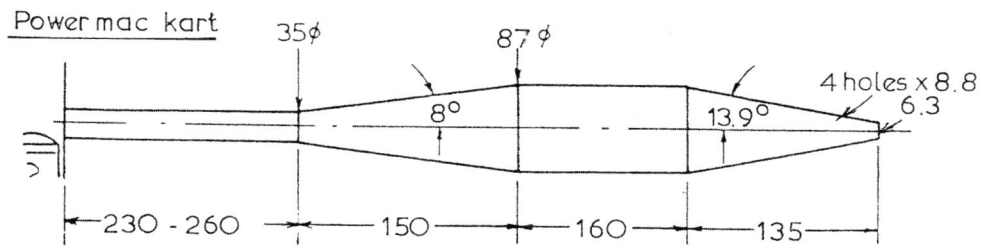

KSI kart

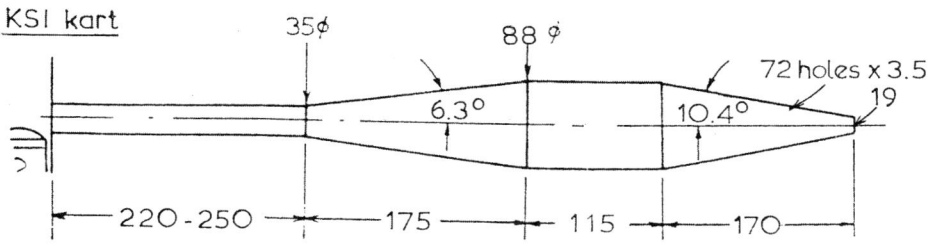

Figure 4.2 Expansion chambers with parallel-wall header pipes.

It is important to note that the above rule for calculating the header length can work only if the exhaust port is of a standard diameter for that particular size of engine. If the engine has a port size outside that shown in Table 4.2, then it will be necessary to work out the length based on a standard port diameter. Also, keep in mind that the calculated length of the header is the measurement from the piston to the end of the header pipe. Therefore, when you fabricate the pipe, remember to subtract the length of the exhaust port.

Table 4.2 Standard exhaust port diameter

Cylinder size (cc)	Port inside diameter (mm)
62–80	30–32
100	34–37
125	37–40
175	42–46
250	44–48
350–500	45–50

Over the past 25 years the majority of manufacturers have realised the benefits to be gained from a tapered-wall header and have gone over to this design. In the first instance it increases chamber volume, which effectively broadens the power range. Secondly, its diverging walls reduce flow resistance both in and out of the cylinder. The burnt gases flow out easily with a minimum of turbulence and any fuel mixture which has spilled into the exhaust is rammed back into the engine more efficiently. Thirdly, and this is the most important reason for the justification of tapered headers, the shallow taper allows the exhaust gas to expand and cool more gently. This results in less loss of kinetic pulse energy than if the gases were to expand rapidly and abruptly on passing from a straight header into the diffuser section of the expansion chamber. With more pulse energy available, a stronger evacuation wave to scavenge the cylinder, and a stronger wave to ram spilled mixture back into the cylinder, is produced.

The taper of the header is normally between 1.15° and 1.5°. However, some manufacturers have used tapers as shallow as 0.8° and as steep as 2.3° in certain circumstances. If the diffuser taper is very shallow (ie 2.8° to 3.25°) then a 0.8° header taper is at times in order. Conversely, if the exhaust port and flange is very long (ie 75mm to 100mm) a steeper 1.7° to 2.3° header may be necessary. (Figure 4.3).

To work out the physical size of a diverging header pipe of a particular length and taper, we use this formula:

$$D_2 = \left(\frac{LH \times 2}{\text{Cot } H}\right) + D_1$$

where D_2 = header pipe major inside diameter, D_1 = header pipe minor inside diameter, LH = header pipe length minus the length of the exhaust port and flange, and Cot H = cotangent of header pipe's angle of taper.

We will assume our road racer has a cylinder volume of 125cc and an exhaust port and flange 70mm long by 40mm id. The total length of the header will be (40 x 6.5) – 70 =

The exhaust system

Yamaha TZ750C pipe.

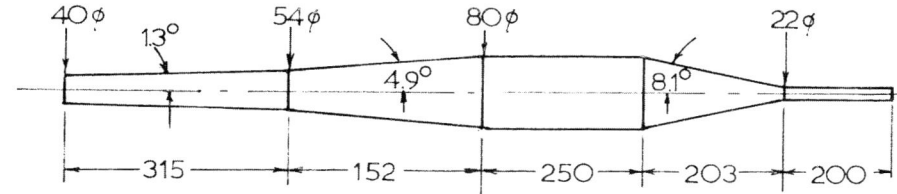

Morbidelli 125 production racer pipe.

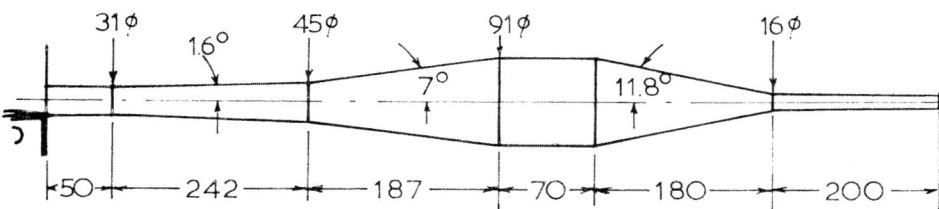

Figure 4.3 Expansion chambers with tapered header.

190mm. The taper of the pipe will be 1.5°, which has a cotangent of 38.19 (from Table 4.3).

$$\text{Therefore } D_2 = \left(\frac{190 \times 2}{38.19}\right) + 40$$

$$= 50\text{mm}$$

Table 4.3 Useful Cotangents

Angle	Cotangent	Angle	Cotangent	Angle	Cotangent
0.8	71.62	6	9.5144	11	5.1446
1	57.29	6.5	8.7769	11.5	4.9152
1.25	45.83	6.75	8.4526	12	4.7046
1.5	38.19	7	8.1443	12.5	4.5107
1.75	32.73	7.25	7.8712	13	4.3315
2	28.64	7.5	7.5958	13.5	4.1653
2.5	22.90	7.75	7.3498	14	4.0108
3	19.08	8	7.1154	15	3.7321
3.5	16.35	8.5	6.6912	16	3.4874
4	14.30	9	6.3138	17	3.2709
4.5	12.71	9.5	5.9758	18	3.0777
5	11.43	10	5.6713	19	2.9042
5.5	10.39	10.5	5.3955	20	2.7475

The next stage of the expansion chamber, the diffuser, is, as we mentioned earlier, a

Two-stroke performance tuning

wave inverter. The duration and intensity of the inverted return wave is determined by the diffuser taper. A shallow taper returns a wave of long duration and low intensity. This has the effect of cutting maximum power but, beneficially, it boosts mid-range power by allowing the expansion chamber to stay 'in tune' with the engine over a larger rpm band. (Figure 4.4).

Conversely, a steeply-tapered diffuser reflects a pulse of high intensity and short duration. Maximum power will be increased but at the expense of narrowing the power band. In applications such as road racing this may be acceptable if the machine has a close ratio 6-speed gearbox and the rider has the necessary riding skill to ride a bike with a narrow power range and a sudden rush of power.

In Table 4.4, you can see the sort of angles that I recommend for diffusers. Some people build expansion chambers with larger diffuser tapers, but I tend to value good mid-range power and a wide, easily managed power band much more than all-out power.

The length of the diffuser is determined by the diameter to which it expands, which should normally be 2.5–2.8 times the exhaust port diameter. If you have room on the bike, and you wish to experiment, you may be able to spread the power range by making the diffuser taper to 2.9–3.1 times the port diameter. This move will at times suppress maximum power, but the improvement lower down the scale usually

Figure 4.4 The angle of diffuser cone taper affects the intensity and duration of the return wave.

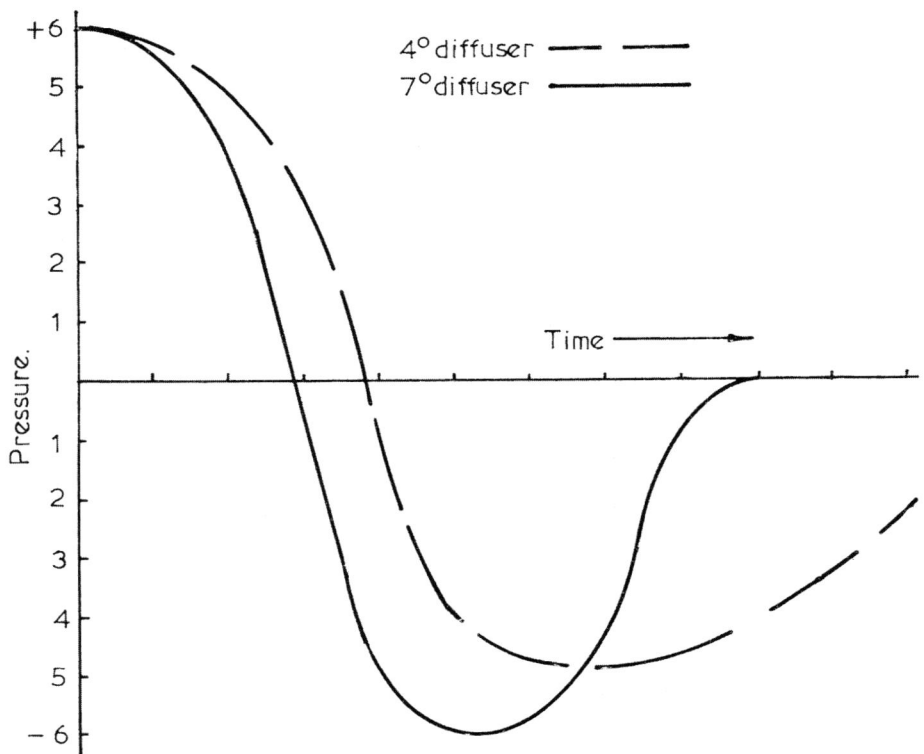

Table 4.4   Diffuser tapers

	Diffuser angle (°)				
	Road race			Motocross & Enduro	
Cylinder size (cc)	single stage	two stage	three stage	single stage	two stage
50–80	6.5 to 7	4.5 & 7	4 & 6 & 8	3 to 3.5	3 & 5
100–125	6.5 to 7.5	4.5 & 7.5	4.5 & 7 & 9	4 to 4.8	3.3 & 6
175	6.5 to 7.5	4.5 & 7	4.5 & 7 & 10	3.5 to 4.5	3.5 & 6
250	7 to 7.5	4.5 & 7	4.5 & 7 & 10	4 to 4.5	4 & 7
350–500				4 to 5	3.5 & 6

compensates. On many bikes, it is difficult to find room for a large diameter exhaust, so you may have to be satisfied with a less than perfect chamber. Most bikes have trouble catering for an exhaust in excess of 110mm diameter, so this could restrict the diffuser outlet to something closer to 2.2 times the port diameter in bikes over 350cc.

Going back to our previous example, we calculate the diffuser size based on a 7° taper (ie 14° divergence) and 2.5 times the port diameter (ie 40mm x 2.5 = 100mm).

The formula is:

$$L_D = \left(\frac{D_3 - D_2}{2}\right) \times \cot D$$

where L_D = diffuser length, D_3 = diffuser major inside diameter, D_2 = header pipe major inside diameter, and $\cot D$ = cotangent of the diffuser's angle of taper

$$\text{Therefore } L_D = \left(\frac{100 - 50}{2}\right) \times 8.1443$$

$$= 204\text{mm}$$

Today, instead of relying on a single taper diffuser, we are using multi-stage diffusers. Generally, a two or three section diffuser is utilised, although some tuners and manufacturers are turning to the use of four stage diffusers. For motocross and enduro bikes, I usually work with a two-stage diffuser as relatively shallow tapers are involved. Road race engines and 125 motocross engines with near road race porting require a three stage diffuser, due to the need for steeper angles of taper to pick up peak power.

A multi-section diffuser allows the exhaust gas to expand and cool more gradually, which means there is less loss of kinetic pulse energy than if the gases were allowed to expand more rapidly in a single taper diffuser. With extra pulse energy available, the expansion chamber can do a better job of scavenging exhaust gases out of the cylinder and drawing up the fresh fuel/air charge through the transfer ports.

In Table 4.4 you will note the diffuser angles which I have found to work well when multi-stage diffusers are utilised. For example, a twin cylinder road race 250 would use a diffuser with the first section tapering at 4.5°, the second section at 7° and the third section at 9°. Just how long each section should be is a secret most two-stroke tuners keep to themselves. However, I will tell you this: motocross and enduro

Two-stroke performance tuning

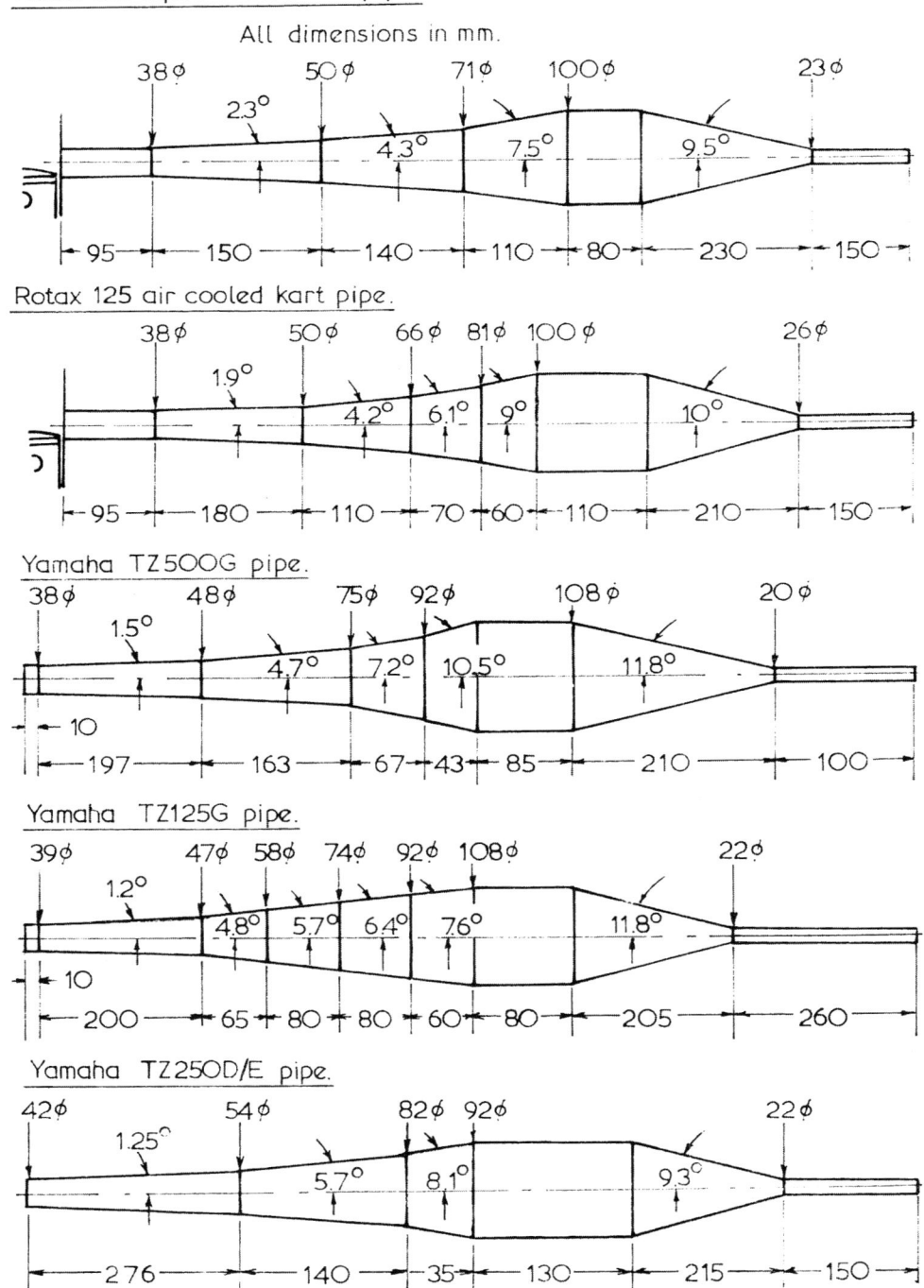

Figure 4.5 Expansion chambers with multi-stage diffusers.

The exhaust system

bikes using a two-stage diffuser will usually require the first section to be 200 to 240mm long; road race bikes with a two-stage diffuser will usually require the first stage to be 140 to 160mm long and, with a three-stage diffuser, the first stage will be 110mm to 140mm long. If, after building an expansion chamber with a diffuser like this, you find that the engine is too peaky, then increase the length of the first stage of the diffuser. This will broaden the power band. Conversely, if the engine lacks peak power, decrease the length of the first diffuser section and lengthen the second stage. In Figure 4.5 you will note the diffuser designs which a variety of manufacturers are using.

The parallel belly section of the expansion chamber naturally has the same diameter as the diffuser outlet, but we cannot calculate its length until we have arrived at a suitable size for the rear baffle cone. After that, the mid-section fills the gap to give the chamber its correct tuned length.

As noted earlier, the baffle cone (or reverse cone) reflects a wave of like sign to stuff the fuel/air charge back into the cylinder. A flat plate could do the same job but the wave duration would be so short that this would only occur over a very narrow rpm range. A cone, on the other hand, extends the duration of the pressure pulse, although reducing its intensity. This serves to broaden the engine's useful power band.

Again, the actual taper of the baffle cone affects the pulse time/intensity factor just as in the case of the diffuser. A short, sharp baffle will increase maximum power, but the motor will lose out in the mid-range; also it will tend to cut dead at just a couple of hundred revs past maximum power rpm. A shallow taper baffle reduces top end power, but the engine will develop more power lower down the rpm range and it

Figure 4.6 The taper of the baffle cone affects the shape of the power curve.

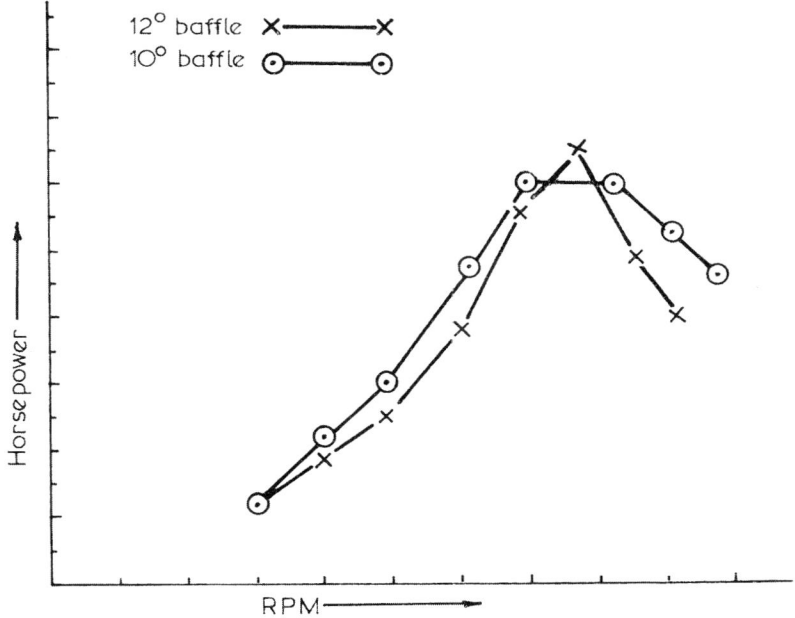

will rev on well past the maximum power engine speed. This broadens the engine's effective power range and means that the bike's speed down the main straight will be less affected by changing wind conditions during a race. If a head wind blows, the engine will have enough power below maximum to fight against it and, if a tail wind comes in, the shallow baffle will enable the engine to over-rev and pick up some more speed. (Figure 4.6).

Table 4.5 indicates the baffle cone tapers with which I prefer to work. These angles have proved to give a good power range without suppressing maximum hp excessively.

Table 4.5 Baffle tapers

Cylinder size (cc)	Road race	Baffle angle Motocross & Enduro
50–80	10.5–12	8.5–9.5
100	10.5–12	9–10
125	9.5–12	8.5–10
175	10–12	8–10
250	10–12	7.5–10
350–500		9–11

In our example, we will assume that an 11.5° baffle taper is used. The formula is:

$$OL_B = \frac{D_3}{2} \times \cot B$$

where OL_B = overall length of baffle cone, D_3 = baffle major inside diameter, and $\cot B$ = cotangent of the baffle's angle of taper.

$$\text{Therefore } OL_B = \frac{100}{2} \times 4.9152$$

$$= 246 \text{mm}$$

You will note that this calculation gives us the overall length of the cone, without allowing for a part of the top to be cut off where the stinger will be attached to bleed pressure from the chamber. By halving this figure, 246mm, we can determine the mean reflection point of the baffle and then go back and calculate how long the belly section has to be to give us the tuned length of 758mm which we originally worked out.

$$L_p = L - \left(L_H + L_D + \frac{OL_B}{2}\right)$$

where L_p = length of parallel section, L = tuned length of chamber, L_H = length of header pipe including the port, L_D = length of diffuser, and OL_B = overall length of baffle.

Suzuki RM125B/C pipe.

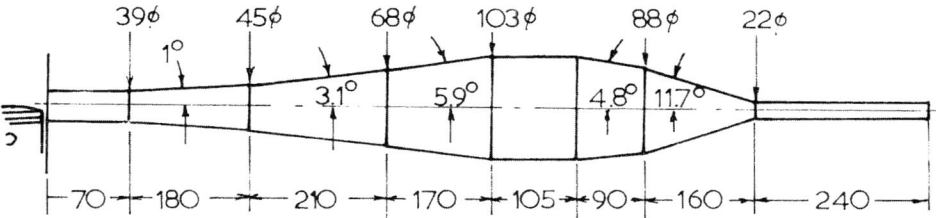

Figure 4.7 Expansion chamber with multi-stage baffle.

$$\text{Therefore } L_p = 758 - \left(260 + 204 + \frac{246}{2}\right)$$

$$= 758 - 587$$

$$= 171 \text{mm}$$

In an effort to broaden the power band on motocross bikes and small displacement road race bikes, tuners and manufacturers have been experimenting with two-stage baffles over the past few years (Figure 4.7). Unfortunately the results have not been as promising as expected. However, small gains have been made by some tuners working with 62cc and 125cc size cylinders. If you wish to do some experimentation along these lines, start out using a baffle which has the first stage tapering at 4.5° to 5.5° over a length of 70 to 90mm, and the second stage tapering at 12° to 14°.

The last section of the expansion chamber, called the 'stinger', is in reality a bleed pipe. Its function is to restrict gas flow out of the exhaust and create back pressure by slowly bleeding off the exhaust gas. This serves to assist the positive pulse wave in pushing any spilled fuel/air charge back into the motor.

Table 4.6 indicates the stinger dimensions which I have found to be most successful. You may find that a minor reduction in pipe diameter will raise the power

Table 4.6 Stinger dimensions

Cylinder size (cc)	Stinger length (mm)	Inside dia. (mm)
50–80	100–205	17–19
100	100–230	19–21
125	100–265	22–24
175	100–270	25–27
250	120–280	26–28
350–500	120–285	27–29

Note – When using this table, first select an intermediate size in both length and diameter and work from there. Do not start with the smallest diameter and greatest length. Most road race expansion chambers will require a stinger not more than 150mm long and of the smallest diameter indicated for each engine size.

Two-stroke performance tuning

All dimensions in mm.

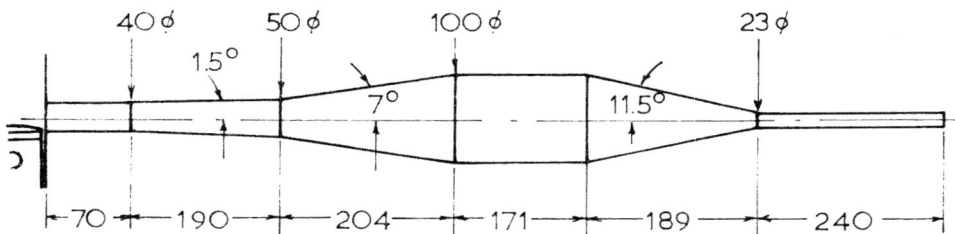

Figure 4.8 Typical 125/250 road race expansion chamber dimensions.

output, but do be careful. A stinger pipe smaller in diameter or longer than specified could easily result in engine overheating and seizure. Therefore, I would suggest you make only small changes and then thoroughly test the pipe before going any smaller. Engine overheating, in its early stages, is indicated by the presence of oil burnt dark brown under the piston crown. On the next stage the burnt oil turns black, until finally 'death ash' appears. After this, the piston can be holed at any time.

In the example we have been using right the way through, we would select a stinger 23mm in diameter (id) and about 240mm long. The entire expansion chamber would look like that shown in Figure 4.8.

The addition of the stinger shortens the baffle cone; its final length can be calculated using the same formula as for the diffuser cone:

$$\text{Thus } TL_B = \left(\frac{D_3 - D_4}{2}\right) \times \text{Cot B}$$

where TL_B = true length of baffle cone, D_3 = diffuser major inside diameter, D_4 = stinger inside diameter, and Cot B = cotangent of the baffle's angle of taper.

$$\text{Therefore } TL_B = \left(\frac{100 - 23}{2}\right) \times 4.9152$$

$$= 189\text{mm}$$

To draw out the shape of either the diffuser or baffle cone on a piece of sheet metal is quite difficult. The cone's dimensions, when rolled out flat, can be worked out geometrically, or the sizes can be calculated mathematically. I prefer the latter method.

Assuming we were going to fabricate the baffle cone for this expansion chamber, we would calculate its size, before being rolled into a cone, in this way.

Looking at Figure 4.9, you will note we only know the inside diameter of the baffle's inlet and outlet, and its length through the centre. The next dimension we must calculate is the length of the cone when measured along the tapered wall, dimension AC. We use the formula:

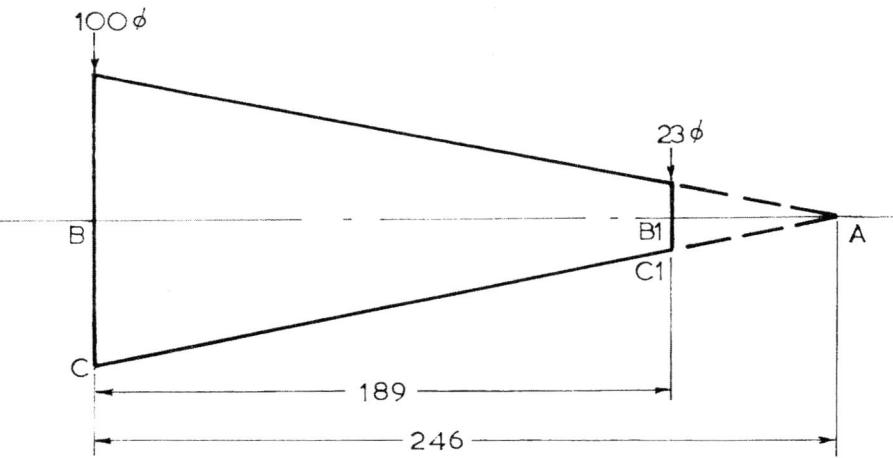

Figure 4.9 Baffle cone dimensions.

$$AC^2 = AB^2 + BC^2$$

where $AB = 246$mm, and $BC = 50$mm (ie half the inside dia)

$$\begin{aligned}AC^2 &= 246^2 + 50^2 \\ AC &= \sqrt{246^2 + 50^2} \\ &= \sqrt{60516 + 2500} \\ &= \sqrt{63016} \\ &= 251 \text{ mm}\end{aligned}$$

Now calculate the length of AC_1 using the same formula:

$$AC_1^2 = AB_1^2 + B_1C_1^2$$

where $AB_1 = 57$mm (ie $246 - 189 = 57$), and $B_1C_1 = 11.5$mm (ie half the inside dia).

$$\begin{aligned}\text{Therefore } AC_1^2 &= 57^2 + 11.5^2 \\ \text{Therefore } AC_1 &= \sqrt{57^2 + 11.5^2} \\ &= \sqrt{3249 + 132.25} \\ &= \sqrt{3381.25} \\ &= 58\text{mm}\end{aligned}$$

The next calculations we make are to work out the circumference of the baffle cone's inlet using the formula:

$$C = \pi d$$

where π = 3.1416, d = diameter of baffle inlet or diffuser outlet.

$$C = \pi \times 100$$
$$= 314 \text{mm}$$

The final calculation is worked on the formula:

$$\theta = \frac{\text{Arc}}{r}$$

where r = the dimension AC, Arc = C the circumference of the baffle inlet or diffuser outlet, r = 251mm, and Arc = 314mm.

$$\theta = \frac{\text{Arc}}{r}$$
$$= \frac{314}{251}$$
$$= 1.2510 \text{ radians}$$

The result, 1.2510, is in units called 'radians'. To be of any value to us we must convert it to an angle by multiplying by 57.3° (note 1.0 radians equals 57.3°). This gives the answer $\theta = 71.7°$.

The final check which we should make involves quite a simple calculation. We multiply the baffle cone (or diffuser) taper by π x 2. In this example the baffle taper is 11.5° so the answer is 72.2°. If all of our calculations have been correct, this figure should be very similar to the angle θ, which we calculated to be 71.7°.

Now that all these calculations have been made, we have the dimensions and angles necessary to draw the cone on to the piece of sheet metal from which the cone is to be cut. Actually, I recommend that you first make a template on a piece of stiff cardboard, and then transfer the shape to the sheet metal, as it is much easier to draw accurately on cardboard than on steel.

On your piece of cardboard, mark a cross (+) near one corner. Then adjust a pair of compasses to draw an arc, the radius of which will be equal to dimension AC, in this case 251mm. Using the cross as the centre, scribe the arc. Next adjust the pair of compasses to 58mm (dimension AC_1) and draw another arc, using the same centre (+) as previously. Then draw a straight line through the centre (+) approximately parallel with the edge of the cardboard. Position a protractor on the centre (+) and mark off the angle θ, calculated to be 72°. Draw a line from the centre through this point and you have the outline of the template. This is shown hatched in Figure 4.10.

Once you have a template you can go ahead and cut out the metal required for the baffle cone. (Note a similar series of calculations are required to get the diffuser cone, or tapered header pipe shape, drawn on to a template.) If you have not had any experience with sheet metal, then I would suggest that you take your templates to a sheet metal fabrication shop and get them to cut and roll the cones. When the cones are formed, you can weld all the pieces together, and you have your super home-made expansion chamber.

The exhaust system

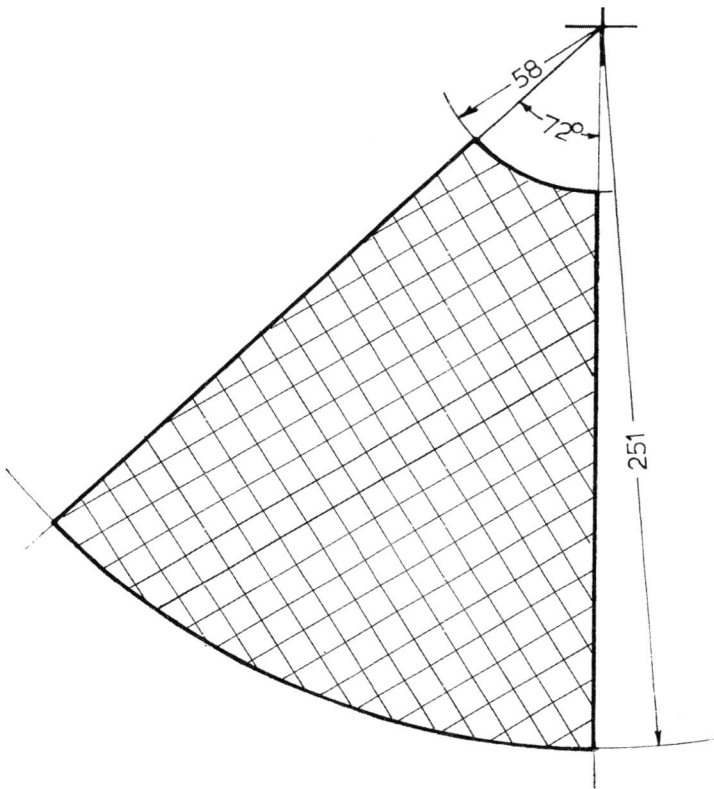

Figure 4.10 Baffle cone template.

Figure 4.11 Smooth header joints enhance exhaust flow out of the cylinder and do not weaken pulse wave activity.

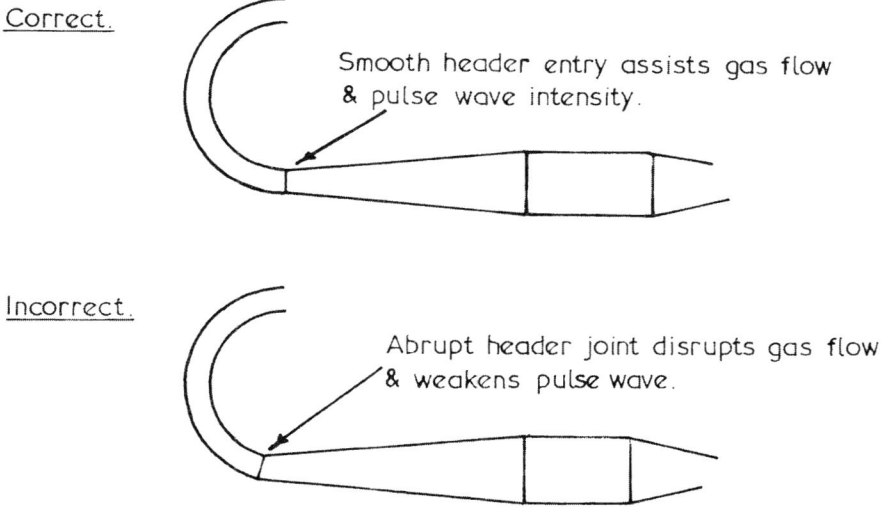

Actually, it is a little more difficult than that. Exhausts for go-karts, and single and twin cylinder road and road race bikes, are fairly straightforward, with the exception of the header pipe. Building a system for a motocross or enduro bike must be one of the worst forms of torture yet devised by man.

The main point to keep in mind, when you set about making the pipe fit, is to keep the header pipe entry into the diffuser cone smooth (Figure 4.11) and avoid the temptation to flatten the chamber anywhere. If the header entry is not smooth, gas flow out of the cylinder will suffer because any abrupt change in direction here, where gas velocity is very high, creates turbulence. Flattening the chamber doesn't upset gas flow, although the sonic pulse wave's effectiveness is reduced as any change in cross-sectional shape lowers the pulse's wave energy. Remember that an expansion chamber's basic function is to maintain and preserve the wave energy at a high level, so you will not want to do anything to cause a loss of this energy which is used to evacuate and recharge the cylinder.

While changes in cross-sectional shape affect pulse wave energy, abrupt bends or turns anywhere in the system do not. Sonic waves are willing to follow even the most tortuous curves without any reduction in pulse energy. Therefore, you can cut and resection the cones and parallel belly section to make the chamber fit your bike. The pulse waves will not know whether the system is straight, or bent all over the place.

There are two ways to cut and re-section an expansion chamber to form a bend: the easy way and the right way. (Figure 4.12). When you cut a cone and reweld it the easy way, you will lose out in two ways. First, the length of the cone is reduced, altering the tuned length of the exhaust, and secondly wave energy will be lost as the exhaust gases will no longer be expanding at a progressive 7° (or whatever the cone taper is) along the inner wall of the cone.

The correct method to form a bend is to make the first cut right through the cone and then take the second cut only as far as the cone's centre. When the cone is re-sectioned, the piece which is cut out is welded into the outside of the bend. This technique keeps the length of the cone through the centre the same as it was when straight. There is still some change in the taper of the walls and in the cross-sectional dimensions, however, which will have a small effect on the pulse wave.

When you get the exhaust right it can have a dramatic effect on not only the maximum hp the engine will produce, but also on the width of the power band. On the other hand, get the exhaust wrong and you can kill a good engine or else finish up with a power band which is not compatible with your riding style or your selected area of competition.

The porting of the Yamaha RZ350 street bike is fairly ordinary except for the exhaust power valve, and a pair of tiny 28mm carburettors would appear to be a major impediment to making serious power. The 'clean exhaust' RZ350 leaves the factory with a pair of pipes which really are not true expansion chambers. They look like proper chambers but delving inside tells the true story. Up to the belly section the shape is familiar two-stroke stuff but where you would expect to find the baffle cone and stinger you find instead that the belly section continues on and contains first a coarse screen CAT converter and then further along a fine screen CAT converter. After that comes the muffler section, but there is no baffle reflector cone or stinger. When dyno tested with the air box and filter removed, the power output was just under 40hp, with a good spread of torque all the way from 6,000 up to 9,500rpm. (Figure 4.13).

The exhaust system

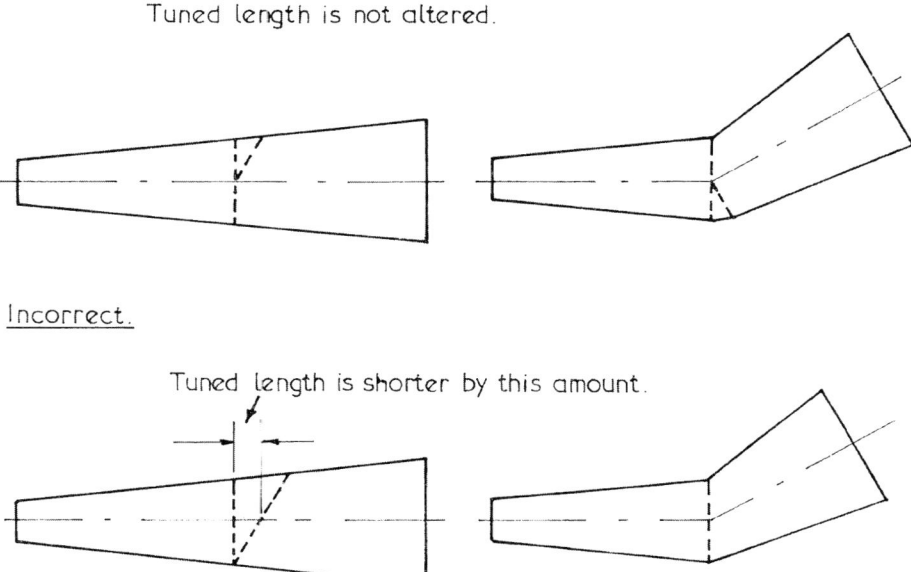

Figure 4.12 To maintain the designed length expansion chambers must be correctly cut and resectioned.

When the 'dirty' factory Euro pipes without CATs were fitted power leapt up about 25 per cent right through the range. This big jump in power was not due to getting rid of the exhaust flow restriction (which is pretty small anyway) brought about by the CAT converters. Rather, it was the product of proper expansion chambers improving the engine's ability to effectively scavenge the cylinders of exhaust gas and then draw in a larger quantity of fresh fuel/air mixture.

The third power curve is for a pair of chambers built for maximum road performance on stock porting and carburation. They are also suitable for production class racing. Note that these pipes, while making almost 10 per cent extra in maximum power, still provide a good wide power band. However to achieve that extra hp the bottom end has dropped away, and while the torque curve is flatter, maximum torque is less. Additionally this exhaust inserts another consideration into the performance equation; namely engine reliability. The engine can now easily spin 1,500rpm faster than Yamaha ever intended, and if that potential is fully utilised engine life will be extremely short. Run this engine continuously above 9,000rpm and piston and crankshaft life can drop to as little as 200 to 300 miles. Piston life can be doubled by taking more care with clearances and with the fitting of hard, higher silicon content pistons. If the crank is lightened, trued, balanced and Loctited as discussed in Chapter 7, it can be expected to live for at least 1,000 miles running continuously in the 9,000 to 11,000rpm range.

However, along with the occasional win, expansion chamber experimentation can be time consuming and frustrating. Such was the situation during intense development of a Kawasaki KX125 motocrosser. Initial modification produced an

Two-stroke performance tuning

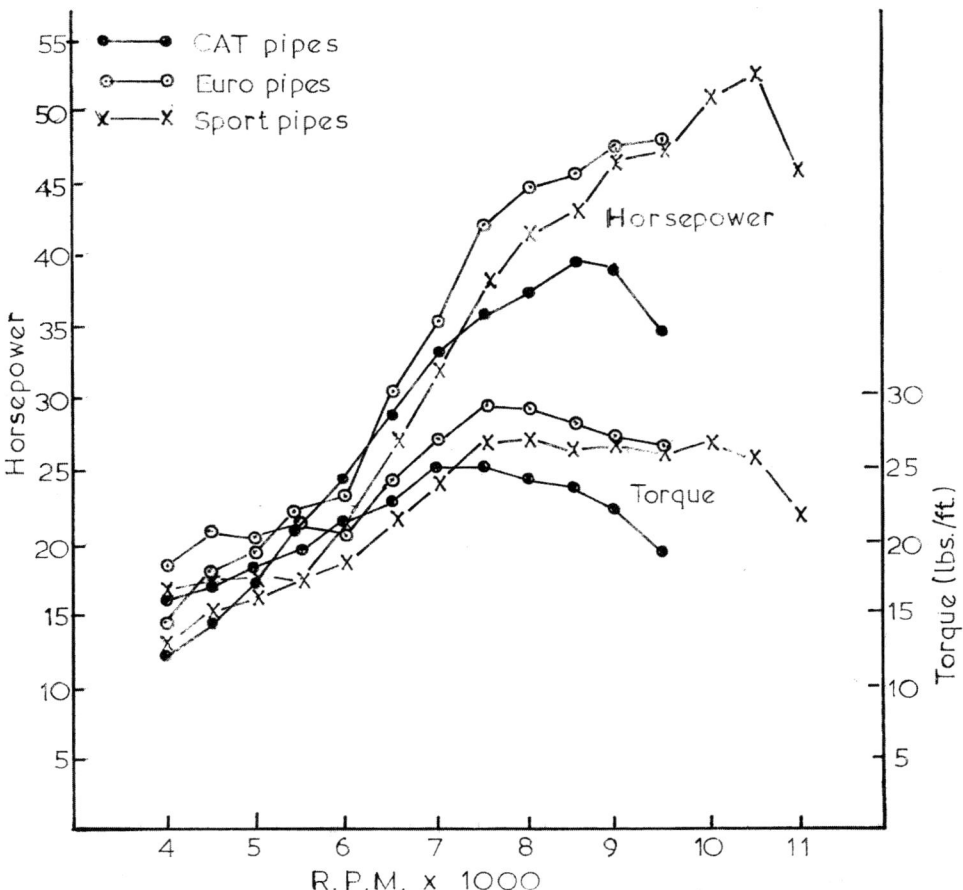

Figure 4.13 Yamaha RZ350 power curves with three different exhausts.

impressively wide power band (Figure 4.14) but the bike lacked top end speed when compared with some of the elite competition. It got off the turns quicker than the other bikes but was being overhauled on any long straights. Getting more peak power should have been reasonably easy. The engine had more than enough 'grunt' in the mid-range so a loss here to gain some top-end was not an issue. All that was required was to build a pipe which would move the current torque curve 500–700rpm to the right. Several attempts were made – all were failures. Then a chamber was obtained which was supposed to be a copy of what the works team were running. It produced a power curve which looked like it belonged on a road racer. Peak power was up 3hp but everywhere in the mid-range the engine was down 4-6hp.

Apart from illustrating the difficulties encountered during expansion chamber development the two Kawasaki power curves also provide a couple of other valuable lessons. Looking at the broad-range pipe power curve you can see a very definite dip at the 11,000 and 11,500rpm points. A dip like this can indicate a tuning problem, or it can result when the power curve is very broad at the top with the expansion chamber working well. If it is a tuning problem the dip usually indicates either carburation richness or leanness. When it is a lean problem the fuel flow meter in the dyno cell would show a significant drop in flow at those points. Such a reduction in fuel flow indicates not excessively lean jetting, but rather that fuel pickup in the carburettor is poor because the 'signal' to pull up fuel into the air stream is weak. Sometimes the lean-dip can be reduced in amplitude or else moved sideways, either to the left or the right on the power curve, by 'adjustments' to the expansion chamber. What we are attempting to do is change the pipe so that the carb receives a stronger signal so as to draw more fuel into the intake air. The adjustment usually involves taking a little bit of length out of the belly section and reducing the baffle cone angle. Of course, if the fuel flow meter did not register a drop in fuel flow, then we would have to conclude that the problem was with rich carb jetting. This isn't always so, but it is often the case.

The second thing we can appreciate from the Kawasaki power curves is the significant influence which an expansion chamber can have in moving the power band around. From the same cylinder barrel we can produce a motocrosser or a road racer power band just by swapping expansion chambers; of course, with appropriate carb

Figure 4.14 Kawasaki KX125 power curves.

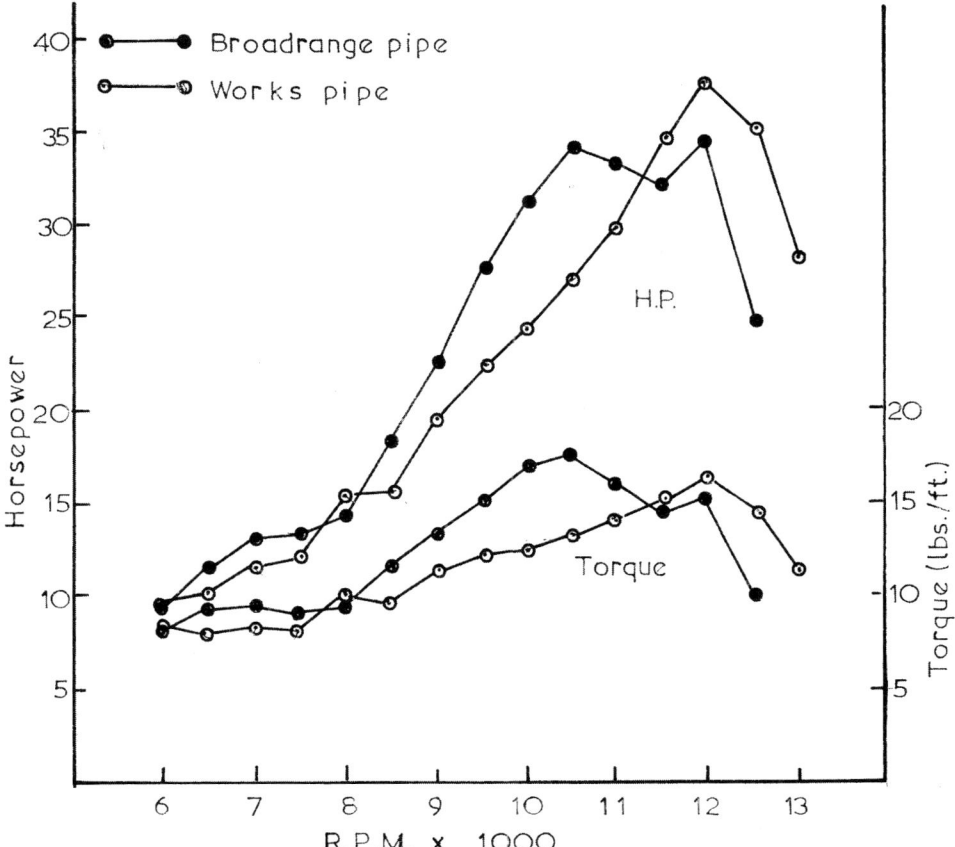

Two-stroke performance tuning

jetting. Therefore just as racers gear their machines to suit each track so likewise they should develop a couple of expansion chambers; one a narrow-band high hp pipe and the other a broadband pipe. On multi-cylinder engines you can go a step further by using the narrow range pipe on one cylinder, and the wide range pipe on the other. In addition to providing a more suitable hp/power band width compromise this can also be a good move to tame an engine which is too 'pipey' and causing the rider problems by getting on to the power band and inducing wheelspin in the wrong part of a corner. Just be sure to identify clearly the pipes and double check that you are re-jetting the appropriate carburettor.

Especially with fixed gear karts a good selection of expansion chambers is vital. Here you not only have to find a pipe which will give the best lap times for qualifying, you also must select a suitable race pipe to help when slowed by heavy traffic during the race.

When it comes to kart header pipes I feel you need to have at least two the same length but with a 1mm difference in bore. Then to adjust the tuned length you need a range of extension tubes cut in 5mm increments. Note that I said 5mm increments; anything more 'coarse' could be costing you a lot of power. Just adding or subtracting 5mm to the header length can affect hp to the tune of as much as $2^{1/2}$ per cent at various points in the power band. In fact some expansion chamber designers are now saying that they are currently working header length changes as fine as 3mm and finding the same sort of power gains as they were with 5mm increments.

To help you understand that I'm not overstating this matter, have a look at what happened during testing of a Suzuki RGV250 gearbox kart. The previous expansion chambers had been written off in an accident and now on a new pair of pipes, hand built to the same specification, hp was down in a number of places. This is always a problem with hand built pipes as no matter how much care is taken the actual dimensions will vary somewhat from pipe to pipe. To make matters worse you can end up with the errors balancing each other out in one pipe, while in another pipe all the errors may go one way, stacking up against you. If you do have both pipes stacking up against you, you can expect to see up to 5 per cent loss in power in places. The easy way to tune this sort of loss out of the system is to run the engine on

Table 4.7   Suzuki RGV 250 dyno test

rpm	Test 1	Test 2	%change
8,000	32.5	32.7	–
8,500	41.6	41.7	–
9,000	48.7	49.7	+2.1
9,500	54.2	53.6	–1.1
10,000	57.4	59.0	+2.8
10,250	61.1	61.3	–
10,500	58.9	59.4	+0.9
11,000	51.8	52.4	+1.2
11,500	42.6	42.2	–1.0

Test 1 – New expansion chambers.

Test 2 – As above but with header length of one chamber reduced by 5mm.

The exhaust system

Exhaust damage like this can cause a 5 per cent power loss.

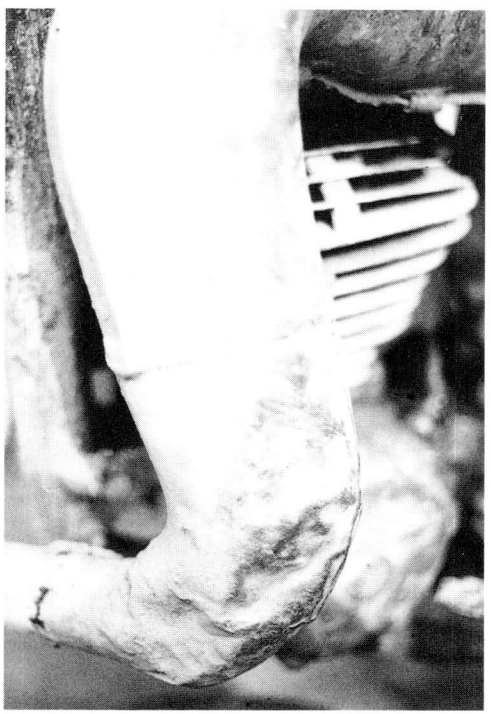

the dyno and then one cylinder at a time add 5mm, then 10mm to the header length; followed by taking 5mm, then 10mm out of the header length. Table 4.7 shows the results of adjusting the header length. One pipe worked best as originally fabricated, but the other pipe worked best overall with 5mm removed from the header length. As you can see, the average power change was less than 1 per cent but in one place the change is almost 3 per cent. That may not seem a lot but if both pipes were deficient

Table 4.8   Aprilia RS250 dyno test

rpm	Test 1 (hp)	Test 2 (hp)	%loss
7,500	31.0	32.7	5.2
8,000	30.9	32.4	4.7
8,500	39.4	41.5	5.0
9,000	46.2	49.1	6.0
9,500	51.2	54.3	5.7
10,000	57.4	60.7	5.5
10,500	59.5	63.2	5.9
11,000	49.7	53.6	7.3
11,500	41.0	42.6	3.8
11,750	37.1	39.2	5.4

Test 1 – Dented expansion chambers.
Test 2 – New expansion chambers.

by that amount you could be way behind the competition and at a loss to understand why.

The situation is similar with what appears to be fairly minor dents either from accidents or from flying stones. An Aprilia RS250 street production racer was way down in power. Whereas it had previously been a regular front runner it was now a bottom mid-fielder. Were the other competitors all suddenly cheating? That would be unlikely, some maybe, but not half the field all at the same time. On the dyno, after a rebuild the engine was still showing no improvement. Both exhausts had dents from 'offs', plus bumps from other competitors. When new pipes were fitted the lost 6 per cent power reappeared (Table 4.8).

Looking at the four Rotax 250 road racer pipes in Figure 4.15 you would be hard pressed to spot the difference between the VSK113 and the VSK148 design. In fact the only difference is that the later VSK148 has a belly 10mm shorter than the VSK113, but in terms of power output that 10mm makes a considerable difference as shown in Table 4.9. However it is not always small changes that produce the required result. The pipe which the VSK113 superseded, the VSK109, is obviously a very different expansion chamber. It is much bigger in diameter and the diffuser cones and baffle cone have a steeper taper. Not so obvious is the fact that the tuned length is approximately 120mm longer than that of the VSK113. In combination these elements mean that the VSK109 is about 9hp down on the two later exhausts, plus the engine runs out of steam 1,000rpm earlier.

Table 4.9   Rotax 250 exhaust comparison

rpm	VSK109 (hp)	124LC/2 (hp)	VSK113 (hp)	VSK148 (hp)
8,500	33.9	29.8	31.9	32.7
9,000	40.5	37.3	38.1	39.9
9,500	42.0	40.9	39.2	39.8
10,000	44.7	46.8	45.4	45.1
10,500	51.4	50.8	51.2	50.2
11,000	57.7	55.9	54.9	54.0
11,500	59.4	59.6	58.0	58.6
11,750	54.9	61.1	60.7	61.3
12,000		61.9	62.6	64.7
12,500		59.7	65.5	68.6
12,750		56.4	58.2	59.8

The other pipe, which I dubbed the 124LC/2, was one I originally developed for the Rotax 124LC kart engine. It predates all of the above factory Rotax designs, but in spite of it being an older design it was given a try on the 250 because of the engine's poor performance on the factory's latest design at that time, the VSK109 pipe. The engine instantly picked up 1,000rpm and about 3hp at the top end over the VSK109. However below 11,000rpm power was down everywhere, except at 10,000rpm.

Again comparing the 124LC/2 pipe with the first successful factory design, the VSK113, there is not much difference in the tapers and lengths. The picture changes considerably though when you check the dyno figures. Up to 12,000rpm the 124LC/2 and VSK113 are near identical, but at 12,500 and 12,750 the Rotax factory design is

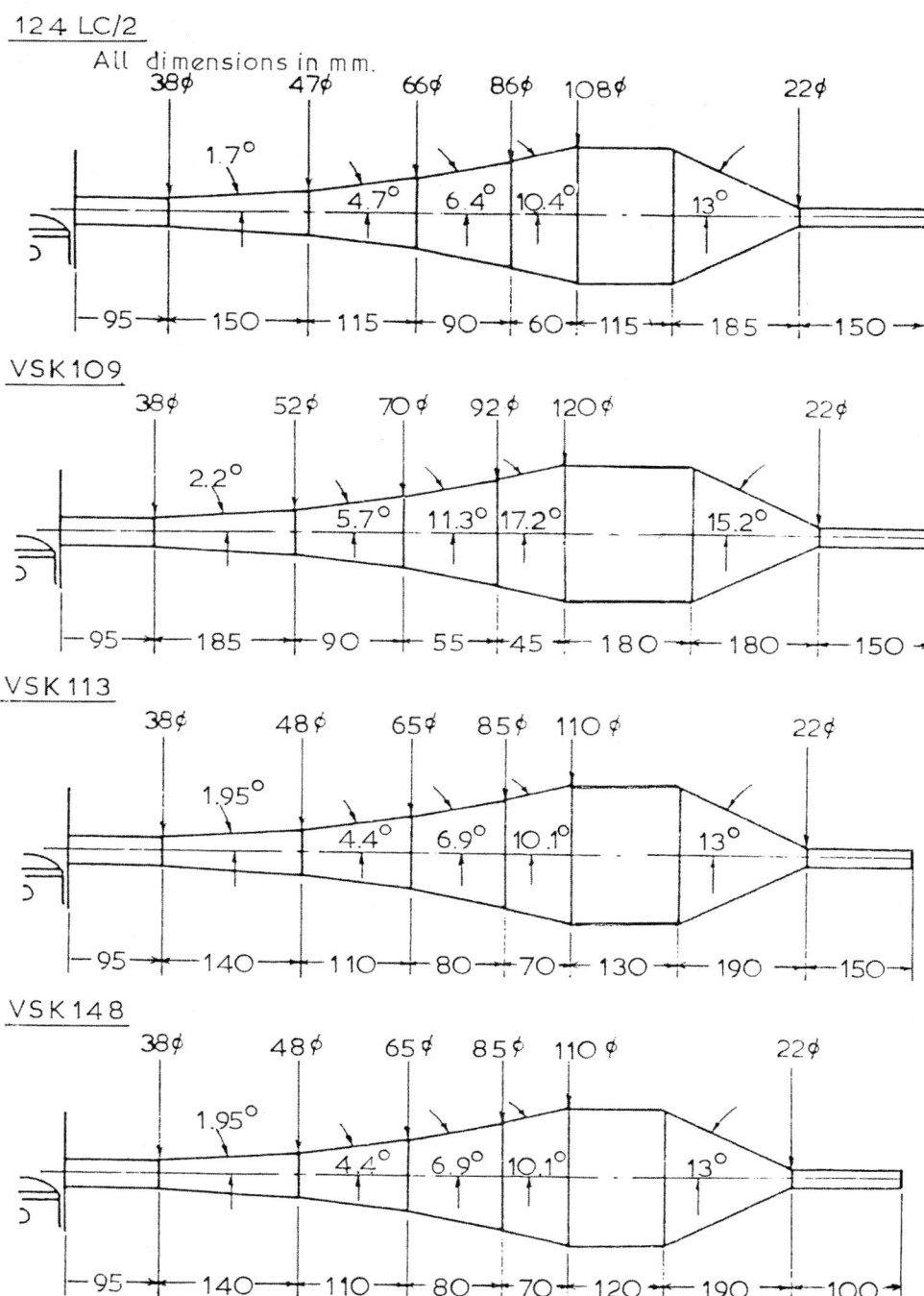

Figure 4.15 These expansion chambers for the Rotax 250 tandem produce dramatically different power outputs.

6hp and 2hp stronger respectively, which again highlights how subtle changes can make a considerable difference.

The next thing you must think about seriously is an effective muffler. Most racers hate the idea of mufflers, seeing them as something which adds weight and cuts power; just the opposite of what is desirable. Yes, they do add weight but on the subject of power loss you can be assured that it is minimal. At some engine speeds you won't see any change, while at others it will only be a loss of 0.3–0.4hp per cylinder if the muffler is designed as in Figure 4.16. Such a muffler is quiet enough for race circuits and motocross tracks but I feel very strongly that every effort should be made to silence motorcycles to a whisper when they are being used on public roads, on private and government land, deserts and reserves, etc. The noise of two wheelers in these areas is upsetting a lot of people and driving wildlife back into more remote parts. This is bad for you and all other off-road riders, because every day a few more gates, which were previously always open to motorcyclists, will be closed and locked. If you want your freedom to ride in places where you are currently permitted to ride, then you are going to have to make sure that the people who live around these areas are free of the noise of your motorcycle.

There are several choices open as to the type of silencer which can be fitted. A few of the aftermarket mufflers are quite good, and relatively trouble-free, providing they are properly mounted to the bike. If you wish, you can make your own detachable muffler. It should be of similar dimensions to the one shown in Figure 4.16. To reduce weight and increase the silencing efficiency, consider fabricating a muffler from aluminium instead of steel tube. Steel tends to resonate more readily than aluminium and this reduces the effectiveness of a steel silencer. Riders operating very close to housing estates should seriously endeavour to reduce noise to the minimum by making their muffler double walled. Sliding another piece of tube over the muffler reduces

Figure 4.16 A detachable muffler constructed to these dimensions will have a minimal effect on power output.

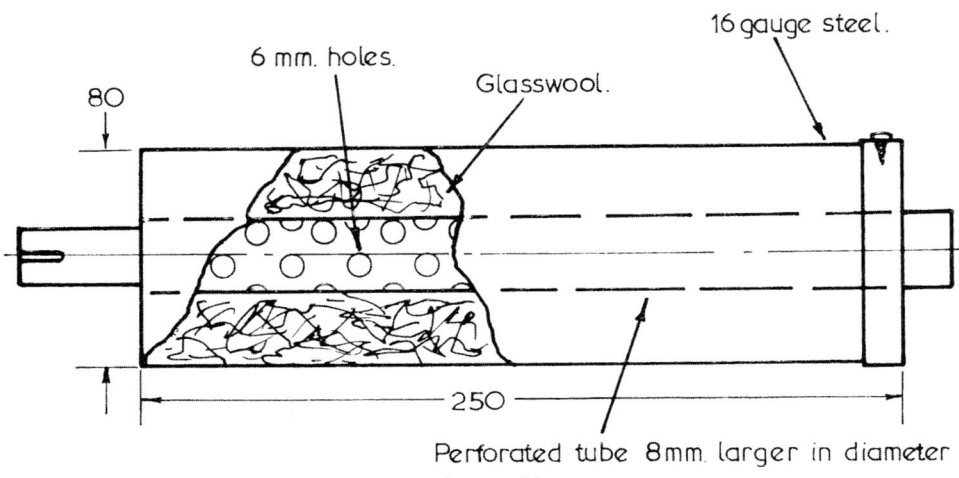

resonance from the inner wall, hence less noise is emitted from the body of the muffler.

Mufflers for road race bikes and karts may be fabricated as part of the expansion chamber. (Figure 4.17). As the allowable noise level is reasonably high, usually around 96dbA, this type of silencer does not have to be so elaborate. Preferably, the main body of the silencer should be 16 gauge aluminium.

Whatever type of silencer you choose to use, make sure that it can be easily dismantled. Then, when the fibreglass sound-absorbing material is choked with oil and carbon, it can be easily replaced with new fibreglass to restore the muffler to its original silencing capacity.

It is not of much use if your exhaust has a good quiet muffler, but the expansion

Figure 4.17 Mufflers may be fabricated as part of the expansion chamber.

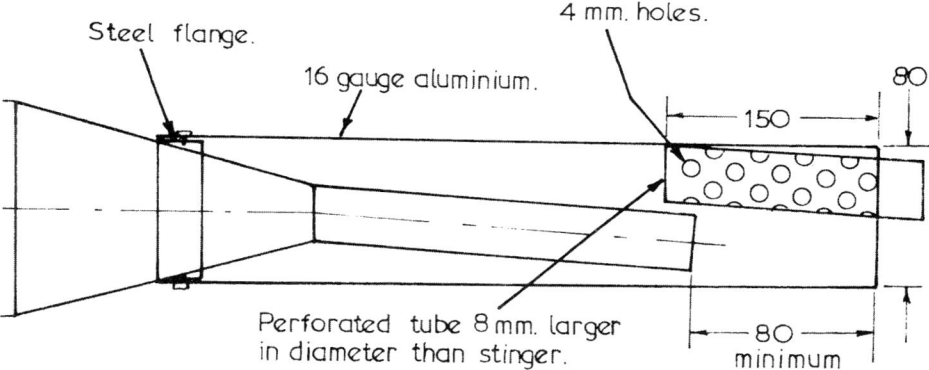

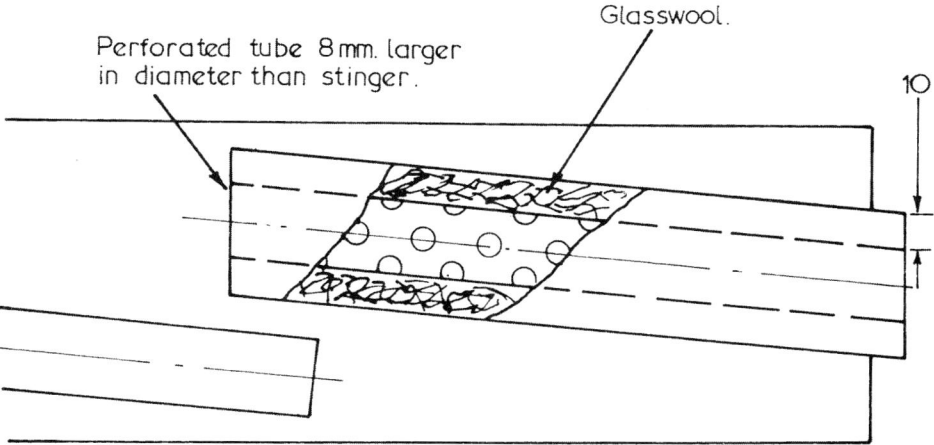

Two-stroke performance tuning

chamber is cracked and almost falling off because of incorrect mounting. Figure 4.18 shows the mounting arrangements that it is advisable to use. You will note in both instances, where the exhaust is attached to the machine by welded brackets, the large area over which the load is supported. This reduces the possibility of the chamber or the bracket cracking. If the mount does start to crack, a length of 6mm rod can be welded into the corners where the mount angles away abruptly from the chamber.

The third method of mounting is the system used by most go-karters. A saddle is fabricated out of fairly heavy steel, usually 4mm thick by 13mm wide, to support the chamber which is held in place by two springs.

The exhaust flange and joint is also a mount of a kind, as well as being a gas and water-tight connection. Usually, the header pipe is secured to the exhaust flange by either a system of springs or a large flange nut. Whatever method is used, make sure the header pipe will not slip out of the flange in the middle of a race. This means if a flange nut is the method of attachment, it must be tight and properly prevented from

Figure 4.18 Suitable expansion chamber mounts.

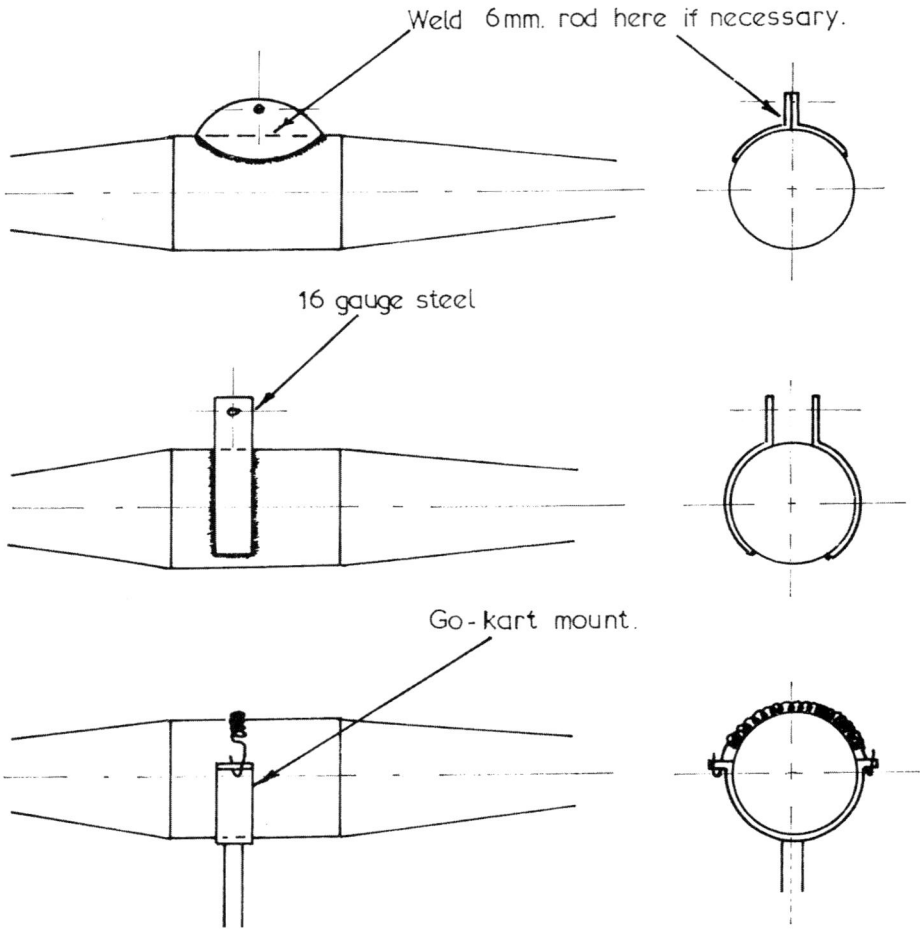

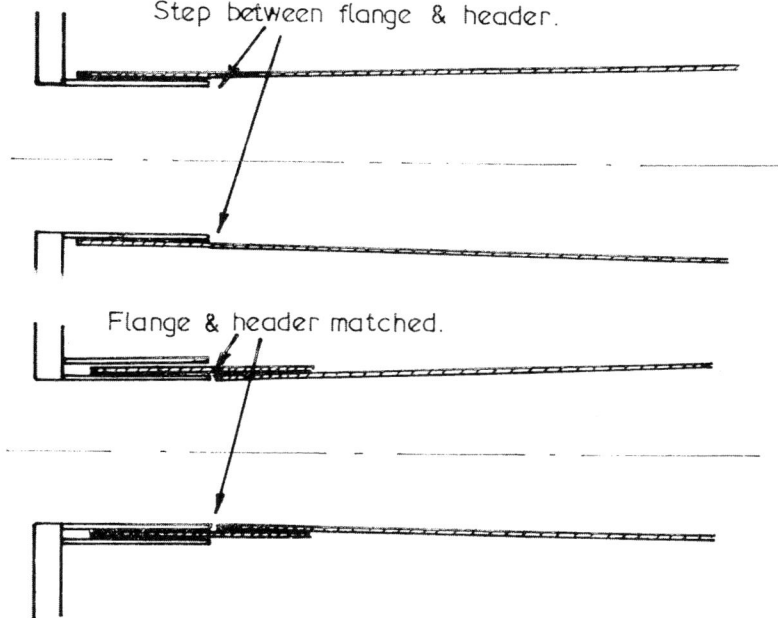

Figure 4.19 The lower slip-joint header connection is preferred as it does not leak, nor does it deter gas flow or pulse wave activity.

loosening by the attachment of two springs from the header pipe to the nut.

Most two-strokes use a flexible slip joint connection with the header pipe held into the exhaust flange by two or three springs. The thing to watch here is that the header and flange should overlap by at least 25mm. This reduces exhaust leakage and makes it much harder for the header pipe to jump out of the flange.

Figure 4.19 illustrates the two most common types of flexible slip joints. The simpler joint is used by the majority of motorcycle manufacturers for economic reasons. As you can see, if the exhaust port is made the correct diameter, the header pipe is too large, or if the manufacturer makes the header pipe the correct size, the exhaust port ends up too small. The joint I prefer is more complicated and can make the expansion chamber difficult to get on and off the bike, but it does keep the port and header pipe sizes matched and it reduces leakage at the joint to a minimum.

Many people don't worry too much about exhaust flange leakage, yet every effort should be made to prevent this, as any leakage reduces the pressure in the chamber, upsetting sonic wave speed and changing the tuned length of the system. If the leak is serious enough, the chamber will cease to evacuate the cylinder properly. Instead, the suction pulse wave draws air into the chamber through the leaky joint. Then the pressure wave will push this air up the exhaust port, and into the cylinder. The fuel/air ratio is upset, power is lost and, in severe cases, piston burning and seizure is the end product.

If you are an off-road rider, you don't want an exhaust flange leak for another reason. A leak that gas can find its way through will also often allow water through too. Every off roader, sooner or later, stalls or bogs in a creek, and when this occurs

Two-stroke performance tuning

Exhaust slip joints should be properly sealed with a suitable jointing compound. Ordinary Silastic silicone RTV will work up to 230°C while Permatex Hi-Temp RTV is for use up to 345°C.

you don't want water to seep into the motor to worsen the situation in which you find yourself.

All exhaust flanges leak to some degree, but slip joint flanges appear to be the worst, until carbon begins to build up and seal the joint. Unfortunately, racing engines have the barrels and chambers removed too frequently for the carbon to seal properly, so it is necessary to use some sort of exhaust putty to stop leakage. I use Silastic RTV silicone or Permatex Hi-Temp RTV to seal all exhaust joints.

Two-stroke exhausts are not a fit-and-forget proposition, as they require fairly frequent maintenance to clean out carbon build-up. Depending on the shape of the system, a layer of carbon 1–2mm thick may form in the header pipe after just a few race meetings and the diffuser and baffle cones usually get a good coat of oily goo at the same time. The carbon and gum deposits upset gas flow by constricting the internal dimensions of the header pipe and by creating turbulence in the system. These effects adversely alter performance, therefore the system must be regularly cleaned out.

Some people use a hot caustic soda solution to scour the chamber internally, but I feel the most effective method is to burn the system clean with either an oxy-acetylene or propane gas torch. This may be done by heating the outside of the system to a red heat and occasionally directing the flame down the header pipe to ignite any oil gum present. If the chamber is made of 16 or 18 gauge steel, it is an idea to tap the chamber quite vigorously with a piece of broom handle as this will help free the carbon clinging to the walls. Wait until the chamber cools a little before you do any banging, otherwise the walls could easily be deformed.

Chambers made of 20 or 22 gauge material are much more difficult to clean, as the heating will often cause the chamber walls to deform slightly, and banging with a stick while the chamber is still hot, to loosen baked on carbon, is definitely out. About

The exhaust system

all you can do is not use an intense flame, but wait until the chamber is quite cool before attempting to free the burned carbon ash. You can tap the chamber lightly with a rod or you can throw a few rough stones or a short length of chain inside and give it a good shake.

There is something you should remember, and that is to give your exhaust a burn-out in a well ventilated area, and preferably outside your shed. The oil in the chamber gives off foul fumes, which will soon have your eyes watering and your throat hurting, so be warned.

The other part of exhaust maintenance involves a careful inspection for any sign of cracks. These usually appear in the mounting bracket proper, or around where the brackets are welded to the main body of the chamber. Chambers constructed of very light gauge material will often crack along seams in the parallel belly section, as this section tends to pulse in harmony with the system's sonic waves.

Chapter 5

Carburation

The basic requirement for any carburettor is that it meters the fuel and air in such proportions as to be easily combustible and hence enable the engine to produce good power over a wide rpm band. Usually, the mixture we want is around 1:12 or 1:13, ie 1lb of petrol (gasoline) for every 12 or 13lb of air. Such a blend is just right for full throttle operation, but for other conditions, such as starting or light load operation, the fuel-air requirement is different (Table 5.1). Therefore, the carburettor has to 'sense' the engine's operating conditions accurately and adjust the fuel-air mix accordingly. If the carburettor is not able to do this, flat spots and engine surging will result, spoiling performance. For this reason we have to be very selective as to the type and size of carburettor we choose for our particular engine.

Table 5.1 Fuel/air requirements

Running condition	Mixing ratio (by weight) Fuel: Air
Starting	1: 1–3
Idling	1: 8–10
Low speed running	1: 10–13
Light load ordinary running	1: 14–16
Heavy load running	1: 12–14

To understand more fully what we should be looking for in a carburettor to suit our engine's requirements, we need to go back to basics and get to know how a carburettor works. Nearly all two-stroke carburettors are a slide throttle type employing a fuel inlet system, an idle system and a main running system.

The inlet system consists of the fuel bowl, float and needle valve (or needle and seat). The fuel, before passing to the metering systems, is stored in the fuel bowl, and it is maintained at the correct level by the float and needle valve.

If the fuel is not at the correct level in the fuel bowl, the fuel metering systems will not be able to mix the fuel and air in the correct proportions, particularly when accelerating, cornering and braking. A high fuel level will richen the mixture, causing excessive fuel consumption and poor running. On the other hand, a low fuel level can be even more serious due to the leaning effect it has. This may result in flat spots when accelerating out of turns, or, in extreme cases the engine could overheat and seize.

A high fuel level may be due to an incorrectly adjusted float, or a needle and seat which is not seating properly and shutting off the fuel when the float rises to the correct level. This can be caused by excessive wear to the needle and/or seat, or by a speck of dirt which may prevent the needle closing fully.

A low fuel level condition may also be the result of an improperly adjusted float, but it could be due to the fuel lines or the needle and seat being too small to flow sufficient fuel to keep the fuel bowl full.

The fuel bowl is always vented so that the fuel is being mixed according to the outside air pressure. Without a vent, accurate metering would not be possible, as fuel vapours would build up pressure in the bowl and displace fuel out through the main and idle metering circuits.

The float is usually made of brass stampings soldered together into an airtight assembly, but it may also be formed from plastic or a closed cellular material. Brass

Figure 5.1 Carburettor idle system operation.

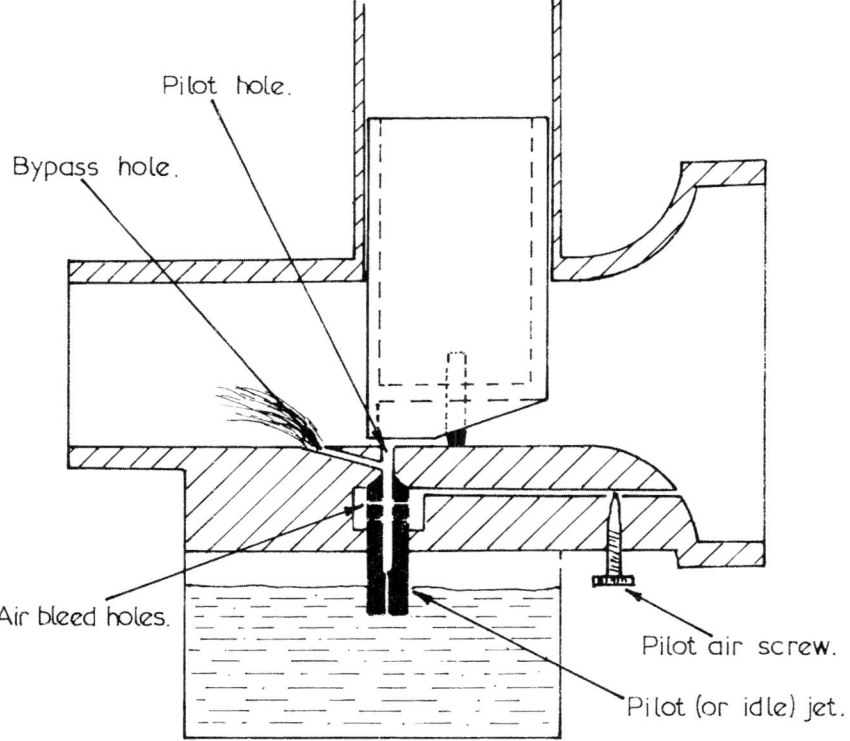

floats are resistant to all types of fuel except nitromethane, and generally plastic or cellular floats are not damaged by most common fuels. However, if you intend to use a fuel other than petrol or methanol, it is wise to check with the carburettor manufacturer regarding possible float damage.

The needle and seat is usually made of steel, although some needles may have a Viton coating on the tip. The Viton promotes excellent sealing and acts as a shock damper, but it should not be used with fuels containing any alcohol or nitro.

The needle valve controls the flow of fuel into the bowl, so some thought must be given to increasing its flow capabilities if an alcohol fuel is being used. Generally, it is not possible to obtain larger needle and seat assemblies for this purpose, so it will be necessary to modify the seat by very carefully drilling the fuel discharge holes oversize. I would recommend you use a pin vice to hold the drill when you do this.

The idle system provides a rich mixture at idle and low speeds when not enough air is being drawn through the carburettor to cause the main system to operate. (Figure 5.1).

When the throttle slide is nearly closed, the restriction to air flow causes a high vacuum on the engine side of the throttle slide. This high vacuum provides the pressure differential necessary for the idle system to operate. The normal air pressure (14.7psi) acts on the fuel in the float bowl, forcing it up through the idle jet and into the air stream.

To emulsify the fuel as it passes through the idle jet (pilot jet), and to provide a fine adjustment of the idle mixture strength, an air bleed circuit and an air adjustment screw is incorporated in the idle system. Turning the air screw out (anti-clockwise) decreases the air bleed restriction and leans the idle mixture. Conversely, turning the

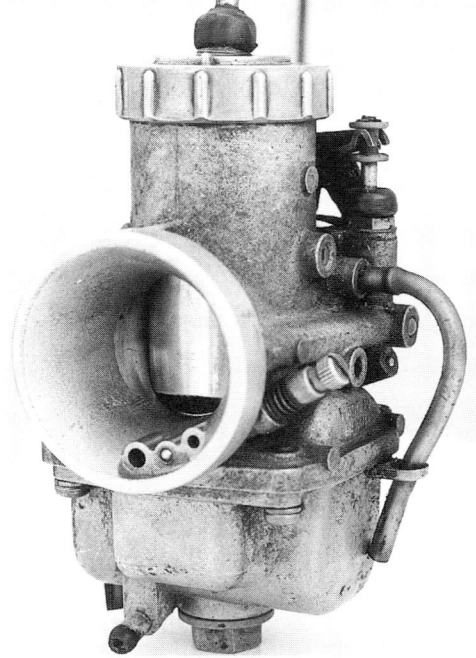

The round slide Mikuni carburettor.

Carburation

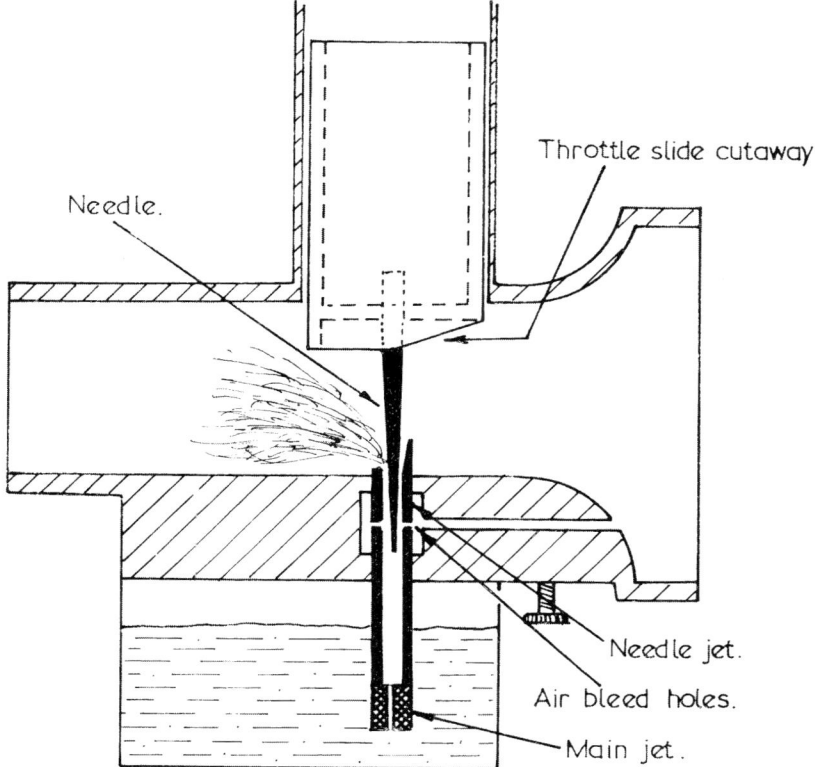

Figure 5.2 Carburettor main metering system operation.

screw clockwise richens the mixture by reducing the amount of air passing through the air bleed circuit.

Many carburettors have the two-hole type idle system illustrated in Figure 5.1, which gives better low speed and mid-range throttle response than the simpler one-hole type. At very small throttle openings, the by-pass hole, and not the pilot hole, actually provides the engine with fuel. Instead, the pilot hole acts as a supplementary air bleed to further atomise the fuel after it passes through the idle jet. Then as the throttle opens wider the pilot hole, too, begins to spray fuel. This serves to supply the engine with additional fuel until the air speed through the carburettor increases sufficiently to start the main system flowing fuel into the air stream.

As the throttle is opened wider and engine speed rises, so the air speed through the carburettor increases, bringing the main fuel delivery system into operation (Figure 5.2). If a carburettor of the correct size has been selected, this should happen at about quarter throttle as the idle circuit reaches the limit of its fuel flow capability at that time. A serious flat spot would result if the main system was not operational and supplying the additional fuel required to prevent a lean mixture.

In the two-stroke engine, a partial vacuum is created in the crankcase by the upward movement of the piston. Because atmospheric pressure is higher than the pressure in the crankcase, air rushes in through the carburettor to equalise the pressure difference. On its way into the crankcase, the air has to speed up as it passes through the carburettor bore. This speeding up of the air reduces the pressure below atmospheric inside the carburettor bore, allowing normal atmospheric pressure (14.7psi) to force fuel up through the main jet and past the needle into the air stream. The pressure differential existing here is often referred to as the 'signal' of the main metering system, because it starts and stops fuel flow.

No fuel discharges from the main metering system until air flow through the carburettor produces a pressure drop or signal of sufficient intensity for the atmospheric pressure, acting on the fuel in the fuel bowl, to push fuel up through the main jet and past the metering needle to be discharged into the air stream. Pressure drop (or vacuum) within the carburettor varies with the engine speed and throttle opening. Wide open throttle and peak rpm give the highest air flow, and consequently the highest pressure difference between the fuel bowl and the needle jet's discharge orifice. This, in turn, produces the highest rate of fuel flow into the engine.

To compensate for various engine displacements and engine operational speeds, a range of carburettors with a variety of bore diameters are available to create the pressure drop necessary to bring the main fuel circuit into operation. A small carburettor will provide a higher pressure difference, at any given rpm and throttle opening, than a larger carburettor, assuming engine size is the same. This is a very important aspect of carburation, which partly explains why the biggest is seldom the best. If the signal being applied is too weak (due to the bore diameter being excessively large), this could delay fuel discharge in the main system, causing a flat spot due to excessive mixture leanness. Also, fuel atomisation will not be good at any engine speed. Poor atomisation spells poor combustion and reduced horsepower, so, if you must err when selecting a carburettor, err on the small side, as the effect on performance will be far less devastating.

In the main fuel system, metering is controlled by the main jet, the needle jet and the needle.

The main jet controls fuel flow from the fuel bowl into the needle jet. An increase in the main jet diameter enriches the mixture, but there are other aspects worthy of consideration. The shape of the jet entry and exit, as well as the bore finish, also affect fuel flow. Carburettor manufacturers measure the flow of every jet, and then number the jet according to its flow characteristics, not according to the nominal bore diameter. For this reason, jets used to meter petrol should not be drilled to increase their size if you desire accurate fuel metering. An engine burning alcohol does not require such accurate metering, unless fuel consumption is a consideration, so jet drilling may be in order when large jets are not available.

In the case of Mikuni carburettors in particular, there is another problem. Due to its popularity, outsiders are manufacturing main jets to suit these carburettors but, unfortunately, many of the jets fall far below the level of quality and flow parity of the genuine Mikuni item. I have seen many jets improperly drilled. Some had deep spiral grooves in the bore, others had steps and tapers in the bore. When jets like that pass through quality control undetected, then obviously their flow checking must be either non-existent or at best haphazard, so I would suggest that you use only genuine jets from the carburettor manufacturer.

At most throttle openings, the volume of fuel introduced into the air stream is controlled primarily by the taper of the needle and the diameter of the needle jet. As

Carburation

illustrated in Figure 5.2, it is the clearance between the needle and needle jet which regulates fuel flow between quarter and three quarters throttle, although the main jet does have some influence. Above three quarter throttle the main jet mainly controls fuel metering, but the needle/needle jet does exercise partial control.

To assist in atomisation and high speed fuel flow metering, many carburettors also incorporate an air bleed circuit in the main metering system. The air bleed aids fuel vaporisation by introducing air into the fuel before it enters the air stream. When the fuel is thus broken down into smaller particles, combustion speeds up because there is more fuel surface area exposed to the combustion flame.

The air bleed also acts as a compensator or air corrector in adjusting high speed fuel flow requirements. Uncorrected, the carburettor would deliver a fuel/air mixture which would become richer as air speed through the carburettor increased. The reason for this is that the pressure drop in the carburettor is in direct proportion to the air speed through its bore and, in turn, fuel flow is in direct proportion to the pressure drop thus created. However, the actual mass of air passing into the engine does not remain in proportion with air speed, so the mixture strength would become excessively rich with increases in air speed through the carburettor if uncorrected.

Fuel flow, in both the idle and main circuits, is also influenced by the throttle slide cutaway size at throttle openings between one eighth and half throttle (Figure 5.3). Increasing the cutaway size reduces airflow resistance and leans the mixture. This occurs because air flow into the engine is greater (due to reduced air flow resistance) but fuel flow does not increase proportionately, as the larger cutaway reduces the pressure drop within the carburettor bore.

The Mikuni Powerjet carburettor, originally designed for snowmobile racing engines was adopted by Yamaha for some of their bikes in 1976. These carburettors have, in addition to the normal metering systems, another separate system with its own power jet and delivery tube, hence the name Powerjet. The delivery tube hangs in the air intake, in front of the throttle slide, and is connected through a metering orifice to the float bowl. Up to three-quarter throttle, the Powerjet carb operates just like any

Figure 5.3 The size of the throttle slide cutaway exercises its largest influence on the fuel mixture up to quarter throttle.

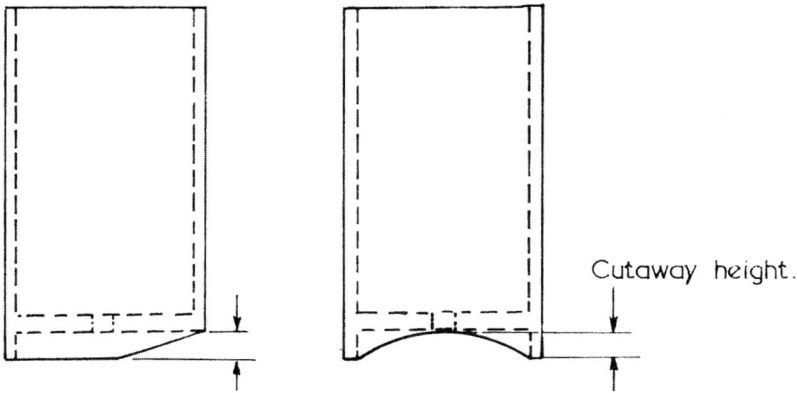

other Mikuni but, as full throttle is approached, the air moving past the tip of the Powerjet delivery tube creates a depression great enough to allow atmospheric pressure to push fuel through the power metering jet and up the delivery tube into the airstream.

The Powerjet therefore has the effect of enriching the mixture past five eighth throttle providing air velocity is high enough to create a vacuum of sufficient intensity to discharge fuel. It is, in effect, a load sensitive enrichment system which allows for more accurate (ie leaner) fuel metering at part-throttle operation. This ensures clean acceleration and smooth running out of turns, yet gives proper full throttle mixture richness for high power and effective engine cooling. The effect is an improvement in part-throttle performance and reduced fuel consumption. Additionally, some two-strokes have a tendency to lean out as they pass beyond full power rpm with conventional carburettors, but the Powerjet Mikuni is able to cure this. Normally, when a Powerjet carburettor is used, the main jet size will be around 70–100 smaller, depending on the size of the power jet fitted to the carburettor (Tables 5.2a and 5.2b).

Table 5.2a Mikuni Powerjet jetting comparison for Yamaha IT175

	Yamaha IT175F	Yamaha IT175G
Carburettor	34mm VM	32mm VM Powerjet
Main jet	360	210
Power jet		82.5
Needle	6F21	6F21
Needle jet	P-8	P-8
Cutaway	2.0	2.0
Pilot jet	70	60

Table 5.2b Mikuni Powerjet jetting comparison for Yamaha TZ250

	Yamaha TZ250D/E	Yamaha TZ250G
Carburettor	34mm VM	34mm VM Powerjet
Main jet	320 to 360	230 to 270
Power jet		60 to 80
Needle	6F9	6DH3
Needle jet	0–2	N–8
Cutaway	2.5	2.0
Pilot jet	70	60

However one of the big problems with early round slide carburettors such as the Mikuni VM and VM Powerjet was air turbulence around the throttle slide. At small throttle openings the intake air rushes down the carburettor throat and heads through the throttle slide cutaway. At that point, because the cavity in the bottom of the traditional round slide is so much larger than the small airway path at the front of the slide, the intake air loses velocity and direction which reduces its ability to create sufficient vacuum to pick up and atomise fuel from the idle circuit. So rather than finely atomised combustible fuel entering the air stream, globs of fuel are collected. Globs of fuel do not burn very efficiently which results in a lean stumble. If the lean

Carburation

The cavity in the bottom of the round throttle slide fitted to Mikuni VM carburettors causes fuel metering problems at smaller throttle openings.

stumble is minimised with richer jetting the mixture then swings to being too rich when the air speed increases at higher rpm.

The round slide goes on causing problems with turbulence as the throttle is opened wider. However this is not such a concern as far as metering goes but rather it becomes a high speed air flow problem because as the air passes from the front of the carb to the round slide recess in the carb it eddies, thus limiting air flow (Figure 5.4).

Figure 5.4 Eddies caused by the round slide cavity disrupt air flow and fuel metering.

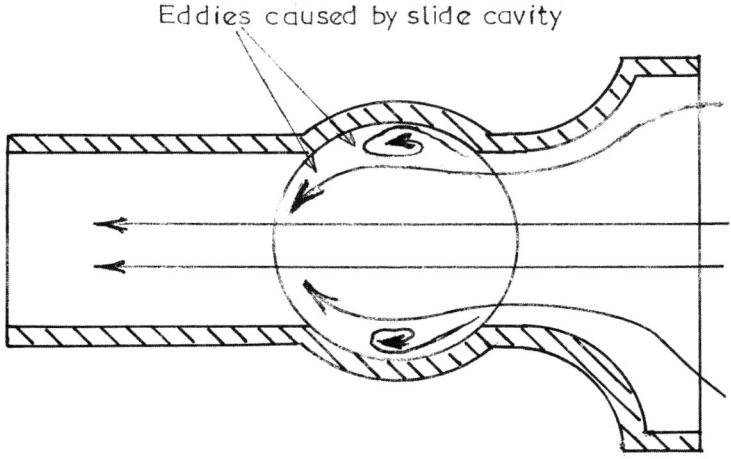

Then, to achieve acceptable high speed air flow tuners fitted bigger bore carbs. These produced even more turbulence at small throttle openings than the carbs they replaced, hence low speed and mid-range metering was even poorer. The consequence of this was a lack of acceptable metering until mid-range rpm were reached making the bikes peaky and difficult to control. In turn this peakyness stifled the adoption of more aggressive expansion chambers.

All of that changed though when the Bill Edmondston-designed flat-slide Lectron carburettor was introduced. This was followed by another Edmondston carb, the E.I., and in 1981 the Mikuni flat slide appeared on some production bikes.

The Lectron with its superior air flow was an immediate hit with road racers. However at small throttle openings it too had metering woes. First the metering circuit was very simple and second, at part throttle, the flat guillotine slide allowed air to flow around the slide. This reduced air flow velocity under the slide and over the fuel pickup area, led to part throttle stumbling.

The E.I. carb advanced a step further. It still had a relatively simple metering arrangement. There were two models, the 'Silver' or the 'Blue'. Generally the Silver

Looking down the bore of a round slide carburettor it can be clearly seen how the slide recess disrupts high speed air flow.

was recommended for reed engines while the Blue was for other types. The Blue had an air bleed circuit similar to a Mikuni. A drilled passage metered air from the air horn into the needle jet. This served to introduce air into the fuel to help atomisation and it reduces signal strength at wide throttle opening to prevent excessive high speed richness. Its effects were adjusted by changing air bleed jets. However the big improvement was in the design of the throttle slide. Now, instead of being like the blade of a guillotine, it was H-section shape. The change eliminated leakage around the slide so part throttle air velocity over the needle jet was higher, providing improved metering accuracy and fuel atomisation. Now riders reported easier starting, 'clean' low speed and mid-range carburation and superior power at full throttle.

Before producing their flat-slide carburettor, Mikuni tried a couple of other types, both being an improvement in some ways on the round-slide VM but inferior to the Lectron. The first type, the round-slide 'smooth-bore' attempted to create a smooth bore at wide open throttle so as to improve high speed air flow. In this area it succeeded but in so doing, the hollow slide created an even greater part throttle metering problem than the VM. The next attempt was the round-slide type ZC. This time around Mikuni decided to tackle the idle circuit to main circuit transition metering deficiency of the VM. The round slide with its high speed air flow problem was retained, but instead of having a hollowed out bottom with a cutaway at the entry side, the slide now had a flat bottom and no cutaway whatsoever. With the slide almost closed the inflowing air was now forced to flow over the needle jet. To ensure the needle jet received full crankcase vacuum a big slot was cut into the rear side of the slide. With the previous VM type idle system plugged the needle controlled fuel flow at all throttle openings, just as on the Lectron. The ZC provided better throttle response in road racers, but still the air flow deficiency remained.

This was solved though when Mikuni made their flat-slide carburettor. It too has an H-section slide and filler blocks bolted into the carb bore to ensure a smooth air flow path. With the production of the flat-slide carb Mikuni now had a product which flowed well and metered accurately. However, the flat-slide carb was an expensive item to manufacture so they produced a budget round-slide carb a little later. They called this the R-bottom. Like Mikuni had determined with the earlier ZC type the idea was to improve fuel metering right through the range, from idle to maximum revs. The round-slide cavity in the carb body still reduced its air flow potential but the new style throttle slide improved fuel metering. The R-bottom slide has a bottom which is pretty much filled in but with an oval channel leading from the cutaway to the rear of the slide. This channel preserves air flow directly over the needle jet, increasing air flow velocity and improving fuel atomisation.

During all of these developments the designers at Keihin weren't sitting on their hands. In the course of time they produced the PJ oval slide and the PWK flat-slide carburettors. Then in 1996 when the '97 model Honda CR250 was released it appeared with a new type of carburettor, the Keihin Power Jet. This carb featured an electronically controlled solenoid valve to shut off fuel flow to the powerjet in the upper rpm range. This served to lean the mixture about 10 per cent which primarily allowed the engine to over-rev 500–800rpm, and in some later applications there was also a peak hp increase.

The ability of any engine to over-rev and produce effective power well beyond the power peak is very important, but many two-strokes basically drop dead soon after max power rpm. For one thing the ability to over-rev allows shorter gearing to be utilised without losing top speed. This produces better acceleration out of the turns. An over-rev without power loss means less gear changes, or the option of having a choice between two gears in some corners. It also widens the effective power band so that on going up through the gears acceleration is improved as the engine can be maintained in the upper reaches of the power band.

One old myth that has become even more strongly entrenched with the introduction of the Keihin Power Jet carb is that two-strokes need to run rich down in the range to produce good torque and they must be lean at the top to make hp. This is complete rubbish! Horsepower is a product of torque so if an engine is producing a high torque figure in the mid-range, then the hp at those revs will also be high. Conversely if an engine is making good hp at the top end, then it must also be producing a high torque number. What is true however is that to make good bottom and mid-range hp a two-stroke has to be slightly rich, and to make good top end power it has to be in comparison, a bit leaner.

This is all a product of the way two-strokes operate. At lower rpm the expansion chamber isn't very efficient at filling the cylinder with fuel/air mixture. Additionally it is not very good at pulling all of the exhaust gas out of the cylinder. Hence at lower engine speed the cylinder is only partly filled with fuel mixture and what mixture is in there awaiting the ignition spark is diluted by residual exhaust gas. Now to make good power we have to ensure that every oxygen molecule is used up in the combustion process. However the fact that the cylinder has been only partially filled and the presence of a lot of inert exhaust gas means that at 'normal' fuel/air ratios there can be too much space between fuel and oxygen molecules for every oxygen molecule to share in making power. What we have to do is slightly oversaturate the cylinder gases with fuel to ensure that a higher percentage of 'happy' fuel and oxygen unions occur. Then, when combustion takes place there will be fewer 'unmarried' oxygen molecules left over that will not be able to contribute to making power. Sure, with the mixture a little rich fuel will be wasted because some molecules can't find an oxygen partner, but what we are looking to do is to turn every oxygen molecule into a power producer.

At the top of the power curve the situation changes. Now the expansion chamber is very effective in evacuating the cylinder of exhaust gas and in pulling up a cylinder full of fuel/air mixture. Also, when air speed through the carburettor is high a much larger portion of fuel will be atomised into combustible droplet size as opposed to lower rpm where a considerable percentage of fuel droplets will be too large to burn at the correct time to make power. Therefore comparatively speaking we reduce, lean, the amount of fuel introduced into the air stream. If we keep the mixture rich at this point we have a relatively cool and relatively slow burn. This makes life easier for the piston which is why in desert racing and endurance events we would usually jet a touch rich. However, normally what we are after is maximum power right up to the rev limit of the engine. For that to be achieved we have to match exactly the right number of fuel molecules to the correct number of oxygen molecules.

When we waste space in the cylinder with an excess of fuel molecules, space that could otherwise have been occupied by oxygen molecules, the engine will not be achieving its full hp output potential. This is where the benefit of the electronic power jet Keihin carburettor becomes evident. At the rpm programmed into the electronic ignition box it closes an electrical circuit to activate the solenoid valve which shuts off fuel to the power jet. Depending on the relationship between the size of the main jet and the power jet the fuel mixture will be leaned about 10 per cent. However this can be easily changed for desert racing for example by using a slightly larger main jet and a smaller power jet. Also, the engine revs at which the solenoid valve operates can be reprogrammed when a replacement reprogrammable ignition box such as the Vortex is fitted. Vortex also have replacement spark boxes available with a power jet solenoid driver to facilitate the adoption of the Keihin Power Jet carb on to bikes with ordinary carburettors.

From the foregoing do not assume that your engine will make more top end and have over-rev capability if a Keihin Power Jet carb is fitted. As I pointed out in the preceding chapter, some two-stroke expansion chambers can produce a very weak signal which reduces the carburettor's ability to introduce sufficient fuel into the air stream at various points in the power band. If that weak signal lean out is down the rev range, away from max torque and max hp engine speeds the power jet shut down at high rpm may help by allowing you to get the bottom end richer without going too rich at the top. However if the weak signal lean period is at the top of the power curve the problem will be worsened and you could cook the engine when the power jet circuit closes off.

Having discussed the way a carburettor works, it should be obvious that not all carburettors are suitable for high performance applications. Some are just too small; others do not have jets and needles that are readily obtainable; some have metering systems which allow acceptable performance on stock machines, but are too crude to provide the correct fuel/air mixture in a racing engine; and others have totally inadequate float/needle valve systems, so that flooding, hard starting and poor running is always a problem.

Obviously the choice is yours when it comes to selecting a carburettor, but do give heed to my advice regarding the size to use. Always remember that the carburettor meters the fuel according to the signal being received in the fuel bowl. No matter how well designed, a carburettor too large for the engine produces a weak signal, consequently the metering system cannot function correctly.

To illustrate this, you might like to try a simple experiment. Fill a container with water and draw the water up through a drinking straw and then through a piece of 3/4in hose. Did you notice how much more sucking you had to do to get the water flowing up the piece of hose? That very same principle applies to fuel flow through a carburettor's metering system. The small carburettor being sucked on by the engine (like you sucking on the straw) gets the fuel responding quickly to the requirements of the engine.

The other advantage of using a carburettor of the right size is this: a high air velocity through the bore lowers the air pressure (ie creates a partial vacuum), which makes the fuel more volatile and easily vaporised. You probably already know that water will change from liquid to gas (steam) at a lower temperature on the top of a mountain than at sea level, because the air pressure is lower. Likewise, with petrol, vaporisation improves with a decrease in air pressure. Additionally, the high air speed in itself assists in breaking up the fuel and vaporising it. Properly atomised fuel blends more evenly with air filling the cylinder and power goes up due to the improved combustion which results.

In Table 5.3, you can see what size carburettors are recommended for various applications. Generally speaking, less experienced riders and all those operating moderately modified engines would use the smaller size carburettors listed for each engine size and application. Rotary valve engines, radically modified engines, and more experienced riders, could use the larger size carburettors. The increase in maximum power will not be very great (maybe 1hp), but the engine will rev perhaps 500–800rpm harder with the larger carburettor before falling off the power band.

Table 5.3 Recommended carburettor sizes

Cylinder size (cc)	Carburettor size (mm)		
	Enduro	Motorcross	Road race
50–62		24–26	27–29
80		27–29	29–32
100		30–32	32–34
125	30–32	33–36	*35–38
175–200	33–38		**32–34
250	36–38	37–40	40–42
350–500	36–38	38–42	40–44

* Street sports bike modified for road race use 32–34mm.
** Refers to street sports bike modified for road race.

To give you some idea of the dilemma which you face when selecting an appropriate carburettor bore size consider a couple of conflicting examples. The first example involves Yamaha RD250 LC and RZ350 street bikes modified for a production race class and a Honda CR500 modified for flat track racing. With revised porting and new expansion chambers the two Yamaha's made 46hp and 58hp on their tiny standard carbs; 26mm on the 250 and 28mm on the 350. You would expect that bigger 32mm or 34mm carbs, would allow these engines to rev harder and make more power, but no, right up to maximum engine speed there was only 0.2 to 0.5hp difference. The situation was similar with the Honda. Running a modified barrel and expansion chamber it produced 65hp on its standard 38mm carburettor. We were looking for more power and figured that a bigger carb would help. A 42 allowed the engine to run 350rpm higher but peak power was unchanged. The 40 and 44 carbs lost hp (up to 0.5hp) in places and gained less than a hp in other places.

The opposite result was produced when a Rotax 250 tandem was being set up for a road race kart. Because the kart and driver were both in need of substantial dieting, plus as the Rotax gear spread was a bit wide for the tight circuits where this combination would be competing, it was decided to fit smaller 36mm flat-slide Mikuni carbs, in place of the usual 38s. The idea was that these would carburet more cleanly at lower rpm as air speed and thus signal strength would be higher. A stronger bottom end, would, it was reasoned, help cover some of the gear gaps and the excess weight the engine had to push around some very tight and quite hilly tracks. Figure 5.5 illustrates how erroneous that calculated guess was, as the small carbs resulted in the

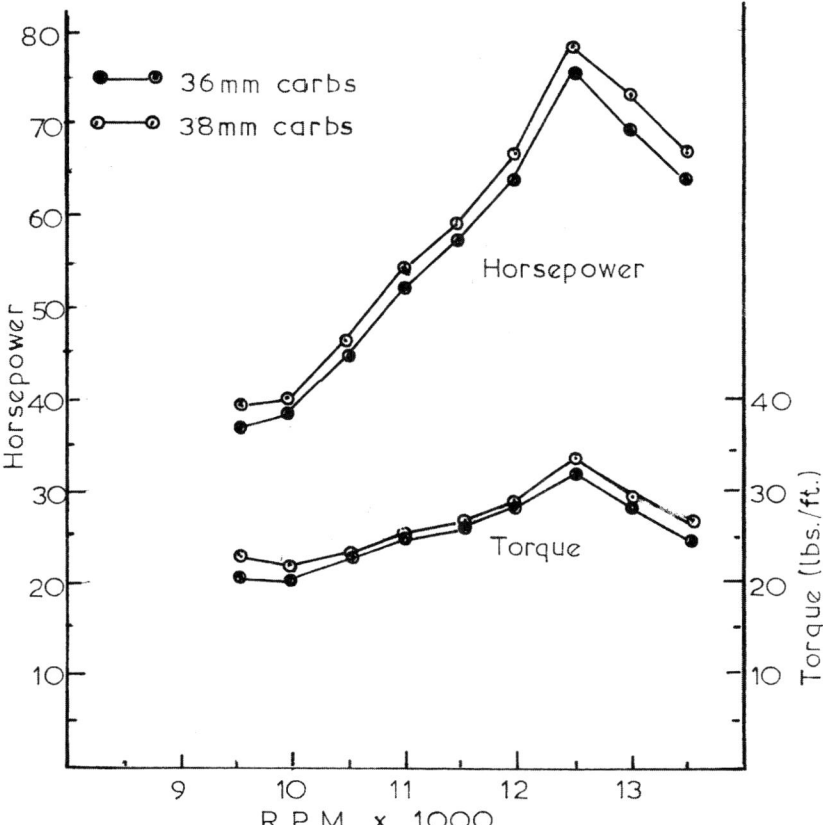

Figure 5.5 Rotax 250 power curves illustrate the effect carburettor size has on horsepower output.

engine losing substantial power at both the top and the bottom of the power curve. Between 10,000 and 12,000rpm the bigger carbs are 1.0–1.5hp stronger, while at 9,500rpm and at 12,500 to the rev limit they are 2.4–3.0hp stronger. All this of course shows the need for time on the dyno.

Looking through the recommended size table, you will note some odd carburettor sizes. You cannot buy a 35mm Mikuni for instance; but you can bore a 34mm VM carburettor out to 35mm. In fact, you can save yourself quite a lot of money if you do this, rather than buy a new carburettor to improve performance. Actually a 32 or 34mm VM Mikuni bored out to 35mm will flow very close to the same amount of air as a 36mm carb because the bored carburettor is more streamlined inside. This occurs because the smaller carburettor finishes up with a bore diameter very close to the throttle slide chamber size, so there is less disruption to the air as the carburettor bore cross-sectional size changes less.

When a carburettor is being bored, the utmost accuracy is required, otherwise the carburettor will be ruined. If the new bore is offset to the left or right of centre, then the engine may refuse to idle due to air leaking past the side of the throttle slide.

Likewise, if more than 0.25–0.35mm is removed from the floor of the bore, the engine will also not idle as it will be impossible to adjust the slide down low enough to restrict air flow into the engine. Therefore, if you are boring a 34mm Mikuni to 35.3mm, which is the maximum oversize by the way, I would suggest that the new bore be offset by 1mm so that the boring tool removes very little from the carburettor floor. A 32mm Mikuni can be bored to 33 or 34mm in a similar way.

When the 32mm Mikuni is bored to either 33 or 34mm, it is necessary to machine the top of the throttle slide to reduce its height. If this isn't done, the slide will run out of travel before the throttle is fully opened.

Mikunis use two types of main jets. The hex head jets are flow rated in cc per minute. Jets from size 50 to 195 are available in steps of 5, and sizes 200 to 500 are in steps of 10. The round head main jets are aperture sized. The largest jet available is a 250, which has an aperture size of 2.50mm.

The needle jet has a code to identify its size. For example, a 159 P-8 needle jet is a 159 series jet which fits 30–36mm spigot mount Mikuni carburettors (Table 5.4). The letter-number combination shows the fuel hole size. The letter denotes the size in increments of 0.05mm, and the numbers signify size increments of 0.01mm. Therefore a P-8 jet would have a hole size of 2.690mm. There is one exception to this: the -5 needle jet is 0.005mm larger than a -4 jet (Table 5.5).

Table 5.4 Mikuni needle jet application chart

Series No.	Type	Main jet	Sizes available	Carb type
159	P	Hex	O-0 to R-8	30–36mm spigot
166	P	Hex	O-0 to R-8	38mm spigot
171	P	Hex	O-0 to Q-8	30mm flange
176	B	Hex	N-0 to Q-8	30–36mm spigot
183	B	Hex	N-0 to Q-8	38mm spigot
188	P	Hex	O-0 to Q-8	32mm flange
193	P	Hex	N-0 to Q-8	24mm flange
196	P	Round	O-0 to Q-8	30–36mm spigot
205	P	Hex	O-0 to Q-8	34mm flange
211	P	Hex	N-0 to Q-8	30–36mm spigot
249	P	Hex	N-0 to Q-8	24–28mm spigot
224	P	Hex	Z-0 to CC-5	40–44mm spigot

Note – 'P' type needle jets are intended for use primarily in two-stroke piston port engines. 'B' type needle jets have bleed holes and are normally used in four-stroke and rotary valve two-stroke engines.

In Table 5.4 you will note there are two types of needle jets available: the 'P' type and the 'B' type. The 'P' type have a single air bleed hole (Figure 5.2) and are used in most piston-ported two-stroke engines. The 'B' type have several air bleed holes. They are for use in rotary valve two-strokes and four-stroke engines. In the case of the primary ('P') type, air that comes from the air bleed orifice is mixed with fuel which has already been metered by the needle and needle jet. The bleed ('B') type, on the other hand, is designed to hold air in the body section of the needle jet, so in this case fuel and air (actually a frothy fuel/air mixture) is metered by the needle and needle jet.

Table 5.5 Mikuni needle jet sizes

Size	Dia (mm)	Size	Dia (mm)	Size	Dia (mm)
N-0	2.550	P-2	2.660	R-4	2.770
N-2	2.560	P-4	2.670	R-5	2.775
N-4	2.570	P-5	2.675	R-6	2.780
N-5	2.575	P-6	2.680	R-8	2.790
N-6	2.580	P-8	2.690	Z-0	3.150
N-8	2.590	Q-0	2.700	Z-5	3.175
O-0	2.600	Q-2	2.710	AA-0	3.200
O-2	2.610	Q-4	2.720	AA-5	3.225
O-4	2.620	Q-5	2.725	BB-0	3.250
O-5	2.625	Q-6	2.730	BB-5	3.275
O-6	2.630	Q-8	2.740	CC-0	3.300
O-8	2.640	R-0	2.750	CC-5	3.325
P-0	2.650	R-2	2.760		

The metering needles are identified by a code such as 6DH2. The first number indicates the needle series. The following letter/s indicate the needle taper. If there is one letter, the taper is uniform along the length of the needle, but if there are two letters, this indicates that the taper changes midway along the tapered section. The first letter indicates the upper taper and the second letter the lower taper.

Starting with letter A, which has a meaning of 15 minutes of arc, each letter in sequence denotes an additional 15 minutes to the angle between the two sides of the needle. Therefore a DH taper has an angle of 1°0′ on the top, and 2°0′ on the bottom taper (Table 5.6).

Table 5.6 Mikuni needle tapers

Letter	Taper	Letter	Taper	Letter	Taper
A	0°15′	J	2°30′	S	4°45′
B	0°30′	K	2°45′	T	5°0′
C	0°45′	L	3°0′	U	5°15′
D	1°0′	M	3°15′	V	5°30′
E	1°15′	N	3°30′	W	5°45′
F	1°30′	O	3°45′	X	6°0′
G	1°45′	P	4°0′	Y	6°15′
H	2°0′	Q	4°15′	Z	6°30′
I	2°15′	R	4°30′		

The number after the letters is a manufacturing code which indicates how far down the needle the taper starts and/or the initial needle diameter, eg: needles marked 6DH2 and 6DH3 have the same taper, but 6DH3 is the richer needle as the taper starts 22.0mm from the top of the needle, whereas the taper begins 28.0mm down with the 6DH2. Tables 5.7 indicate the dimensions of the more common Mikuni needles.

Table 5.7a Mikuni series 4 needles
To fit some 26mm carburettors and 22 and 24mm flange mount carburettors

Needle	X	Y	10	20	30	40	50
4D3	50.3	25.3	2.511	2.511	2.421	2.253	2.100
4D8	50.3	22.8	2.519	2.519	2.381	2.211	2.000
4E1	50.3	28.0	2.515	2.515	2.345	2.127	1.924
4DG6	50.3	24.0	2.518	2.518	2.405	2.119	1.850
4DH7	50.3	23.0	2.518	2.518	2.386	2.098	1.790
4F15	50.3	26.5	2.512	2.512	2.400	2.120	1.881
4J13	50.2	24.0	2.513	2.513	2.230	1.800	1.400
4L6	50.3	24.5	2.515	2.515	2.178	1.660	1.190
4F6	50.5	25.3	2.514	2.514	2.406	2.145	1.876
4L13	45.1	25.0	2.518	2.516	2.339	1.842	
4F10	50.2	24.5	2.513	2.513	2.385	2.135	1.877
4J11	41.5	21.3	2.512	2.506	2.188	1.776	
4P3	50.5	25.0	2.510	2.506	2.436	2.284	2.122

Note – X is the overall length of the needle in mm.

Y is the dimension from the top of the needle to the start of the taper.

The numbers 10, 20, 30 etc. indicate the needle diameter in mm at a point 10, 20, 30 etc. mm from the top of the needle.

If the needle is identified as a 6DP5-3 in the bike owner's handbook, this would indicate that the needle is a 6DP5 and that it is fitted standard with the circlip in the third groove, counting the top groove as number one.

The throttle slide cutaway size is indicated by a number stamped under the slide, eg: a 2.0 signifies a 2.0mm cutaway. Smaller engines usually need a 1.5, 2.0 or 2.5 cutaway, and larger engines a 2.5 or 3.0.

The idle jet (pilot jet) is available in sizes 15 to 80, in steps of 5. Fine adjustment of the idle mixture is by means of the idle air screw which richens the idle mixture when turned in (clockwise).

The float level is adjusted with the fuel bowl removed and the carburettor inverted (Figure 5.6). With the float tongue contacting the needle valve, the distance 'A' should be equal to the specified float level. Usually, this will be 20 to 35mm depending on the carburettor type. This measurement is made with the fuel bowl gasket removed.

Some Mikuni carburettors have the float level adjusted to dimension 'B'. In this instance the level is usually around 9 to 10mm; again this varies from model to model.

Many tuners begin tuning carburettors by trying to determine the correct main jet size. This procedure is correct if the engine has not been extensively modified and the standard carburettor is still being used. However, if you find large changes in the size of the main jet do not seem to be having very much influence on the half to full throttle mixture strength, then you can be fairly certain that a larger needle jet is required.

When an engine has been extensively modified, I prefer to begin testing (after ensuring that the float level is correct) with the main jet removed. If the engine will

Table 5.7b Mikuni series 5 needles

To fit 26–32mm spigot mount and 28–34mm flange mount carburettors

Needle	X	Y	10	20	30	40	50	60
5D6	59.3	27.5	2.515	2.515	2.460	2.290	2.120	
5FJ9	59.2	35.0	2.517	2.517	2.517	2.364	2.021	
5D120	59.1	28.2	2.520	2.520	2.479	2.311	1.980	
5F3	58.0	27.4	2.519	2.519	2.419	2.135	1.863	
5EH7	57.6	28.5	2.517	2.517	2.473	2.210	1.848	
5E13	57.5	29.5	2.515	2.515	2.484	2.197	1.803	
5EJ13	57.8	26.5	2.519	2.519	2.431	2.210	1.766	
5DL13	60.2	32.0	2.515	2.515	2.515	2.362	1.922	1.463
5EJ11	60.3	28.5	2.515	2.515	2.515	2.241	1.839	1.420
5EL9	60.3	27.0	2.517	2.517	2.441	2.221	1.780	1.248
5FL11	60.3	28.2	2.518	2.518	2.438	2.175	1.740	1.256
5EP8	60.2	33.0	2.513	2.513	2.513	2.245	1.780	1.120
5FL14	58.0	28.0	2.520	2.520	2.440	2.170	1.735	
5FL7	58.0	28.0	2.518	2.518	2.440	2.170	1.735	
5DP7	57.6	26.4	2.512	2.512	2.440	2.259	1.580	
5J6	58.0	27.5	2.518	2.518	2.340	1.890	1.450	
5L1	58.0	27.0	2.518	2.518	2.330	1.811	1.297	
5C4	55.1	24.0	2.516	2.516	2.448	2.310	2.179	
5F18	58.0	27.0	2.521	2.521	2.515	2.257	2.006	
5J9	58.0	27.0	2.522	2.520	1.432	1.996	1.505	
5F12	51.5	23.3	2.021	2.021	1.882	1.631	1.375	
5D1	53.5	27.6	2.510	2.510	2.496	2.338	2.169	
5DP2	60.3	32.4	2.515	2.514	2.513	2.418	2.067	1.418
514	60.0	27.0	2.514	2.509	2.442	2.071	1.690	1.332
5D5	57.6	30.0	2.513	2.513	2.510	2.366	2.205	

Note – X is the overall length of the needle in mm.

Y is the dimension from the top of the needle to the start of the taper.

The numbers 10, 20, 30 etc. indicate the needle diameter in mm at a point 10, 20, 30 etc. mm from the top of the needle.

just run at part throttle, but floods as the throttle is opened, then the needle jet is large enough. However, if you find that the engine keeps going at three-quarter to full throttle you can be sure a larger needle jet is needed. Note that this test should be done with the needle lowered to the No. 1 (ie lean) clip position.

After you have determined a needle and needle jet combination that is too rich, you can then try various size main jets until you find one that allows the engine to run reasonably well at full throttle. Don't worry about throttle response or acceleration for the time being. Carry out this test with the needle raised to the middle clip position.

Next we have to find what size idle jet (pilot jet) is required. Start these adjustments by backing out the idle speed screw until the throttle slide is completely closed, and then turn the screw back in until the slide just barely breaks open. Having done that, close the idle air screw completely and back it out 1 to 1½ turns. Start the

Two-stroke performance tuning

Table 5.7ci   Mikuni series 6 needles
To fit 30–38mm spigot mount carburettors

Needle	X	Y	10	20	30	40	50	60
6H1	62.3	37.5	2.510	2.510	2.510	2.412	2.041	1.696
6DH2	62.3	28.0	2.511	2.511	2.466	2.295	2.000	1.660
6F9	62.3	28.9	2.516	2.516	2.475	2.210	1.949	1.678
6CF1	61.5	29.5	2.512	2.512	2.429	2.240	1.974	1.710
6FJ6	62.3	35.2	2.505	2.505	2.505	2.376	2.040	1.606
6DH3	62.3	22.0	2.512	2.512	2.458	2.286	1.948	1.607
6J3	62.3	36.7	2.515	2.515	2.515	2.359	1.912	1.456
6L1	62.3	37.0	2.512	2.512	2.512	2.335	1.826	1.313
6DP5	62.3	32.1	2.518	2.518	2.518	2.372	1.834	1.141
6N1	62.3	37.0	2.514	2.514	2.514	2.278	1.672	1.058
6DP1	62.3	28.9	2.511	2.511	2.476	2.312	1.748	1.075
6F3	60.5	34.2	2.512	2.512	2.512	2.313	2.050	
6DH4	62.3	25.5	2.520	2.520	2.440	2.258	1.915	1.575
6J1	64.0	36.2	2.517	2.517	2.517	2.339	1.919	1.495
6DH7	62.2	28.5	2.516	2.516	2.505	2.316	2.009	1.688

Note –   X is the overall length of the needle in mm

Y is the dimension from the top of the needle to the start of the taper.

The numbers 10, 20, 30 etc. indicate the needle diameter in mm at a point 10, 20, 30 etc. mm from the top of the needle.

Table 5.7cii   Mikuni series 6 needles
To fit 30–38mm spigot mount carburettors

Needle	X	Y	Z	10	20	30	40	50	60
6F5	62.3	38.1	19.0	2.515	2.456	2.454	2.364	2.098	1.840
6F4	62.3	32.0	19.4	2.515	2.442	2.436	2.206	1.939	1.678
6F8	62.3	34.0	21.5	2.512	2.512	2.386	2.214	1.945	1.688
6FJ11	62.3	36.0	18.7	2.519	2.481	2.481	2.367	2.030	1.610
6F16	59.1	36.7	18.5	2.519	2.489	2.489	2.372	2.104	
6DH21	52.3	30.1	16.5	2.515	2.470	2.465	2.328	2.024	
6F16	64.6	31.2	18.4	2.520	2.404	2.400	2.201	1.941	1.679

Note –   X is the overall length of the needle in mm.

Y is the dimension from the top of the needle to the start of the taper.

Z is the dimension in mm from the top of the needle to the pronounced taper point.

The numbers 10, 20, 30 etc. indicate the needle diameter in mm at a point 10, 20, 30 etc. mm from the top of the needle.

engine and attempt to obtain a smooth 1,000–1,500rpm idle, by juggling the air screw and the idle speed screw in turn.

If you are tuning a multi-cylinder engine and you can get the engine to settle down to a good idle at this point, you should then synchronise the throttle slides so

Carburation

Table 5.7d   Mikuni series 7 needles
To fit 40–44mm spigot mount carburettors

Needle	X	Y	10	20	30	40	50	60	70
7FO6	72.3	29.0	3.005	3.005	2.951	2.680	2.415	2.140	1.876
7H2	72.3	28.9	3.005	3.005	2.928	2.575	2.230	1.868	1.507
7J2	72.3	28.8	3.005	3.005	2.904	2.460	2.010	1.569	1.125
7F2	73.0	43.5	2.515	2.515	2.515	2.515	2.312	2.040	1.703

Note – X is the overall length of the needle in mm.

Y is the dimension from the top of the needle to the start of the taper.

The numbers 10, 20, 30 etc. indicate the needle diameter in mm at a point 10, 20, 30 etc. mm from the top of the needle.

that they open and close together. The easiest way to do this is to remove the needle and needle jet and then open the throttle just wide enough for a length of 1/8in bronze welding wire to be slipped down a carburettor throat for the throttle slide to seat on. Then, holding the throttle open just wide enough to keep the welding wire trapped,

Figure 5.6 Mikuni float levelling method varies according to the specific carburettor type.

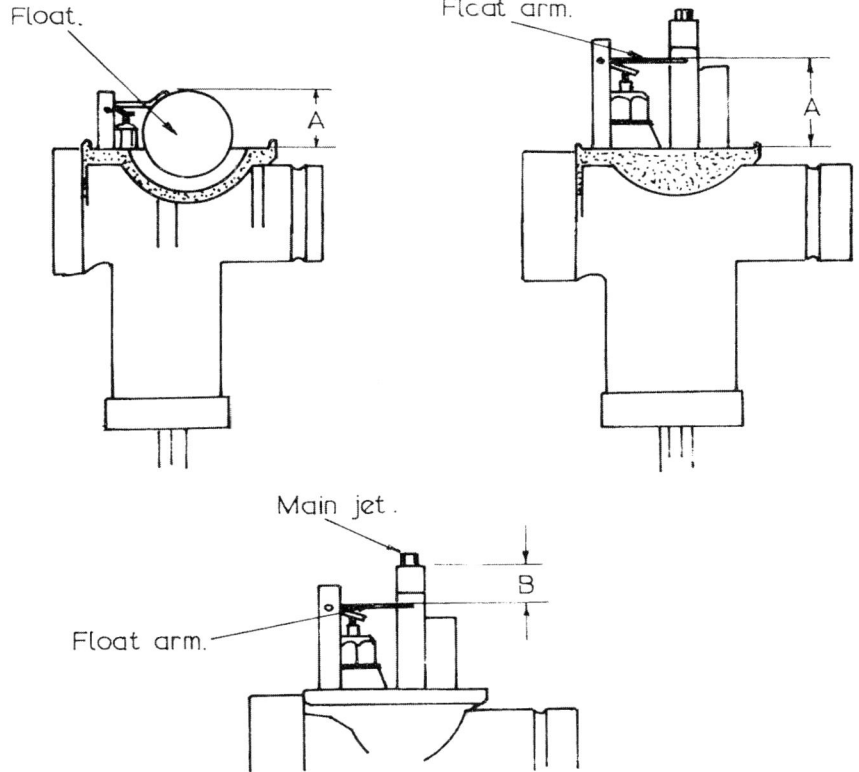

adjust the throttle cable length on the other carburettor/s so that another piece of 1/8in welding rod can just slip under the throttle slides.

If the engine will not idle, it is probable that the idle jets are wrong. Jets that are too small are indicated by an increasing idle speed as the air screws are turned in. Turning the screws in should cause the engine to run rich at some point (usually 1/2 to 1 turn from being fully closed) when the idle jets are of the correct size. An idle jet that is too large is indicated by an ever-increasing idle speed, as the air screws are backed further and further out. Note that the air screws must not be opened more than three turns, otherwise they will vibrate out.

Once the idle has been adjusted, you can test that the slide cutaway is of the correct height. The cutaway influences the mixture between 1/8 and 1/2 throttle, and especially in the range of 1/8 and 1/4 throttle opening. Therefore, if the engine tends to cough and die when the throttle is cracked open, but gains revs as the throttle is closed, change to slides with less cutaway (ie richer). (At times minor leanness at this point can be corrected by raising the needle one or more notches.)

If you find it necessary to change the slides, recheck throttle synchronisation and determine that the idle jet is still of the correct size. A change in idle jet size is generally only required when a large change in cutaway height has been made.

Now take the bike for a run and check that the main jet is approximately correct by testing how it runs at 3/4 to full throttle. If the engine runs well and the plug reads a good colour then the main jet is close enough to begin searching out the correct needle profile and/or needle position.

The needle taper and position controls the fuel/air mixture between 1/4 and 3/4 throttle. To determine if a change is required, test the bike on a smooth and level road

A good selection of carburettor needles and jets should be on hand during carburettor tuning sessions.

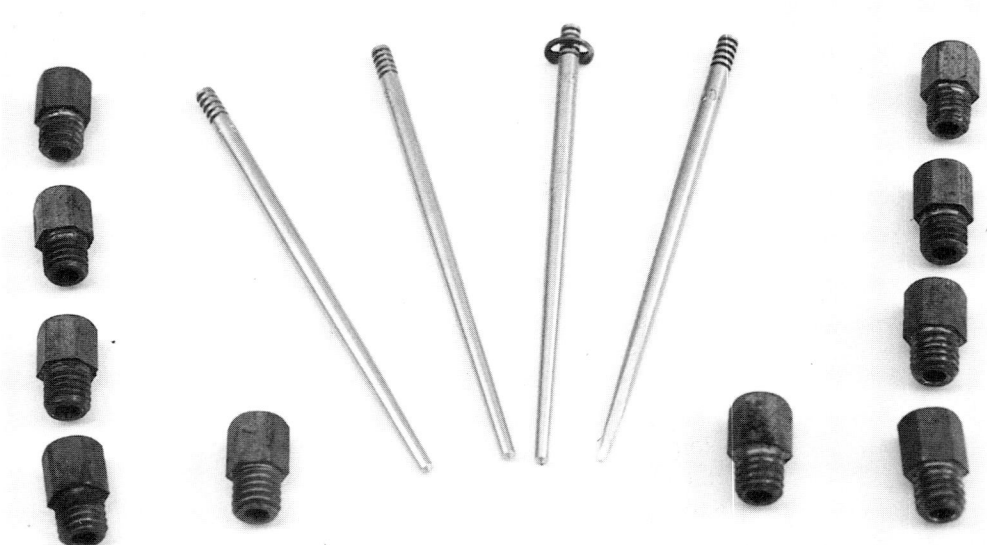

for at least half a mile at ¼ throttle, and then at ½ and finally ¾ throttle. If the engine 'four strokes' and misses at a steady throttle opening, the mixture is too rich, so lower the needle one groove at a time until smooth running is realised. On the other hand, an engine that snatches and surges is running lean, so the needle will require raising.

Table 5.8 Mikuni metering guide

	\multicolumn{6}{c}{Throttle opening position}
	1/8	1/4	3/8	1/2	3/4	full
Slide cutaway	A	A	B	B	C	D
Pilot or idle jet	A	A	B	C	D	D
Pilot air screw	A	A	B	C	D	D
Needle jet	B	A	A	C	C	D
Needle size	C	B	A	A	B	C
Needle position	B	B	A	A	B	D
Main jet	D	D	C	C	B	A
Power jet	D	D	D	D	C	A

These letters indicate metering effectiveness at various slide openings.

A – most effective
B – fairly effective
C – small influence
D – no influence

Next, try steady accelerations from ¼ to ½ throttle and from ½ to ¾ throttle and note whether the engine appears to be lean or rich. Then repeat the test but snap the throttle open each time. You may find that the mixture is lean at ¼ throttle and changes to rich between ½ and ¾ throttle. This would indicate that the needle is too thin between those points, so a change to a needle with a thicker profile, generally at the 30 and 40mm points (see Tables 5.7) would be required. After fitting the new needle you may find it necessary to change the clip position either up or down, depending on the needle thickness at the 10 and 20mm points, to cure the ¼ throttle leanness originally experienced.

Once the correct needle and needle position has been determined, the bike should be tested at ¾ to full throttle to find the right main jet diameter. What we want is a mixture rich enough to avoid piston and engine overheating, but not so rich that the engine is losing power due to poor combustion.

When trying to find the optimum size main jet, it is always preferable to begin testing with a main jet way too large and work down from there, otherwise if you start out testing with a jet too small the engine could easily overheat and seize. As you come down closer and closer to the size required, the engine will perform progressively better. When you reach the point where the engine feels to be running at its best, you should do some careful and serious spark plug reading to ensure that the mixture is, in actual fact, correct.

It takes practice and a proper magnifier of 4X or 6X power to pinpoint correct mixture strength. The things to look for that indicate certain operating conditions are indicated in Table 5.9. You will note that all of the plug end actually exposed to the combustion flame is examined and read, not just the insulator nose as some so mistakenly believe.

Table 5.9 Checking mixture strength by spark plug reading

Spark plug/mixture conditions	Indications
Normal – correct mixture	Insulator nose light tan to rust brown. Little or no cement boil where the centre electrode protrudes through the insulator nose. The electrodes are not discoloured or eroded.
Fuel fouled – rich mixture	Insulator nose black and possibly wet. Steel plug shell end covered with a black soot deposit.
Overheated – lean mixture	Insulator nose chalky white or may have a satin sheen. Excessive cement boil where centre electrode protrudes through insulator nose. Cement may be milk white or meringue-like. Centre electrode may be 'blue' and rounded off at the edges. Earth electrode may be badly eroded or have a molten appearance.
Detonation – lean mixture	Insulator nose covered in tiny pepper specks or maybe tiny beads of aluminium leaving the piston crown. Excessive cement boil where centre electrode protrudes through insulator nose. Specks on the steel plug shell end.

For the plug reading to be accurate, it will be necessary to run the engine at full throttle and maximum speed for at least 6 miles, and then immediately cut the engine dead. If you allow the engine to slow down as you bring the bike to a stop, and then slowly trickle through the pits, the plug reading will be meaningless.

When you find a main jet which gives a good plug colour then, by way of a double check, make a piston crown and combustion chamber reading. If the top of the piston and combustion chamber is dry and coloured very dark brown to black with hard carbon, then the mixture is correct. A wet and black sooty appearance indicates an over-rich mixture, and greyish deposits are a sure sign of dangerous leanness.

Having found the correct main jet size don't be fooled into thinking that you will not have to change it again. Two-stroke engines are very fussy about mixture strength and engine overheating, therefore you will find it necessary to jet the engine to suit different tracks and compensate for changes in atmospheric conditions. Even your ability as a rider comes into consideration. As your skill develops and you are able to hold the throttle wide open for longer distances around a track, you will have to jet richer to cool the engine. Fast tracks with long straights demand larger jets than short, low speed twisty circuits. High altitude running requires leaner jetting, and so forth.

Since the temperature, humidity and barometric pressure all affect air density, it is obvious that the mixture strength, the ratio of fuel to air being introduced into the engine, will vary from day to day and from place to place (because of altitude difference). Under normal circumstances the change in air density is of little or no consequence to the average road rider, but the racing engine tuner, seeking as much power as possible and desiring to avoid engine and piston overheating, has to take the air density into consideration before each and every race, or even during an enduro where large changes in altitude are experienced.

When the air density decreases, this reduces the amount of oxygen inducted into the cylinder, therefore the mixture becomes richer. Conversely, an increase in air density increases the quantity of oxygen entering the motor, so there is a corresponding leaning of the fuel/air mixture. To compensate, it will be necessary to fit richer or leaner main jets.

Remember, when compensating for a change in air density, that the change in density also affects the pressure exerted on the fuel in the float bowl. Therefore a decrease in relative air density (RAD) will automatically lean the fuel/air mixture to a degree, because of the lower air pressure. This means that you don't fit 5 per cent smaller jets when the RAD falls by 5 per cent. I usually reckon that a change in RAD of 12 to 15 per cent requires a 5 per cent change in fuel jet size. Remember, too, that a decrease in RAD is usually due to hot or high altitude conditions, conditions which in themselves reduce the engine's cooling efficiency. To compensate for possible engine overheating it is good to keep the mixture slightly rich when the RAD is low.

If you intend to tune your carburettor taking RAD changes into consideration, you must have a reference point and work from there. After you have tuned your carburettor as outlined on previous pages, you should make a record of the RAD and then experiment with jet sizes at other RADs, according to the percentage difference between the baseline RAD and the RAD on the day you are retuning the carb.

The relative air density can be worked out from Figure 5.7, providing you know the air temperature and the uncorrected barometric pressure. RAD meters are available and these give a direct percentage density reading.

There is another factor involved and, unfortunately, this cannot be read off the RAD graph or meter but, as the humidity affects true air density, we have to take it into account to be completely accurate. The effect of the humidity is quite small, except when both the temperature and relative humidity are high. Water vapour has weight and, as such, combines with the weight of the air to distort the true 'weight', or density, of the air. Think of it in this way: you are the air and your clothing is water

Table 5.10   Humidity saturation pressure and percentage

| Temperature | | Saturation pressure | | Saturation percentage of water |
°F	°C	in Hg	Millibars	
40	4.4	0.247	8	0.83
60	15.6	0.521	18	1.7
70	21.1	0.739	25	2.5
80	26.7	1.03	35	3.3
90	32.2	1.42	48	4.7
100	37.8	1.93	65	6.5

Two-stroke performance tuning

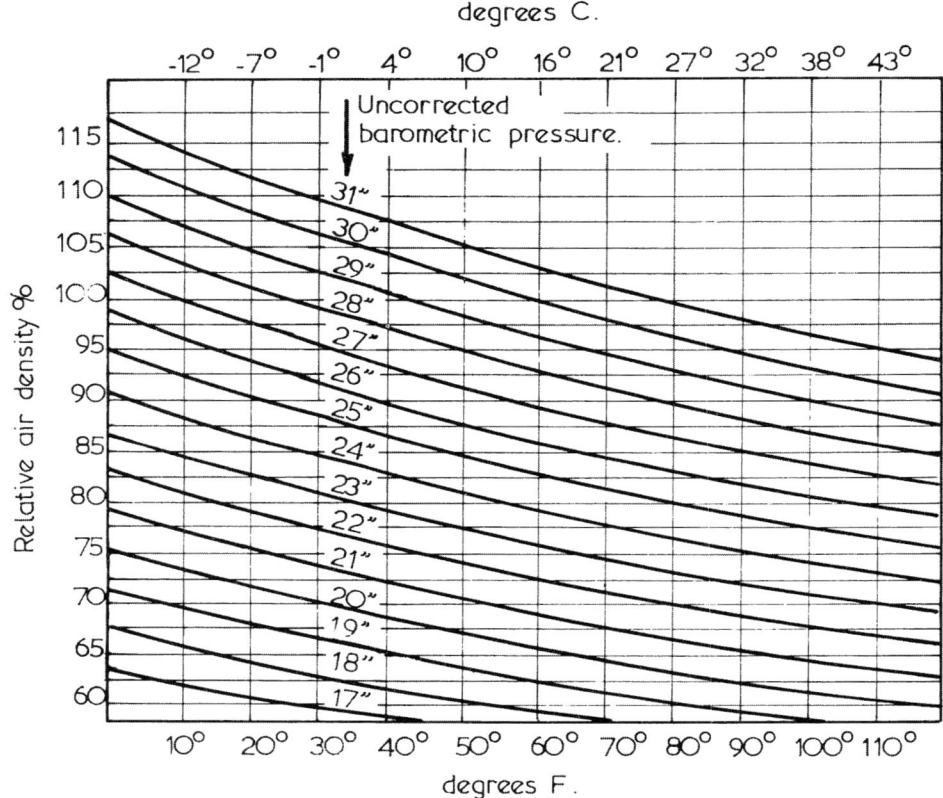

Figure 5.7 Relative air density chart.

vapour, wearing clothes you are going to exert more pressure (weight) on the bathroom scales than your true undressed weight. To find your true weight, you have to subtract the pressure exerted by your clothes. Similarly, when we want to find the true air density, we have to subtract the pressure exerted by water vapour in the atmosphere.

If you take a look at Table 5.10, you can see that the pressure exerted by water vapour at 100°F is 1.93in Hg. If the barometric pressure at the time is 30in Hg the true air pressure is only 30.0 – 1.93 = 28.07in Hg, a decrease of 6.4 per cent. Therefore, in this instance, the mixture could have ended up 6.4 per cent rich if the relative humidity was not taken into account.

Usually the amount of water vapour is less than the amount indicated in the column headed 'Saturation Pressure', as this assumes a relative humidity of 100 per cent. (Relative humidity compares the amount of water vapour present with what the atmosphere is capable of holding.)

To find the true air pressure, use the formula:

$$CAP = UBP - \left(\frac{SP \times RH}{100}\right) \text{ inches mercury}$$

where CAP = corrected air pressure, UBP = uncorrected barometric pressure, ie read straight off the barometer, SP = saturation pressure from Table 5.10, and RH = relative humidity.

Once the corrected air pressure has been calculated, the true relative air density can be read straight off the relative air density graph.

If you are using a RAD meter the percentage reading must be corrected using the formula:

$$\text{Corrected RAD} = \text{RAD reading} - \left(\frac{S\% \times RH}{100}\right)$$

where S% = saturation percentage of water from Table 5.10 and RH = relative humidity

Consequently, when the humidity nosedives two-stroke tuners can run into serious problems, but probably not for the reason you imagine. It must be understood that when the humidity is high not only is there a smaller number of oxygen molecules being drawn into the cylinder, there is also a much larger than usual number of water molecules interspersed between the oxygen and fuel molecules. The effect is that the space between the two increases which slows combustion flame speed because, rather than progressing rapidly from fuel molecule to fuel molecule, the flame front comes up against water molecules which do not contribute to the combustion process. Thus the water suppresses violent combustion and detonation, just as a higher octane fuel or an octane boost additive would. Now if the humidity falls that detonation suppression effect is diminished or even eliminated. It is as though the engine is now firing on a lower octane fuel. Clearly, if mixture richness is not increased and perhaps spark advance reduced to compensate, the engine could be destroyed by detonation.

The one thing you must be sure to do, if you wish to be successful in tuning your engine according to the relative air density, is to keep complete and accurate notes. If you find that your engine works best with 270 main jets with a 90 per cent RAD and 80 per cent relative humidity, be sure that you make a note of the fact in your tuning diary. Then, on each occasion the air density is again 90 per cent and the RH 80 per cent, you will know exactly what size jets to try, providing the tracks are of a similar layout. If the track is a faster one with larger straights, you may have to jet richer. At another location the RAD might be 98 per cent and the RH 63 per cent, so armed with the information in your diary you know that you should try a 280 main jet. May be it will be the correct size, but then it may not be. There are no hard and fast rules here, no two engines respond to RAD and RH changes exactly alike. Usually small displacement, high rpm road race engines in a high state of tune are most affected by a change in air density.

The mixture ratio is also affected if there are any air leaks, so you must be very careful to seal the manifold to the barrel, and the carburettor to the manifold, using the correct gasket and the right type of gasket cement. This may seem a rather small thing, but you would be amazed at the number of inexperienced tuners who use Silastic to seal the manifold/carburettor. Silastic is excellent as an oil sealant, but it is not petrol resistant, therefore it should not be used anywhere in the induction tract. I recommend the use of Permatex No.3 for this purpose.

An air filter also reduces the air density (particularly if it is clogged full of dirt), but in this instance by restricting air flow into the cylinders. Obviously the restriction can be done away with by removing the filter. This may be acceptable for some road racing bikes, but all other types, including road race karts, should be fitted with effective filters.

Long engine life and effective air filtration are very closely related and, contrary to popular opinion, a clean air filter of good design (such as K & N) does not reduce engine power output by very much at all. On road bikes such as the Yamaha RD250 LC/RZ350, however, the air filtration system does have a devastating effect on performance. This is due to the use of a tiny air filter and a very restrictive air box, designed to cut down induction roar and reduce noise pollution.

Many enthusiasts are fooled into thinking that they have gained increased horsepower by fitting 32 and 34mm carbs to their 250 and 350 twin cylinder street bikes. Usually, the improvement in power is more the result of the removal of the stock air box and filter than from the bigger carburettors. As I pointed out earlier, I have found some Yamaha 250 and 350 production racers which do not make any more power with 32 or 34mm carbs and when such big carburettors are fitted to street bikes the result very often is reduced bottom end power, with no increase at the top end.

Table 5.11 indicates the performance increase realised from removing the air box and filter from the rotary valve square-four Suzuki RG500 and fitting a low profile itg brand filter over each pair of carburettors. Each filter was sealed in an air box and fed cool air directly from a ram-air scoop in the fairing.

Suzuki claim that the standard engine produces 95hp at the crankshafts but in this case the bike's owner wanted at least that at the back wheel without losing engine tractability. With that brief in mind the barrels were only very lightly ported. In this state of tune, with standard flat-slide 28mm Mikuni carburettors and standard air filter system, the engine produced 92.4hp at 9,000rpm. Next, 32mm carbs were tried as the engine didn't want to rev-on past 10,000. These weren't connected to the air box ducts, instead a pair of fans blasted cool air to each pair of carbs. The power now shot up to 103.6hp at 10,250rpm. However, below 4,000–4,500rpm the dyno wouldn't hold full throttle and up to 6,500rpm power was way down on the 28 flat slides. For the next test the 28s were again fitted but without the air box and ducting. With the cool air fans operating the engine made virtually the same maximum horsepower and the bottom end came back. After this a lot of time was devoted to building a non-restrictive cool ram-air intake system with a pair of low profile itg filters. In this form the engine made almost 102hp at the back wheel, and in so doing confirmed the benefits of retaining relatively small carbs on street bikes while highlighting the need to unplug the air inlet system.

If you are an off-road rider, you should take a good, careful look at your bike's filtering system. I would estimate that the majority of dirt bikes suffer considerably more wear from inducting dust, mud and water into their internals than they do from very hard, high speed competition riding. We don't want any dust or mud going into the engine at all. This material is very abrasive, prematurely wearing the piston, rings,

Carburation

Table 5.11   Suzuki RG500 air box test

rpm	Test 1 (hp)	Test 2 (hp)	Test 3 (hp)	Test 4 (hp)
4,000	23.4		23.2	23.5
4,500	27.9		27.8	27.9
5,000	29.8	20.8	29.3	30.1
5,500	34.6	24.6	34.7	35.1
6,000	37.7	30.3	37.5	37.3
6,500	44.7	38.7	45.2	45.0
7,000	56.2	51.9	55.8	55.6
7,500	65.2	63.9	65.5	66.1
8,000	71.0	69.6	70.8	69.7
8,500	80.8	80.3	82.6	82.7
9,000	92.4	94.0	94.8	93.9
9,500	88.6	98.5	98.2	98.6
10,000	82.3	101.5	100.8	100.5
10,250	71.5	103.6	102.5	101.8
10,750		92.7	93.3	92.4
11,250		77.1	76.9	75.8

Test 1 – Modified engine with standard 28mm carburettors and standard air box, duct and filter system.

Test 2 – 32mm carbs with cold air supply from fans.

Test 3 – 28mm carbs with cold air supply from fans.

Test 4 – 23mm carbs with new air boxes and itg filters fed by ram-air scoops in the fairing.

cylinder and bearings. Also, we must keep water out of the engine. A few specks of water on the plug will stop the engine dead. If the engine takes a good gulp of water, the piston and head could crack because of the sudden temperature decrease.

The first thing you need is a good air filter, properly oiled. For dry events, I think the K & N range of cotton filters are tops. If there is a lot of mud and water around, I prefer the Uni foam filter as it does a better job of keeping water out. Whatever type of filter you use it must be oiled, but be sure to use a water-resistant oil, not engine oil. Waterproof oils will not stop water getting into the engine, that is the task of the filter, but they will not break down into a soapy mess like engine oils, and most gear oils, when they contact water. When an oil breaks down like this, the air filter operates in a similar way to an unoiled filter: very poorly. Bel-Ray filter oil is good, but expensive. If you are worried by the price, try Castrol ST90. It is a 90 weight waterproof gear oil, about one third the price of Bel-Ray filter oil.

A good, well oiled filter is totally useless if it isn't correctly fitted. Unfortunately many bike manufacturers make it very difficult to do just that and this adds to the problem, as it is often impossible to see or feel if the filter is properly in place or not. About the only thing to do is give the sealing edge of the filter a coat of waterproof grease, and then take your time fitting it, being as careful as possible.

The design and location of the air box has a large bearing on just how much dust, mud and water the air filter is going to have to cope with. By paying special attention to sealing the air box you can reduce the load on the filter considerably. Often there are gaps between the rear mudguard, the side covers and the air box, which let in whatever is being thrown off the back tyre. If you seal up these gaps with duct tape or with rubber strips glued into place, you will be contributing much to longer engine life.

Remember, when you go about sealing off the air box from the mudguard and side covers, that you have to let air into the motor or it won't go. The driest, most dust-free place from which to draw air is from under the fuel tank, so direct your efforts towards getting air into the air box from that area, if possible.

Another point to remember when you seal the air box is that, if water does get in, it has no way of getting out unless you put a reed valve in the bottom of the air box. If you cut a simple hole in the bottom of the air box to let water out it will also allow water in, and in huge quantities. A reed valve type drain allows flow in one direction only, so if your bike doesn't have this style of drain, do fit one.

Don't be fooled into thinking that, now the air box is taken care of, there is no way water will ever again get into your engine and cause it to stop part way through an enduro. Water can, and does, get into carburettors in many different ways. The most obvious way is in the fuel itself. Leaving your fuel drums standing upright in the rain is inviting trouble, and so is leaving the funnel stuck in the bike's fuel tank while you run through the rain to fetch a drum of fuel, which you have been so careful to place inside a nice dry service tent.

Another way water can enter the carburettor is via the float bowl overflow tube. Some enduro riders plug the overflow outlet – in fact, Bing have been leaving overflows off some of their carburettors for years now, relying on excess fuel to drain off past the tickler. I still believe in leaving the overflow tube in place, and running down-hill from the outlet, so as to allow excess fuel to spill out of the fuel bowl. The overflow tube should drain into a catch tank of about 150cc. It will require emptying at each fuel stop.

Ridding the Bing of an overflow hasn't been the answer to keeping water out of this carburettor, as water can still find its way in past the tickler. What you must do, if you have a bike fitted with a Bing, is remove the tickler and cover the hole with a plastic tube terminating high up under the fuel tank. If you need a rich mixture to get the bike started in freezing weather, you will have to lay the bike on its side and let it flood for a couple of seconds.

Water can also get into the carburettor by running down the throttle cable, so make sure that the adjusters at both ends of the cable are taped up. Tape the top of the carburettor too; remember that a vacuum exists within the throat and slide chamber, so water can enter the engine by being drawn through the thread on the top of the slide chamber.

Some tuners wonder if 'pumper' diaphragm-type carburettors offer any advantages over the more normal slide throttle and float bowl carburettor. In motorcycle applications, I would say a definite 'No', but they do have a place on go-kart engines which otherwise require the extra burden of a fuel pump to lift fuel from the fuel tank up to the carburettor. The majority of kart engines are fitted with this type of carburettor which employs engine crankcase pulses to both pump and meter fuel.

Karts with fixed gearing (ie no gearbox) require a carburettor capable of accurate fuel metering over the greatest possible rpm range, and the pulse carburettor fills this need. The 100cc McCulloch engine, for example, utilises a large 35mm bore

Carburation

The Mikuni pumper carburettor employs engine crankcase pulses both to pump fuel to the carburettor and to meter the fuel into the air stream.

carburettor for good high speed air flow, yet the pulse metering enables the engine to run well and accelerate cleanly out of tight turns where engine speed drops to 5,000rpm.

The main difficulty with pulse carburettors is that they work well only on the engine for which have been specifically made. Their air passages, which bleed crankcase pressure into and away from the chamber behind the metering diaphragm, have orifices calibrated to be sensitive to the crankcase compression ratio and the cylinder displacement. Any change in either of these factors will upset fuel metering.

However for that metering system to be able to supply the engine's fuelling requirements it must be correctly adjusted and assembled. Some of the more common problems seen on Yamaha KT100 'S' and 'J' type fixed gear sprint kart engines fitted with the Walbro pumper carb are seized pistons and blown big ends due to lack of attention in this area. Because of the effects of the octane enhancing additives in today's high octane unleaded fuels on the pump and metering diaphragms they require regular replacement, often after each race meeting. Now, because they carry out the task so frequently karters often do not note, or even know, the order in which gasket, diaphragm and cover are refitted. Consequently fuel metering is upset and the engine suffers an expensive blow-up.

Keep a diagram and refer to it each time the diaphragms are replaced or the metering lever height or pop-off pressure is adjusted. The pump diaphragm fits against the carb body, with the gasket next, then the pump cover. The metering diaphragm, which hooks into the metering lever, has the reverse order. The gasket mates to the carb body, then the diaphragm, and finally the cover.

The metering lever height, with the gasket in place, should be 1.4mm below the face of the gasket. If the measurement is larger than this less fuel will be allowed in past the needle valve.

The pop-off pressure has a large bearing on top end mixture strength. If the pop-off pressure is high, say 12psi, then the mixture will lean off at the top end. Conversely if the pressure is adjusted to say 7–8psi then the top end will be richer. I prefer to see this latter figure on tarmac sprint tracks. To check the pressure put a drop of red auto transmission fluid on the needle valve. Then, using a Tillotson pump and gauge, see at what pressure the oil spurts, indicating the pressure at which the needle valve opened. To change the pop-off pressure the needle spring is stretched or shortened.

The other problem with pulse carburettors concerns the tuning of them. I have seen many engines seize because their tuners didn't understand very basic tuning procedure. Maybe it is because these carburettors are so simple to tune that some have been caught out. There are just two screws to adjust: one controls the low speed

There must be correct balance between how far the high and low speed metering needles are open. If the high speed needle is open too far and the low speed is closed to get the mixture back on track the engine will seize when pulling out of turns. Conversely if the low speed needle is open too far and the high speed is closed to compensate then the engine will lock up along the straights.

This piston was seizing even on the inlet side due to a lack of lubrication caused by the low speed needle being closed right off.

mixture and the other the high speed mixture. There is an overlap of functions at medium speed, and herein lies the problem. If the low speed mixture is adjusted too rich, then the high speed screw will have to be leaned right off to allow clean acceleration out of tight corners. This then causes an excessive lean condition at high speed, so that the engine overheats and locks up.

What you have to do is adjust both the low and the high speed screw to the setting recommended by the manufacturer, and then do your fine tuning out on the circuit. McCulloch recommend 1½–2 turns open for the high speed screw and 1¾ turns open for the low speed screw. I have found it preferable to set the screws a little closer than this. Usually ¾ turn open for the low speed and 1½ turns open for the high speed is a good place to begin testing on short tight sprint tracks.

After you have allowed the engine to warm up to normal operating temperature, you can fine tune the mixture adjustment. Turn the high speed screw out until the engine 'four strokes' and then lean it off a little at a time (1/16 turn) until the engine runs correctly at full throttle and maximum rpm on the main straight. Next adjust the low speed mixture for good acceleration out of tight turns. If the engine runs rough and smokes badly when accelerating, turn the low speed screw in to lean the mixture. If the engine falters and misses coming out of turns, open the low speed screw a little to attain smooth acceleration. When you are happy with the way the engine pulls out of corners, you should check the high speed adjustment again to ensure that the mixture is still correct.

It is very important that you adjust the mixture in the way outlined, otherwise you will soon run into trouble. I would say that of the seized kart motors I have worked on 80 per cent have been caused by improper carburettor tuning. Usually, I find the low speed screw is anywhere from four to seven turns open, and the high speed screw closed up to about a quarter to a half turn open.

Many go-kart classes are Box Stock categories which do not permit carburettor modifications. However, if you read the regulations you will generally find a maximum size listed for the carburettor bore, for example, also the rules usually state that all fixing devices (ie nuts, bolts, screws, etc.) are free. To get the best out of your engine you will have to use these rules to advantage, for you can be sure other tuners will. The 100cc Box Stock McCulloch class rules give the maximum carburettor size as 35.66mm for the throttle bore, and 29.08mm for the venturi diameter. Many carburettors have a venturi closer to 28.3mm, so you should enlarge and polish it, using a hand scraper and emery paper. The screw fixing the butterfly to the throttle shaft should also be modified to reduce air flow restrictions. Remove the star washer from under the head of the screw and file the screw down until there is a groove just deep enough to turn the screw back into the shaft. Put Loctite on the screw to prevent it vibrating back out of the throttle shaft. Next file the protruding thread of the screw flush with the throttle shaft.

If the carburettor regulations are free, performance can be improved by heliarc welding the butterfly to the throttle shaft, and then filing the shaft down to a thickness of 1.5mm. When this is done, be sure to adjust your throttle pedal stop so that you can just get full throttle, otherwise you will twist the end off the throttle shaft.

The fuel discharge nozzle causes quite a loss of flow, so this should be filed to reduce its width to a minimum. You will also note a small plug protruding into the airstream of the McCulloch carburettor just in front of the venturi. This can be removed entirely and the hole filled with epoxy, or the plug can be shortened and refitted.

Karts raced on faster circuits may have the venturi diameter increased up to a maximum of 32mm, but do be careful to richen the mixture as this modification can give rise to a dangerous high speed lean-out.

All McCulloch carburettors benefit from an air horn to smooth air flow into the carburettor bore. An air horn, with a nicely rolled radius similar to that illustrated in Figure 5.8, should be fabricated. Without the air horn, flow is reduced by up to 8 per cent.

Most road race karts use motorcycle engines equipped with conventional float chamber carburettors, which of course means that some type of fuel pump must be installed. The pump which I recommend is a pulse type unit manufactured in America by Outboard Marine Corporation. The type 'AY' pump (Part No. 385784), designed for 60–100hp outboard motors, is the most readily available and easily capable of supplying the required fuel volume for even the hottest TZ250 engines. The alternative from Mikuni is their DF 52–48 pulse pump.

To ensure a good strong pulse, the pulse tube connecting the crankcase with the pump must be short (ie no longer than 4in) and of small internal diameter. (Most fittings require ¼in bore tube.) If you find the pump will not supply sufficient fuel (this usually only occurs when pulse tubes longer than recommended, are used), you can fit a small 150cc auxiliary fuel tank to gravity feed straight into the carburettor fuel inlet. The auxiliary tank fills at low rpm and keeps the carburettor full at high speeds when pump pulses are weak.

Unlike automobile carburettors, motorcycle carburettors will not tolerate even very low fuel pressure before they give way to flooding. Being designed for gravity feed operation, their float system is not capable of shutting off fuel flow through the needle and seat at a pressure as low as 1psi. Therefore, when a pulse pump is fitted,

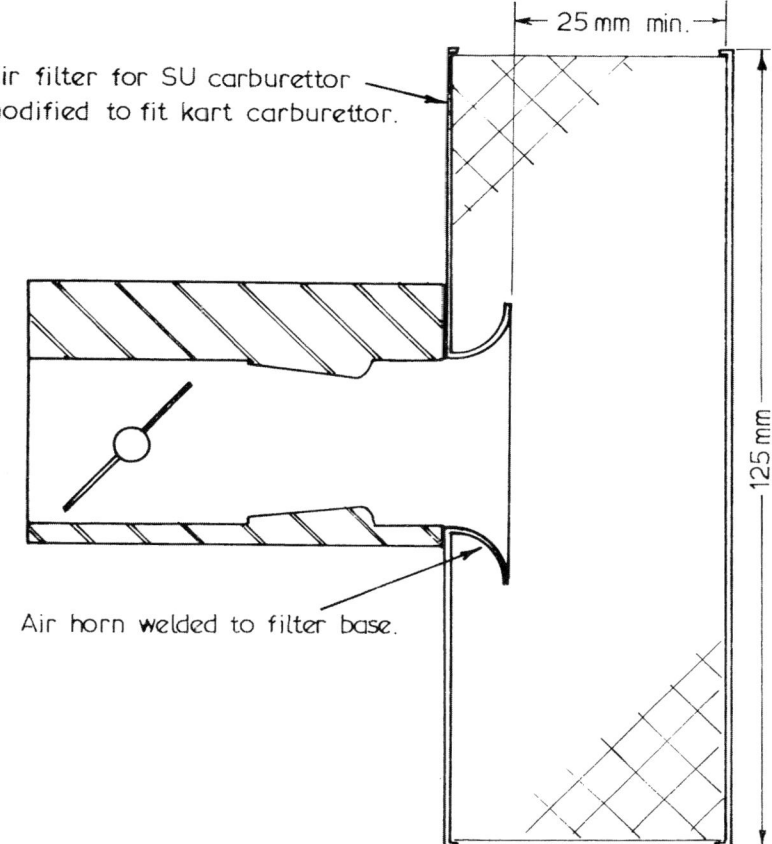

Figure 5.8 McCulloch carburettor air horn and filter improves air flow and ensures that dust, sand and grit does not enter the engine.

you will have to incorporate a fuel by-pass line, and also perhaps a pressure regulator, in your fuel system. (Figure 5.9). The by-pass line should be of the same internal diameter as the fuel line, and installed such that it is at a level approximately 6–12 inches higher than the carburettor fuel inlet. It must tee-off within one or two inches of the fuel inlet and discharge excess fuel into the top of the fuel tank. These precautions will ensure the fuel will not take a path of low resistance and return to the fuel tank, starving the engine.

When a carburettor is being installed, there are a few things you need to watch. The first point you must check is that you are actually getting full throttle. Inexperienced tuners are often fooled to believe the slide is opening fully when they see it disappear from the bore of the carburettor, forgetting that the cutaway is perhaps 2–3mm higher than the back of the throttle slide. What you must do is look right down the bore of the carburettor, using a mirror if necessary, and see if the back of the slide is hanging in the airstream. Alternatively, you can remove the needle and feel with your finger if the slide is opening fully.

Also check that the carburettor, manifold and inlet port are correctly aligned, as

Two-stroke performance tuning

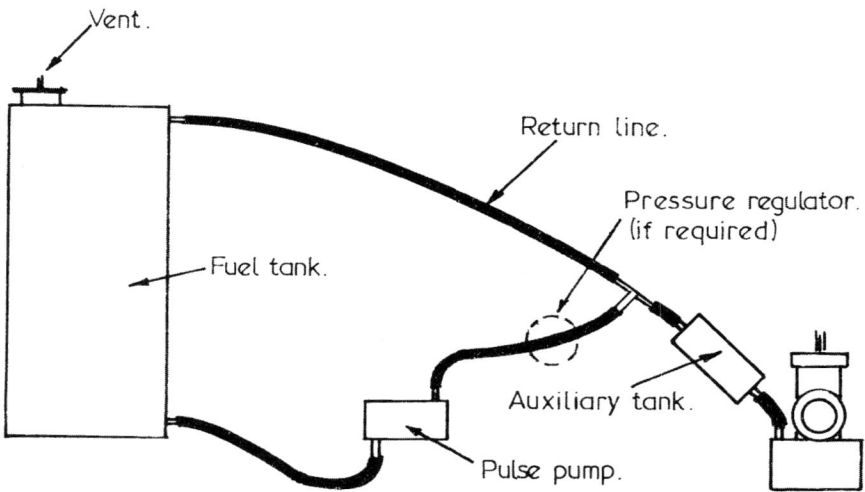

Figure 5.9 Gearbox kart fuel supply system.

misalignment will disrupt air flow. Most inlet manifolds are a loose fit on their retaining bolts, so if care is not exercised the manifold can quite easily be mounted out of line with the inlet port. When you test the alignment of the carburettor and inlet manifold, ensure that the retaining clamp is fully tightened as this tends to distort the neoprene material of which the manifold is made. Without the clamp tensioned, the carburettor and manifold often appear to mate perfectly, but once tightened the neck of the manifold may protrude into the air stream. Any offending material should be cut away with a very sharp knife or razor blade. Then dress the surface with a high speed grinder or a chainsaw file.

Each time the carburettor is refitted to the engine, ensure that it is not tilted off centre. Some manufacturers have a locating notch moulded in the inlet manifold to facilitate correct carburettor positioning, but most leave it up to the tuner to sight through and line up. A carburettor, when tilted off vertical, will be leaned-out, as this lowers the fuel level in the float bowl.

When it comes to fuel, basically we want to use the best available to ensure maximum horsepower without detonation. The type of fuel that you are permitted to use will be laid down by the body governing your particular branch of competition, so take the time to read the rules carefully and don't blindly follow the interpretation which fellow competitors have given them. For example, fuel regulations may state that 'only pump commercial premium unleaded petrol is permitted'. Some racers take that rule to mean that only 96 octane unleaded from the local service station pump is allowed. Others figure that 'pump commercial' means any high octane unleaded from a service station pump anywhere in the world. Consequently at great expense, some purchase 98 octane unleaded from Europe while others purchase 100 octane unleaded from Japan as both are available from service stations. A few racers have adopted a more liberal understanding of this fuel rule. For some racing car classes 100 octane unleaded petrol is specified. As it is available from a commercial pump at larger race circuits some competitors feel that they are operating within the rules by using this 'pump commercial' fuel. Clearly the competitor using 'pump' 96 octane fuel will be at a disadvantage in comparison with those choosing 'pump' 100 octane if the latter take full advantage of the added octane levels to increase ignition advance and/or the compression ratio.

The situation is similar when the rules specify 'racing unleaded 100' fuel. In many parts of the world that means 100 octane by the Research test method. However in America they rate their fuel differently. As shown in Table 5.12 their 'racing unleaded 100' has a Research Octane Number (RON) of 104–106, and using this fuel along with a higher compression gives a clear power advantage.

Table 5.12 Fuel characteristics

Fuel	Specific gravity	RON	MON	Fuel/air ratio (lb/lb)	Heat energy (Btu/lb)	Latent heat of evap	Weight (lb/gal)
Acetone	0.79			1:10.5	12,500	225	8
Avgas 100/130							
'green'	0.69	105–110	100–102	1:12.9			7
'blue'	0.71	105–110	100–102	1:12.7			7
Benzol	0.88	105–110	95–100	1:11.5	17,300	169	8.7
Ethanol	0.79	108–115	90–92	1:6.5	12,500	410	8
Ether (diethyl)	0.71				15,000	153	7
Methanol	0.79	105–115	89–91	1:4.5	9,800	472	8
Nitromethane	1.13			1:2	5,000	258	11.3
Nitropropane	1.05				6,700		10.5
Petrol							
premium unleaded	0.74	*96	*85–86	1:12	19,000	135	7.4
premium leaded	0.73	96	86	1:12.5	19,000	135	7.3
racing leaded (USA)	0.73	112–114	102–104	1:12.7			7.3
racing unleaded (USA)	0.75	104–106	94–96	1:13.2			7.5
racing unleaded 100	0.75	100	90–92	1:13.0			7.5
Propylene oxide	0.83				14,000	220	8.2
Toluol (methyl benzine)	0.87	120–124	110–112	1:9.8			8.7
Triptane	0.69	110–112	100–102				6.9
Xylene	0.86	117–118	115–116				8.5

  * *Premium unleaded in some lands has a higher rating (often 98 RON, 87 MON).*

In other classes the fuel rule may simply state that any commercial fuel with a specific gravity of less than 0.75 is permitted but no alcohol blending is allowed. For many competitors this means a choice between either 'green' or 'blue' Avgas 100/130. The old green-coloured variety has more lead (4.5–6gm/gal) than the newer blue low lead type (2–2.7gm/gal). Tested by the Research method the octane rating of both is 108–110; by the Motor method it will be 100 minimum, and up to 102. Both are often marketed as 'leaded racing 100'.

Another variety of Avgas, 115/145, is no longer available except to the military and operators of old fighter aircraft. Avgas 112/160 used to be available in some parts of the USA and is now marketed there as 'leaded racing 108', thus reflecting its Pump

Octane Number. It has a Research number in the 112–114 range and a Motor number of 102–104. Additionally it would be legal where a maximum SG of 0.75 was specified.

Apart from an octane advantage over Avgas 100/130 this latter fuel has a fuel consumption edge which is significant in endurance type events. With a specific gravity of 0.73 as compared with 0.69 for green Avgas it has more energy in every litre so the mixture has to be leaned off, which enables each tankful to go another 5–6 per cent race distance.

A similar thing can also be achieved with green Avgas 100/130 by blending in 33 per cent toluol (toluene, methyl benzine) ie a 1:2 mix ratio. This produces a fuel with an SG of 0.75 and Research numbers of 109–116 and Motor octane of 102–104.5, making it very similar to American 'leaded racing 108'.

However the same result will not be achieved with blue Avgas 100/130. Because it has a lower lead content the fuel refiners have had to blend in high-octane hydrocarbons to bring its octane rating back up. Therefore blending in 24 per cent toluol will bring the SG up close to the limit and reduce fuel consumption by 5 per cent but the octane levels may not increase at all.

This highlights the problem we have today in not being able to reliably increase the octane levels of ordinary road fuels by the addition of octane boosters. In the past a fairly reliable method of increasing the octane rating of petrol was to add up to 33 per cent toluol. This percentage would raise the octane rating of ordinary leaded pump petrol (0.2gm lead/litre) by about 6 numbers Research and 2.5 numbers Motor respectively. However as lead levels have decreased, government regulations permitting, refiners have substituted octane boosting hydrocarbons to bring octane levels back up, so a reliable octane increase cannot be guaranteed by their addition.

The situation is similar with another very effective octane booster, MMT (methyl cyclopentandienyl manganese tricarbonyl), which is the base product in both 104+ and 104+ Super brand octane boosters. Whereas a can of this stuff, which at normal concentration will treat 83 litres of petrol, raises the Research number of low octane unleaded (91–93 RON) by about 3–4 and 4–6 respectively for 104+ and 104+ Super, with premium leaded and premium unleaded (95–98 RON) this boost effect decreases with 104+ to around 0.5–1 and with 104+ Super to a 1–2 octane increase.

The message is that at any time you blend a fuel other than Avgas, which is highly regulated and a known quantity, you are at the mercy of the fuel companies, so what worked with one batch of fuel may not work with another. Also, when switching from one type of fuel to another make sure that you know what you are dealing with. Many engines have unwittingly been destroyed by racers switching from 'leaded racing 100' to 'unleaded racing 100'. They saw the 100 and assumed incorrectly that both fuels had the same octane level. The leaded 'racing 100' (Avgas 100/130) gets its 100 rating with the tougher Motor test, whereas the unleaded is rated by the Research method.

Since much confusion exists as to what octane ratings are all about, we will have a look into the subject. Most people realise that we can get more power with a high octane fuel because we can use a higher compression ratio, and perhaps more spark advance, without running the engine into detonation. However, did you know that a change from, say, 97 octane pump petrol to 110 octane racing fuel (100/130 Avgas) would not give a power rise for most motocross bikes? In fact, you could possibly lose

Carburation

Octane boosters and fuel additives are an unreliable method of increasing fuel octane levels. Sometimes the additives work, on other days they do not, depending on how the fuel companies are blending their fuel. This piston suffered damage on one of those days when the fuel additives didn't provide the usual octane increase.

power, if the engine was not changed in any way to take advantage of the octane increase.

To understand why this is so, we have to delve back into history a little to see why a numbering system was introduced to grade fuels, and discover exactly to what the octane numbers refer. Back about the time of World War I, aircraft engines would suddenly self-destruct, through detonation. An engine might run just fine on one batch of fuel, but punch holes in the pistons on the next batch. The fuels seemed to be the same, weigh the same, and may perhaps have come from the same refinery.

The fuel companies tried chemical analysis in an endeavour to achieve parity from one load of petrol to the next, but, in spite of an intensive programme, they were not able to weed out the batches which were prone to produce knock (detonation). Therefore special fuel research engines, with a variable compression feature, were constructed to evaluate and grade fuels. Such a standard test engine (a single cylinder, heavy duty unit) would be warmed up to standard test temperature, run at standard rpm and load, and then have the compression ratio increased until the fuel being tested just produced engine knock. Its anti-knock quality would then be specified as its Highest Usable Compression Ratio (HUCR).

Even with every fuel lab supplied with the same type of test engine, and using the same standard test procedure, it was discovered that the same fuel could test out with differing HUCR numbers in different test laboratories. It was decided that some unvarying standard was needed by which to calibrate the lab engines. Two pure substances were chosen as reference fuels. The high reference fuel chosen was iso-octane, while the low reference fuel was normal heptane (n-heptane).

Now it was decided that a fuel under test would be run in the variable compression engine and its HUCR determined. Then a series of runs would be made with various mixtures of iso-octane and n-heptane until a blend was found which produced knock behaviour identical to the fuel being tested. At this point the quality of the test fuel would be rated in relation to the percentage of iso-octane in the reference fuel mixture which gave identical test results. For example, a test fuel which behaved the same as a mixture of 75 per cent iso-octane/25 per cent n-heptane would be called 75 octane fuel. Using this standard test procedure, fuel of constant quality could be refined and supplied for a variety of applications.

Since that time a number of test procedures have come into use to simulate a variety of engine operating conditions. Motor spirit is usually rated according to the Research or Motor test methods. Both measuring techniques use the same single cylinder, variable compression test engine, but the Motor method employs a greater engine speed and a higher inlet mixture temperature than the Research test. Hence the Motor method is a more severe test, and generally yields octane numbers 6 to 12 less than the Research test. (Table 5.13). This distinction is important, as it informs us that the Motor Octane Number (MON) is more relevant to a racing engine than is the Research Octane Number (RON).

Table 5.13 Octane test comparison

Research octane number	Motor octane number	Pump octane number
92	85	88.5
96	88	92
98	90	94
100	91.5	95.8
105	95	100
110	100	105
113	103	108
115	105	110

Another common number seen on American service station pumps is the Pump Octane Number (PON). This is the average of the RON and MON:

$$\frac{RON + MON}{2}$$

and yields a very creditable rating of a fuel's performance under actual road conditions.

The Supercharge test is applied to aircraft fuels which exceed 100 octane numbers, as the other tests become meaningless at just over 100. The SON (Supercharge Octane Numbers) are significant from 100 to well over 300. Two tests are involved, the F3 and F4 tests, which explains why aircraft fuels have a dual rating such as 100/130. The first number refers to the F3 test, which simulates a supercharged engine running on a chemically-correct fuel/air mixture, as when cruising. The F4 number gives an indication of the fuel's performance rating with an enriched mixture and increased supercharge boost, as would be supplied during aircraft take-off.

Numbers over 100 cannot refer to percentages of iso-octane. They are, in fact, performance numbers devised to extend the scale of anti-knock measurement past that possible with pure iso-octane. As such, they give a rough estimate of the power potential of the fuel when a heavy supercharge boost is applied to a suitable engine. For example, 115/145 Avgas has the potential of increasing an aircraft engine's power by 45 per cent over that possible using pure iso-octane fuel.

The anti-knock properties of hydrocarbon fuels are related to their molecular structures. The paraffins (such as normal heptane and kerosene) are long chains of carbon and hydrogen held together by weak molecular bonds which are easily broken by heat. Iso-octane is a member of the iso-paraffin family. These have a branched chain structure that form stronger bonds to resist detonation better. The cycloparaffins (or napthenes) also have good anti-detonation properties with their hydrogen and carbon atoms well bonded in a ring shape molecule. The aromatic fuels, such as toluol, also have a ring-shaped structure with very strong bonds. This explains why they have such good anti-knock characteristics.

The chemical composition of the fuel determines just how rapidly the fuel will burn and whether it will be resistant to detonation at high compression pressures and temperatures. The fuels with weak molecular bonds break up and burn spontaneously (ie without being ignited by the combustion flame initiated by the firing of the spark plug) at lower temperatures and pressure than fuels with strongly bonded structures. Some fuel additives, such as the aromatics, make excellent anti-detonants because they burn slowly and don't oxidise or burn completely until combustion chamber temperature and pressure is very high. Aromatic fuels therefore inhibit, or slow down, combustion. For this reason a high octane fuel will not increase engine power unless the engine actually needs a fuel which is chemically stable at high temperature and pressure. Obviously, if the engine does not have a compression ratio and spark advance great enough to produce high combustion pressure and temperature, then the high octane racing fuel will not burn completely until too late after TDC, resulting in loss of power.

A similar situation exists with alcohol fuels which are permitted in most categories of speedway and dirt flat-track racing, and also some kart classes. Both methanol (methyl alcohol) and ethanol (ethyl alcohol) require more spark advance to give a big power increase but they are less fussy about an increase in compression ratio. In fact two-strokes will often show more increase in peak horsepower on stock compression whereas a higher compression engine will show a greater increase lower in the power band. On stock compression the power increase could be around 11–13 per cent right across the power band, but on a higher compression ratio this could change to 13–15 per cent at the bottom end and a 9–10 per cent increase at the top end.

Where does this power increase come from, particularly if the compression ratio is not increased? Well, there are two aspects to consider. First, both methanol and ethanol have a very high latent heat of vaporisation ie, it takes a lot of heat to convert them from liquid into vapour. Petrol has a latent heat of evaporation of 135Btu/lb, methanol 472Btu/lb and ethanol 410Btu/lb. This heat, required for proper atomisation, is removed from the crankcase, the piston, the transfer passages, the combustion chamber etc., resulting in an internally cooler engine. This means less expansion of the fuel/air mixture after it leaves the carburettor which allows a greater number of air and fuel molecules to be drawn into the crankcase and then transferred up the transfer passages into the cylinder. A denser fuel/air change means more power.

Secondly, the two cycle engine is a type of heat engine, ie one that burns fuel to cause the expansion of gas, and the subsequent movement of the piston. The more heat produced by the combustion fire, the more pressure there will be exerted on the piston, which gives us a power increase.

Using petrol, the fuel/air ratio for best power (ie the strongest force on the piston) is 1:12.5. With methanol, for example, we can increase the fuel/air ratio to 1:4.5, although I usually prefer a ratio of 1:5.5; less than 1:7 is too lean.

One pound of petrol has the energy potential of about 19,000Btu (one British Thermal Unit is the amount of energy required to raise the temperature of one pound of water by one degree Fahrenheit). In comparison, methanol delivers around 9,800Btu/lb, which means that it produces less than 52 per cent of the heat energy of 1lb of petrol. However, because we are mixing more methanol with each pound of air (1:5.5) than petrol (1:12.5), we are actually producing more heat energy by burning methanol.

To work out how much more heat energy is produced, we have to divide 12.5 by 5.5, which equals 2.27. Next we multiply 9800 by 2.27, which gives us 22,246. This indicates that methanol, in the correct fuel/air proportions, will produce 17 per cent more heat energy than petrol at the correct fuel/air ratio.

$$\left(\frac{22,246}{19,000} \times 100\right) - 100 = 17\%$$

From the above calculation, it can be seen that an engine running on straight methanol will burn more than twice as much fuel (1.8 times as much for ethanol) as one burning petrol. Therefore you must be careful to ensure that the fuel tap, fuel lines and needle valve will flow the required amount of fuel.

This can present some problems, as many carburettors will not flow the required amount of fuel through the standard needle and seat. Often a larger replacement is not available, so you will have to enlarge the discharge holes to increase flow by the amount necessary. At times you will find it impossible to get main jets large enough, so again you will have to resort to some drilling.

Most Mikuni carburettor jets (the hex head type) are classified with regard to their fuel flow rate, the number stamped on the jet standing for the ccs of fuel the jet is capable of flowing in a certain time. If you are changing from petrol to methanol, then you should start testing with jets at least 2.3 times as large, eg change 210 jet to a 480.

The round head Mikuni jets are rated according to their nominal bore diameter in millimetres, eg a round head 250 jet has a nominal aperture of 2.5mm. Again, when changing from petrol to methanol you will have to begin with jets with an aperture area 2.3 times as large. (Aperture area = πr^2).

Keep in mind also when you convert to an alcohol fuel, either neat or blended, that the fuel/oil ratio may have need of adjustment. Straight methanol would require only 80 per cent as much oil, or a 25:1 ratio in many applications, although some

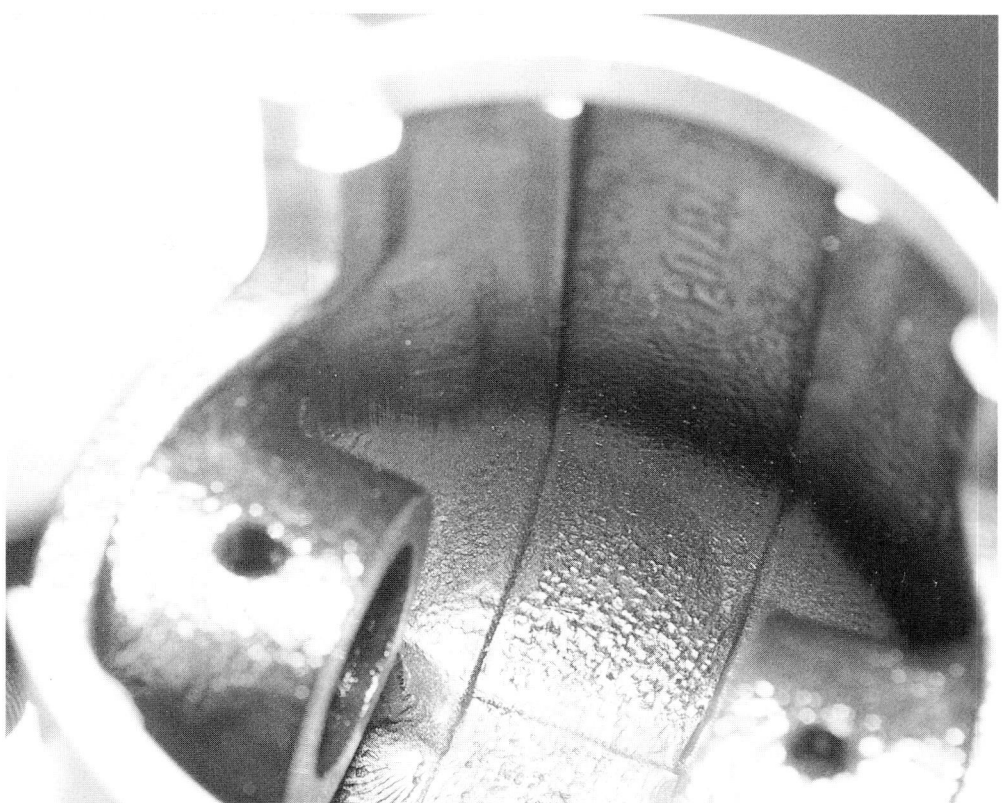

This grey deposit under the piston crown is 'death ash'. Its presence indicates that the piston is overheating due to excessive spark advance or a lean fuel mixture. In this case it was found to be caused by leanness due to the needle and seat not being able to flow sufficient methanol to keep the carburettor float bowl full.

engines will require a 16:1 mix. It is always best to start testing at 20:1 and work from there.

There are other problems involved in the change to alcohol, some of which will affect you and some your engine. Since your life is the most important, we will deal with you first. Methanol is extremely poisonous and, as it is an accumulative poison, it can build up over a period of time and oxidise to form formaldehyde, eventually causing blindness or even insanity. It is absorbed through the skin and lungs, either by direct contact or from the vapours. Inhalation of the exhaust gas can also be dangerous as vaporised methanol is usually present, especially when rich mixtures are being used.

Alcohols are a very effective paint stripper, and they may attack some fibreglass resins. They have a scouring effect on fuel tanks and lines so these should be soaked in alcohol and then drained so that the residue does not find its way into the carburettor when you switch from petrol to alcohol.

Methanol and ethanol will absorb huge amounts of water out of the air, so they must always be kept in an air-tight container. The fuel will also have to be completely drained from the tank and the carburettor to prevent the formation of water-induced corrosion and oxidization. This can be particularly damaging to a carburettor and usually results in blocked metering passages.

After burning alcohol in a two-stroke engine it is most important to run a petrol/oil mix rich in oil through the engine each time you put your machine away after a day's running. If this is not done, you will soon find corrosion and etching of the cylinder wall, crank and piston pin, needle and ball bearings that will lead to premature failure. To prevent this occurrence, I would suggest that you run a half pint of 16:1 petrol/oil mixture through the engine.

In colder climates, starting difficulties may be encountered when pure alcohol is being burned. Some use other more volatile fuels blended in, to help overcome this problem. Usually 5 per cent acetone or a maximum of 3 per cent ether is used. I do not recommend starting aerosols containing ether, due to the possibility of engine damage being caused by detonation. Personally I feel the best method is to remove the spark plug and pour about a half teaspoon of either petrol or neat acetone into the cylinder before you attempt to start the engine.

Alcohol burners demand a good ignition system. Not only does the ignition have to cope with much higher compression pressures, it may also be called on to fire plugs wetted by the very rich mixture being inducted. Alcohol fuels burn much more slowly than petrol, so it will be necessary to experiment with more ignition advance. It is not possible to predict just how much additional advance will be required as there are so many variables involved, but you should begin testing with about an extra 3–5° advance.

Before you advance the spark lead, do make sure that the carburation is fully sorted out. If the engine runs just slightly lean, with added spark lead you could very easily hole a piston. As well as a much larger main jet, it is probable that a different needle profile and a larger needle jet will be required. To correct off idle leanness, a bigger pilot jet and a small 1.0 to 1.5mm slide cutaway may be needed.

When methanol or ethanol is the base fuel, propylene oxide may be added to increase the combustion flame speed. If you decide to use propylene oxide, be very careful to blend in not more than 3–5 per cent by volume and ensure a rich fuel/air mixture of 1:4.5–5.0 is maintained, otherwise mechanical damage may result. Propylene oxide can become explosive if allowed to come in contact with rust particles or copper and its alloys. Therefore it must be stored in plastic or aluminium containers. Once blended with other fuels it is relatively stable.

Acetone is often blended with alcohol to accelerate combustion flame speed, and also to reduce its tendency to pre-ignite when lean mixtures are used. Usually a 10 per cent acetone/90 per cent alcohol blend is all that is required for this purpose, although much higher percentages of acetone may be blended if desired.

In some classes of competition 'nitro' type fuels may be permitted. The first, nitromethane, is not a good fuel as such but if used sensibly it will give two-stroke engines a power boost. Nitromethane's only virtue is that it contains approximately 53 per cent by weight oxygen, so in effect it is a chemical supercharger. In drag car engines it is blended 80–90 per cent nitro to 10–20 per cent methanol, but there is no way a two-stroke engine can hold together with more than a 20 per cent nitro–80 per cent methanol blend. Even then, I would only use nitro in small and rigged single cylinder dirt track engines. To deter detonation, or other engine damage, it is always necessary to lower the compression ratio. If your engine runs reliably at a 17:1 compression ratio on methanol, then you should be able to use a 14:1 ratio with a 20 per cent nitro–80 per cent methanol fuel mix.

As with methanol, nitromethane demands a rich fuel/air mixture. Using a 20 per cent nitro blend, the mixture would be approximately one part fuel to three or four parts air, ie 1:3–4. This means that you will have to increase the main jet size by about 22–25 per cent above that required for pure methanol with a 20 per cent nitro-80 per cent methanol mixture. A 12–15 per cent jet increase will be close for a 10 per cent nitro blend.

Care is in order when handling nitromethane, as it may become explosive. Normally nitro is quite safe, but it may be made shock sensitive by any of the following practices:

a) The addition of hydrazine in fuel blending.
b) The use of caustic soda or any other alkaline for cleaning the mixing drum.
c) The use of 'unpickled' anodised aluminium fuel tanks. After anodising, the tank must be allowed to stand for a few days filled with a solution of 10 per cent vinegar-90 per cent water.

The other nitro fuel, nitropropane, can be blended with methanol (and also 100/130 Avgas) in concentrations up to a maximum of about 12 per cent to produce a horsepower gain in the order of 5–6 per cent. However, extreme caution is called for to avoid engine damage as it also is an oxygen-bearing fuel and will cause a severe lean-out. As a guide increase carb jets by 1.5 times the percentage of nitropropane added. Thus if using the maximum 12 per cent, increase jets 18 per cent. Note that when using nitropropane the acetone content in the fuel should not exceed 3 per cent. As many fuel companies blend in 2–3 per cent acetone to improve methanol's ability to mix with other fuels do not add any acetone without first consulting the refinery chemists. Additionally do not use propylene oxide with nitropropane.

Chapter 6

Ignition

Two-stroke engines of the type being considered in this book rely on an electric spark to initiate combustion of the fuel/air charge which has been inducted into the cylinder. For the engine to operate efficiently, the spark must be delivered at precisely the right moment in relation to the position of the piston in the cylinder and the rotational speed of the crankshaft. Additionally, the spark must be of sufficient intensity to fire the fuel mixture, even at high compression pressure and high rpm.

Today, very few two-stroke engines use a coil and battery ignition system. However, we will consider the operation of this type of ignition first as this will enhance your understanding of the workings of magneto, magneto-type capacitor discharge, and battery-type capacitor discharge systems.

The battery and coil type system relies on a battery, either 6 or 12 volt, to supply the initial electrical energy; a set of points to time the spark, and a coil to intensify the voltage of the electrical energy supplied by the battery so that it is capable of jumping the spark plug gap and firing the fuel mixture (Figure 6.1).

When the points are closed, electric current flows through the coil's low voltage primary winding, and then through the points to earth. The current in the low tension winding produces a magnetic field which surrounds the coil's secondary or high tension winding. As soon as the points open, current flow through the primary stops, and the magnetic field collapses, causing an electric current to be induced in the secondary winding. This creates a high voltage spark (up to 25,000 volts) capable of jumping across the spark plug electrodes to fire the fuel/air mixture.

In Figure 6.1, you will note that a condenser is also included in the primary ignition circuit. Many have the idea that the condenser stops the points from burning, but this is secondary to its main function, which is to drain off electrical energy quickly from the coil's primary winding. This speeds up the collapse of the magnetic field when the points open and increases the high voltage spark intensity. Without a condenser, the electrical energy inducted in the coil's high voltage winding would be too feeble to produce a spark.

Ignition

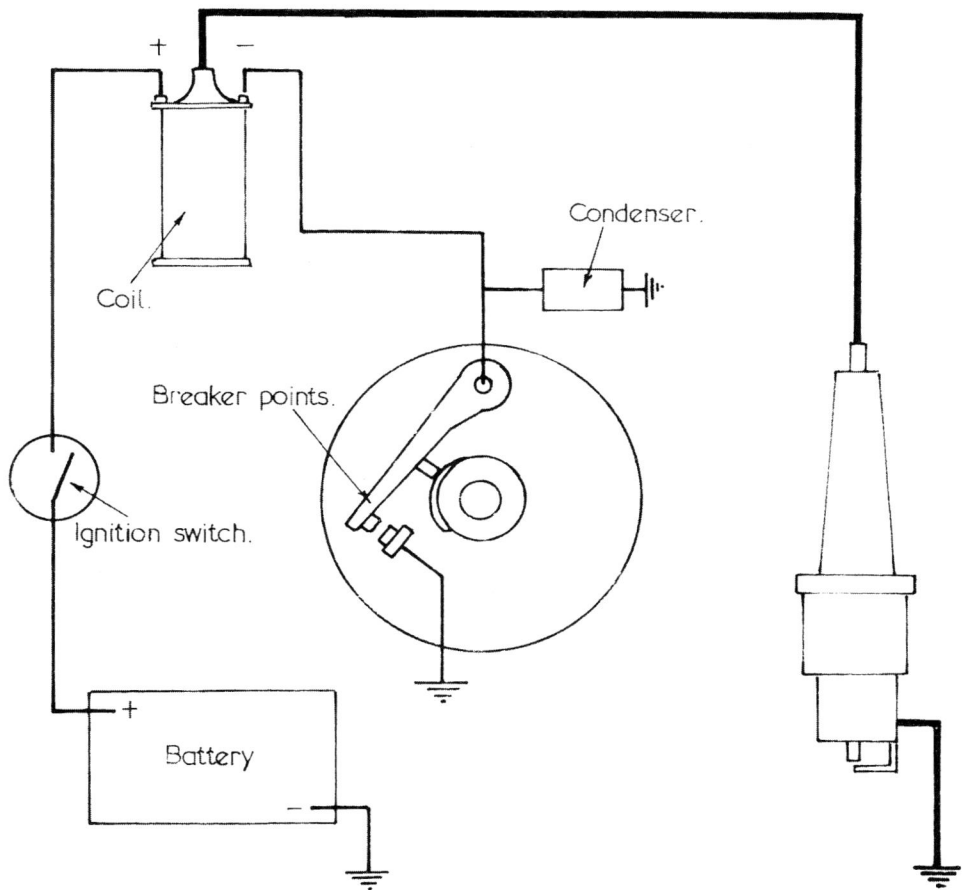

Figure 6.1 Conventional points type negative earth ignition system.

The coil and battery system works very reliably and efficiently. However, due to ignition problems peculiar to two-stroke engines, this type of ignition has largely been replaced by ignition systems which produce a high voltage spark more quickly. The rise time (ie the time from 10 per cent of maximum output voltage to 90 per cent of maximum output voltage) is between 75 and 125 microseconds with a conventional coil and battery system. In that interval a spark plug, surrounded by a conductive petrol/oil mist, will have time to bleed off voltage across the insulator nose, causing a misfire, or at best result in a spark of low intensity. Capacitor discharge and magneto systems both overcome this problem, having a rise time of 20 microseconds and 45 microseconds respectively.

The magneto systems fitted to two-stroke engines have, over the years, often gained a bad reputation. True, in a few instances this reputation was well founded but, generally, a magneto system will give reliable and efficient service if correctly maintained. The most common causes of trouble are burned out and pitted points due to a lack of regular maintenance, and electrical collapse of the condenser. This is usually as a result of the condenser being mounted in a very hot area. For good

Two-stroke performance tuning

service, the condenser must be mounted away from extreme engine heat in a relatively cool location, preferably close to the ignition coil. When this is done, it is a simple matter to connect the condenser to the coil's input terminal rather than run a long wire back to the points. Only when the condenser is collapsing the coil's primary magnetic field quickly will an intense spark result.

With a magneto, the primary current is produced in a similar way as for an alternator, hence there is no need of a battery. The alternating current (a.c.) is not rectified to direct current (d.c.), but passes through the magneto's points as is. This contributes to good point life, as there is little possibility of pitting if the point gap is correct and the points are kept free of oil and grease.

When the points are closed, the primary current passes through the coil's primary winding, producing a strong magnetic field that surrounds the secondary high voltage windings. This magnetic field collapses when the points open, inducing a high voltage current in the secondary which fires the spark plug. In this respect the magneto is very similar in operation to the coil and battery system (Figure 6.2).

Because the magneto primary voltage is not regulated to 6 or 12 volts as with a battery ignition system, the primary voltage increases proportionately with the engine speed. This feature of the magneto ensures that the coil is fully energised (saturated)

Figure 6.2 Points type flywheel magneto ignition system.

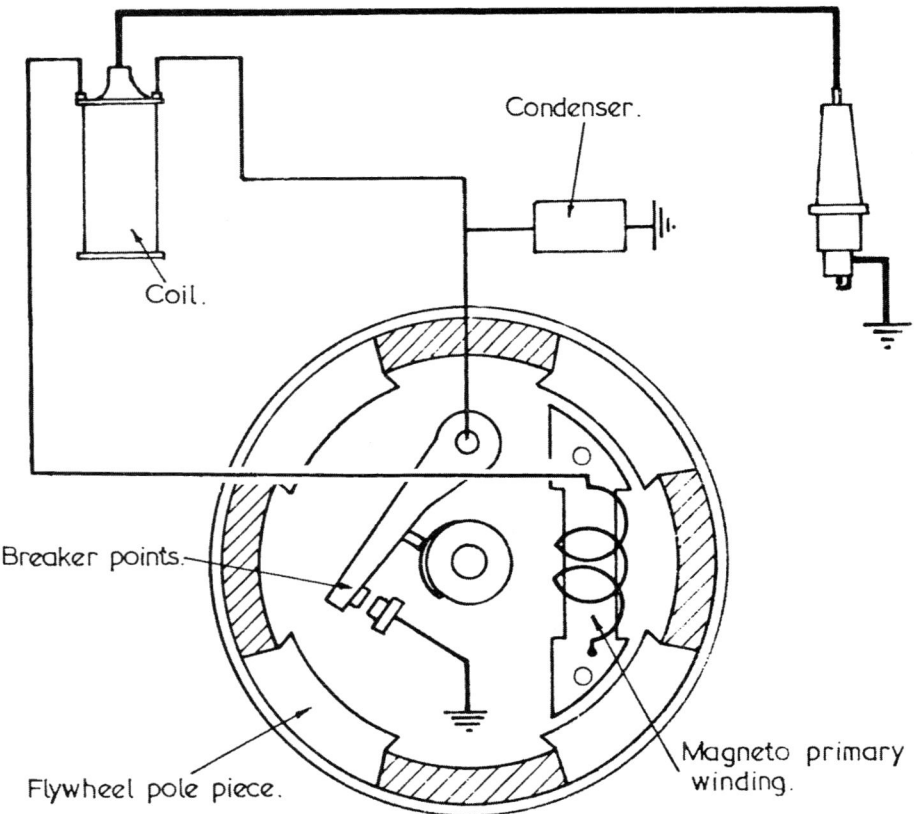

between each plug firing, regardless of how fast the engine is spinning. Also, because the primary voltage increases with an increase in rpm, the secondary voltage does likewise, producing a proportionately bigger spark. For example, if a primary voltage of 8 volts induces 10,000 volts in the secondary, then a primary voltage of 24 volts will result in a secondary voltage of 30,000 volts.

As this time, the majority of two-stroke engines come from the manufacturers with capacitor discharge ignition (CDI). There are two types of CDI systems, the battery type and the magneto type. The battery type requires a battery to supply the system's primary current, while the magneto type generates its own primary current. In all other respects both types function in a similar way.

Basically, the system utilises electronic devices to step-up the output voltage from the battery or magneto to something like 375–400 volts. The primary current, after being boosted to this voltage, is then stored in the storage capacitor. An electromagnetic trigger (usually located in the rotor attached to the end of the crankshaft), on passing close by the trigger coil, induces a pulse which closes an electronic switch called an SCR (silicon controlled rectifier). This allows the storage capacitor to send a surge of power through the ignition coil's primary windings, which induces a high voltage in the secondary to fire the spark plug.

CDI systems will generally give good service if a few basic precautions are taken. Most problems result from poor electrical connections or poor earth connections. To avoid bad connections, clean all the terminals with solvent and, when pushing the connectors together, ensure that they are a good tight fit. Then tape the connectors together so that they cannot vibrate apart. Earth connections are just as important as any other electrical connection. Usually, the connection between the 'black box', for example, and the bike's frame, is quite good. As the earth has to go back to the engine, test to see that the engine is earthed to the frame, using a continuity tester.

When mounting the 'black box' and coils on a sidecar outfit or go-kart, keep them away from engine or exhaust heat. Heat is a killer of electronic components, which is why they are often mounted on a heat sink. Remember that, as well as dissipating heat, a heat sink can also absorb huge amounts of heat if mounted in a hot environment.

With capacitor discharge ignition it is very important to ensure that a spark plug wire will not jump off or be in a position where it could be pulled off accidentally. If this occurs when the engine is running, it is quite probable that the electrical insulation of certain components will break down and cause the system to burn out. Also, keep in mind that you should ensure the plug wires are grounded if you ever flood the engine and it becomes necessary to turn it over vigorously with the plugs removed.

Unfortunately, CDI systems do, at times, fail. Total failure in itself is quite serious, especially if you are in the lead! But more difficult to detect is the type of problem more usual with CDI. Generally, the spark-producing system functions very reliably, but the automatic advance/retard station, which is made up of electronic components, does give trouble. When this occurs, the engine will continue to run, as strong sparks are still being supplied. But if the system is 'locked' in the full advance mode, detonation and engine seizure are a likely result. Less serious is a system functioning fully retarded, as this will cause sluggish performance at lower engine speeds.

The ignition unit advance/retard function can be checked using a strobe timing light. If the timing marks line up at all engine speeds it shows that the ignition is malfunctioning.

There is nothing that you can do to prevent advance/retard system failure, but you can save yourself a good deal of expense and worry if you realise such a problem can exist. Some racers chase their tails for a whole season trying to find lost horsepower, unaware that the CDI is the cause of the trouble. Others spend hundreds of dollars replacing seized pistons and barrels for several successive meetings when all along the trouble has been the CDI stuck on full advance.

If your bike has a CDI with an advance/retard system (check with the manufacturer) you can avoid this sort of unnecessary expense and frustration by regularly checking the timing with a strobe light. Usually you will have to connect the strobe to a 12 volt battery as most bikes do not have a suitable power source. If the advance/retard system is working properly, you will note that the timing marks line up at certain engine speeds and move apart at others. When the system is malfunctioning, the timing marks will not line up at any speed, or in the case of a system which employs a high rpm retard mechanism which has locked on full advance, they will remain aligned at all engine speeds.

To get some idea of how seriously over-advanced an engine can be when a high-speed retard station fails to function correctly, we should examine the Motoplat ignition unit fitted by Rotax to their type 124 air-cooled and liquid-cooled kart engines. Prior to mid-1980, the ignition provided steady ignition timing (ie without any advance/retard). With this Motoplat unit the timing was set at 1.0mm before TDC (14° advance). Later model engines were fitted with an ignition unit incorporating a high-speed retard function. With this type of Motoplat the advance is set at 3.76mm before TDC (27.5° advance). After 5,000rpm the automatic retard gradually reduces the ignition advance with rising engine speed, so that at 11,000rpm the timing has decreased to 1.2mm or 15.5° before TDC. Obviously if the automatic spark retard failed, and the engine was operated with the Motoplat locked on full advance, the engine would quickly fail.

Regardless of the ignition system employed, the spark plug must fire at the correct time if good power is to be made and engine damage avoided. Some two-strokes are timed to spark at 2mm before TDC, others at anywhere from 0.4mm to 4mm. Probably you are wondering why the difference? Well, first, the length of the engine's stroke will affect the amount of advance required. A short stroke engine with 2mm advance will have considerably more advance measured in degrees of crankshaft rotation than a long stroke engine with spark timing of 2mm before TDC. For example, a 125cc engine with a stroke of 60mm and 2mm ignition advance has 18.8° advance, the same as a 125 engine with a 50mm stroke and 1.65mm advance. The advance angle can be calculated using the formula:

$$A = \text{Cos} \frac{(P^2 + R^2 - L^2)}{2 \times P \times R}$$

where A = ignition advance in degrees, R = engine stroke divided by 2 in mm, L = con-rod length (usually stroke multiplied by 2) in mm, T = ignition timing in mm, and P = R + L − T.

As mentioned in the previous chapter, the burn rate of individual fuels differs. Methanol burns slowly, therefore more advance is required as the combustion flame must have progressed just as far by the time the piston passes TDC as it would if the fuel were petrol. To compensate for the slower rate of burning, the flame must be started earlier.

Beside the type of fuel being burnt, there are other factors which influence flame speed. Very rich and very lean mixtures both burn slowly, hence more spark advance will be needed. A mixture close to full power lean burns the hottest and requires the least advance.

An increase in the compression ratio increases the density and temperature of the compressed fuel/air charge. This increases the rate of combustion. Likewise, an improvement in the volumetric efficiency of the engine will have a similar effect. Therefore a change in porting, a new expansion chamber, or a carburettor which allows the engine to breathe better, may require that the spark lead is reduced.

As the engine speed increases, fuel atomisation improves, which means that the fuel will be broken up into smaller particles. Small fuel particles have a proportionately larger surface area, therefore they burn more rapidly, which means less advance is necessary.

The size of the combustion space and the position of the spark plug in that space also influences the advance required. Obviously the further the flame has to travel, the longer it will take to completely burn the mixture. Consequently an engine with a squishless (quiescent) combustion chamber or a large bore diameter will usually require more advance. Also an engine in which the spark plug is offset from the centre of the combustion chamber will increase the distance the flame must travel, hence the need for additional advance.

Two-stroke performance tuning

Any modification which significantly raises the peak hp output of a two-stroke usually results in higher piston crown and cylinder head temperatures. A hot fuel charge burns more quickly, so the advance must be reduced.

Perhaps the most perplexing question for many enthusiasts is why a two-stroke engine has an ignition advance curve which is totally the reverse of that of a four-stroke. A typical four-stroke race engine will start out with about 12° spark advance at idle and finish up with around 32° spark timing at peak torque rpm, extending to perhaps 37° at peak hp revs. However the advance curve in Figure 6.3 is what will be found in many two-stroke race engines. The Yamaha KT-100S kart engine starts out with 27½° advance, rising to a maximum of 28° at 5,000rpm. After that it falls back to 27° at 7,000rpm, then to 22½° at 10,000rpm where it remains fixed, up to maximum engine speed.

The big difference between the two engines is that the two-stroke suffers terribly, both from exhaust gas contamination of the fuel/air mixture and limited cylinder filling at lower engine speeds. Therefore the spark plug must fire a comparatively long time before TDC to give the diluted charge maximum time to burn and expand to force the piston down on the power phase. Then as rpm rise the expansion chamber works more efficiently at evacuating exhaust gas from the cylinder, and correspondingly it draws in much more fuel mixture to more densely fill the cylinder. With less exhaust gas to slow combustion and more tightly packed fuel and oxygen molecules, combustion speeds considerably. Thus spark advance must be reduced to ensure that the pressure peak resulting from this more rapid combustion occurs at 12–15° degrees past TDC. If it were to peak before TDC or even too soon after TDC, combustion pressure would rise too rapidly and the unburned portion of mixture would explode. It is this uncontrolled burn, detonation, which destroys so many race engines.

Figure 6.3 Yamaha KT-100S ignition advance curve with fixed high rpm ignition retard.

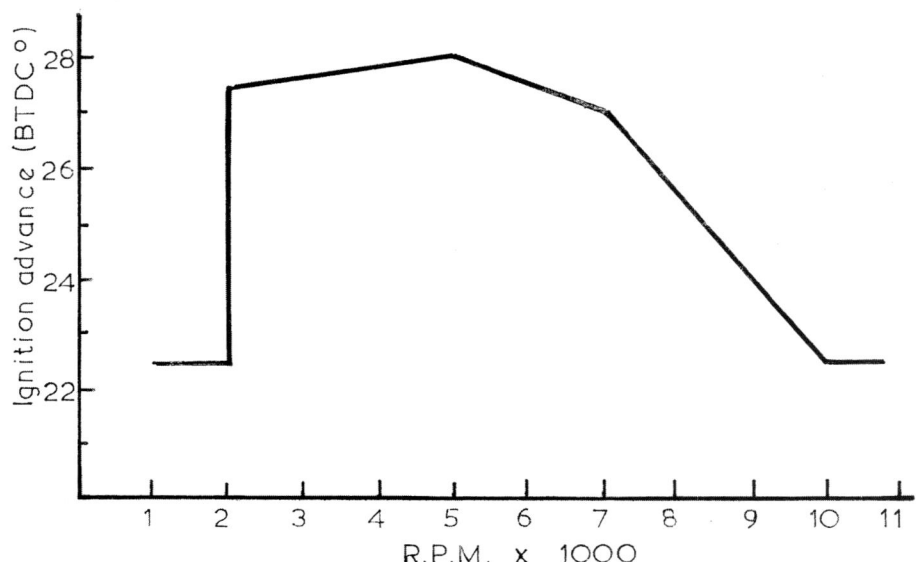

Ignition

With advances in electronics it became possible to go a step further with spark advance. Like four-strokes, two-strokes also experience reduced cylinder filling past the point of peak torque. This means that mixture density is less, therefore to extract maximum horsepower an increase in spark advance is required past maximum torque rpm. Hence rather than fixing the timing at the peak torque/peak hp angle what is really beneficial in many engines is to start adding advance once cylinder filling capacity begins to taper off at the top end of the power range.

Figure 6.4 illustrates just such a spark advance curve. The engine which this spark box was programmed to suit produced maximum torque and horsepower at 11,000 and 11,200rpm respectively. That means that mixture density is at a peak just about 200rpm either side of 11,000rpm; from 10,800 to 11,200. Therefore that is where the ignition is fully retarded, and to give a margin of safety because cylinder fill is quite high at lower and higher rpm the retard is extended 300rpm in both directions ie from 10,500 to 11,500rpm. However once past 11,500rpm mixture density is reducing rapidly so more advance is added so that at 12,000rpm 1° is added, and by 13,000rpm 3° are added. This added spark lead adds power past peak hp revs and broadens the power band by extending the engine's over-rev potential.

Just how much power this additional advance will unleash is hard to quantify. The tendency of tuners is to over-advance the early type ignition slightly at peak torque rpm so as to get close to what the engine really needs to make best power right at the very top of the power band. Thus if the engine made best power at the torque peak of 11,500rpm with 14° advance and it went on to make best horsepower at the rev limit of 12,500rpm with 16° advance the tendency has been to split the difference

Figure 6.4 Modern digital ignition advance curve provides additional ignition advance past maximum horsepower engine speed.

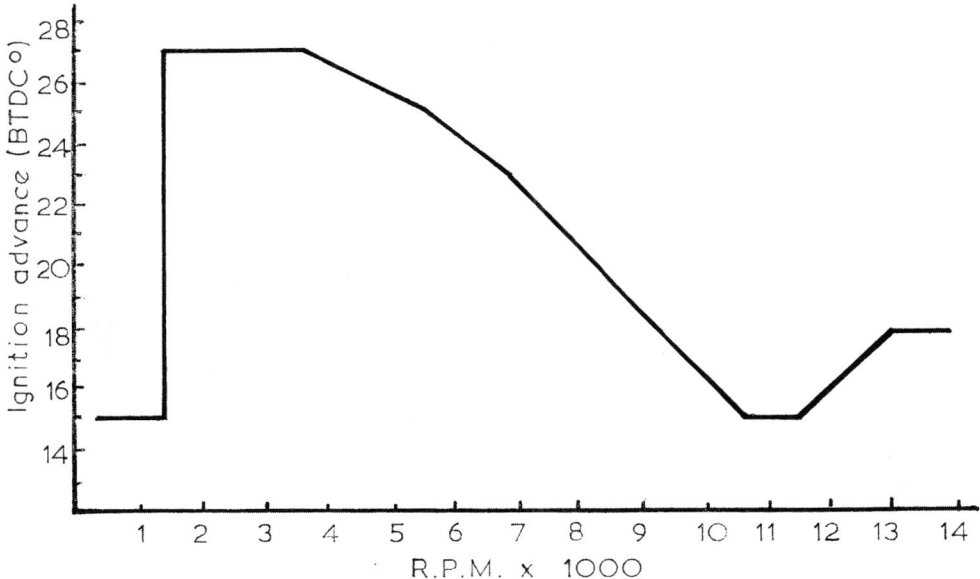

Two-stroke performance tuning

when running on shorter tight circuits. Consequently the engine would be given 15° advance, which is 1° over at 11,500 and 1° under at 12,500.

When an engine has been tuned in such a manner a switch to a programmable ignition box will probably show a 1–2 per cent power rise at the torque peak and a 2–3 per cent increase in hp at the rev limit. However if the engine had been previously tuned as I prefer, with the advance set to give best horsepower at the torque and power peaks, changing to a programmable ignition it will show no increase in maximum hp and torque, but once past that point the increase will be around 3–5 per cent right up to the rev limit. Additionally, as shown in Table 6.1 the rev limit from the aspect of useful power output rather than from a mechanical standpoint, may possibly be moved 200 to 500rpm higher. You will also note a power rise with this example down at the bottom of the power range of between 2 and 3½ per cent. This was achieved because the programmable ignition could have the advance angle optimised to provide the exact ignition timing for every engine speed to suit this engine's modified barrel and expansion chamber, whereas the stock factory ignition could only be adjusted to allow the correct firing angle around the peak torque/peak hp area. Consequently at other rpm the stock advance angle was less than ideal.

Table 6.1 Yamaha TZ125 ignition comparison

rpm	Test 1 hp	Test 1 Torque	Test 2 hp	Test 2 Torque
9,500	22.9	12.7	23.7	13.1
10,000	25.8	13.5	26.3	13.8
10,500	26.8	13.4	27.7	13.9
11,000	31.3	14.9	31.0	14.8
11,500	34.5	15.7	34.8	15.9
12,000	36.6	16.0	36.8	16.1
12,500	37.3	15.7	37.2	15.6
13,000	34.6	14.0	35.4	14.3
13,250	32.4	12.8	33.8	13.4
13,500			27.5	10.7

Test 1 – Stock Yamaha ignition.
Test 2 – Programmable Vortex ignition.

With four-stroke engines it is usual for the modified engine to require considerably more advance than standard. However, from the foregoing you can see that this does not apply in the case of two-strokes. In fact, it is quite unusual for a two-stroke engine to need more advance than that specified by the manufacturer. It is difficult to say just how much advance a modified engine will require, but as they can be easily damaged because of too much ignition advance, I would suggest that you reduce the recommended timing by 20 per cent to begin with. Therefore, if the standard timing is 2.5mm before TDC, start testing with 2mm advance and then increase it in steps of 0.1mm to find if more advance improves the performance. Many enthusiasts have a tendency to over-advance the timing in an effort to pick up every last fraction of performance. My advice is to use the least amount of advance conducive to peak

Ignition

performance. For this reason, I think it is a good idea to back-track to make sure that more advance is better. In this example, if the engine seemed to run better with each 0.1mm increase until you arrived at the standard timing of 2.5mm, drop back to 2.2mm just to make sure that the engine is really performing at its best with 2.5mm advance.

The fact that an engine will tolerate a certain amount of advance does not indicate that that is what makes best power. Nor does it indicate that it will be the correct timing for another track. An engine operating on a short, twisting track where you are on and off the throttle all the time in the lower gears will run best with a little more spark angle than on an average circuit with a broader combination of turns and straights. Conversely, at tracks where you can hold 5th and 6th gear at full throttle for long periods you will have to reduce the timing figure to hold maximum horsepower the full length of the straight. Even if the engine doesn't go into detonation and begin destroying itself the build-up of internal heat will reduce power part-way down the straight. If you run into a headwind, or if there is a rise in the straight the engine will heat even more rapidly and lose power. You may not even notice the loss, but you can be sure that power will be down 4 or 5 per cent, and it can drop off twice that amount before engine damage becomes clearly evident.

If you don't know the signs to watch out for a problem like this can be difficult

A timing dial gauge kit is needed to adjust the ignition timing on two-strokes. When the gauge is not in use keep it in its protective case.

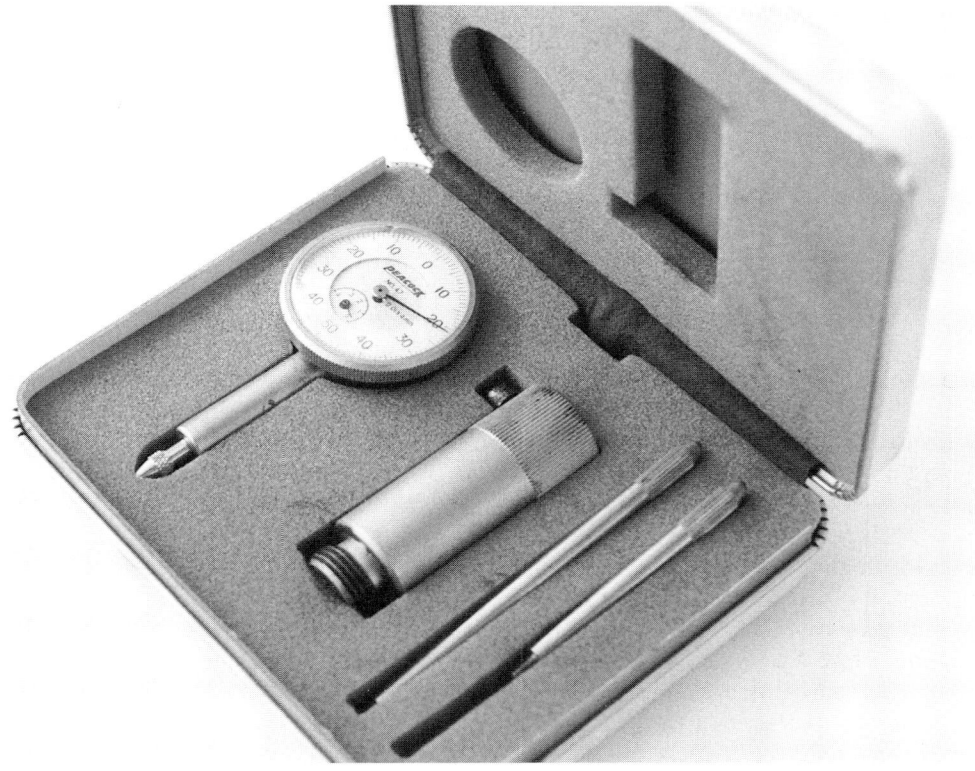

Two-stroke performance tuning

to track down. Such was the situation with a rider who competed quite successfully on a WR200 Yamaha enduro bike. In rough going he could hold his own against the bigger 250 and 400 bikes but once out into the high speed stuff he would be left behind. He concluded that the answer was wilder porting to lift peak horsepower. He also noted the speed difference was significantly reduced during the cooler months and summised that richer jetting may fix the problem when the thermometer rose. It did, but only if the engine was so rich that it was just off four-stroking. After a lot more experimenting he realised his need of expert help. The jetting was brought back to what he had previously found to give the best overall performance. The timing was checked and moved back 0.27mm to the factory recommended position. A thermocouple washer was installed under the spark plug to record cylinder head temperature and the engine was run up on the dyno. A good engine should have made 33–34hp; this one got to 31.8 at 10,000rpm and sent the thermocouple temperature rocketing up to 228°.

This is well into the danger zone. An engine will live to give a flash power reading with the thermocouple indicating between 218° and 230°, but the power will be down and it quickly falls away as the engine dives into meltdown mode. Testing shows a clear relationship between what sort of power you can expect and the thermocouple reading. You can pull maximum power at perhaps 205°C and hold that horsepower level for as long as it takes the temp washer to record 212°C, but as soon as those 5–10 seconds have passed and the temp rushes past about 218° horsepower falls rapidly.

During a few more tries the mixture was richened slightly and the timing was progressively retarded a total of 0.35mm. In this state the engine wasn't so sharp down low but above 8000rpm the horsepower was up over 1hp and at 10,000rpm the dyno reading was now 33.7hp; 1.9hp or 6 per cent stronger than previously. Additionally the engine would now hold the maximum power figure for over 15 seconds and at 20 seconds it was still stronger than the previous flash reading.

In case you are thinking this sort of thing is peculiar to dirt bikes with comparatively small radiators I will now relate the saga of a TZ racer. He had done all the right things – closed the squish clearance to 1.0mm and set the timing to Yamaha's recommended figure. Yet the first time he stripped the engine one combustion chamber and the lip of the corresponding barrel showed signs of detonation sand blasting. He knew what the erosion meant so he backed the timing off 0.4mm in the peppered cylinder and 0.2mm in the other cylinder; but now the bike was slower.

The suggested remedy was to recut the combustion chambers as once the erosion starts the roughness seems to get detonation started even when everything else is okay. Also the squish would be taken down to 0.8mm. With the engine rebuilt and the timing adjusted to stock it was connected to the dyno. A thermocouple washer was placed under each spark plug to enable both cylinders to be independently scrutinised. The engine made 60.4hp at 12,250rpm but while the previously good cylinder was holding at 202–208°C the 'detonator' was heading toward 232° after just a couple of seconds at full load.

When you see something like this you need to check that the fuel flow meter is giving the same flow rate on both carbs – it was. Next the mufflers were removed just in case one was altering exhaust back pressure due to soot build-up. Also the expansion chambers were checked very carefully for cracks or dents as anything like

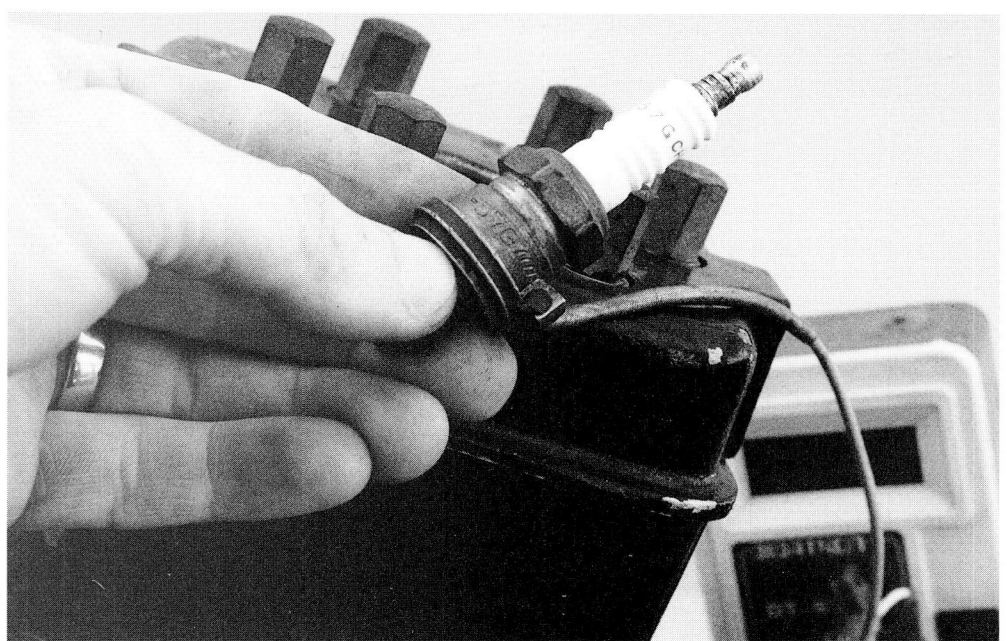

The thermocouple washer is fitted under the spark plug to provide valuable temperature data during track and dyno tuning sessions.

this can alter how efficiently they pull mixture into the cylinder and how much exhaust gas they leave behind, or even push back into the cylinder to slow and cool combustion. There weren't any dents or cracks so the conclusion was that one cylinder had to be burning faster or hotter than the other. This isn't unexpected as production tolerances can mean that a set of ports will flow more mixture than another. Then if that cylinder which is flowing just a touch better is matched with an expansion chamber which is producing slightly stronger pulses you end up compressing a larger volume of fuel/air mixture into the same size combustion space as its neighbouring cylinder. The end result is higher effective compression, which calls for less spark advance.

For the next dyno run the ignition timing on the problem cylinder was retarded a full 0.5mm. As expected the horsepower hardly changed, moving up a touch to 60.7hp. However the plug washer temperature was way down to 201°. On the next two pulls the spark was bumped first 0.15mm and then another 0.1mm. Horsepower had now progressively increased to 62.5hp and finally to 62.9hp, with the thermocouple temperature holding below 215°C after several seconds at full load. The good cylinder was still showing no sign of stress according to the washer temperature. So for the next run the timing on this cylinder was advanced to 0.12mm more than stock. At 12,250rpm peak power now went up to 63.2hp. That doesn't seem much extra, only 0.3hp, but between 9,500 and 11,000rpm the engine was up to 0.6hp stronger. Overall the tuning session netted almost 3hp, or close to 5 per cent. In addition the engine will sustain that power level without engine damage.

Why though, prior to the dyno session, was the bike faster with the engine apparently over-advanced? Well there are two reasons: the good cylinder didn't in fact need to be retarded, it actually worked better with more advance, and secondly on the detonating cylinder the timing was retarded too far. The end result was that the good cylinder was retarded 0.32mm and the detonator was retarded 0.15mm from best sustainable horsepower advance.

The other question begging an answer is, why did one cylinder work best with more than the factory recommended amount of advance? Again this can be due to production differences with that cylinder not filling as well, or having more residual exhaust gas, because of minor differences in port flow or expansion chamber pulsing. Exhausts can also work differently when one pipe gets more air flow over it than the other. Because the gases cool more quickly wave action in that exhaust changes. Another reason can be that the factory test track has a great whopping 1.5km long straight which will keep the engine fully loaded in 5th and 6th gear for perhaps 30 seconds. To avoid losing power half way down the straight it is possible that the factory had found that they had to retard the timing. However the circuits where you run may have only 0.5 to 0.7km straights, so because you will not be heating the engine up so much you can perhaps run more advance and win a horsepower benefit.

When Rotax introduced their digital ignition in 1988 for their kart and road race 125 and 250 engines they made allowance for this by programming in four separate ignition curves. Below 12,000rpm all four curves were pretty much the same but once past 12,000rpm curve No. 4 provided maximum high rpm ignition timing while curve No. 1 gave maximum retard. During development work Rotax found that not only did atmospheric conditions and circuit layout have a large bearing on which ignition curve best suited the engine, but riding/driving styles and whether the engine was installed in a kart or a bike also influenced the outcome.

Digital ignition has progressed considerably since then and many more programs can now be installed in much smaller ignition units. A basic curve such as in Figure 6.4 is sufficient for most dirt bike applications, however a more sophisticated system with at least three ignition curves such as in Figure 6.5 will provide a higher level of performance and maximum engine protection. For road circuit competition we really need several more advance curves, preferably with on-the-move selectability to provide additional retard on very long straights without compromising the advance available on tighter sections. Such selectability also allows for switching curves to give a degree of traction control on corners where the engine is hitting hard in the wrong part of a corner, or it allows for changing track surface conditions or changing atmospherics in longer events.

Naturally when there are several ignition curve options available it can be quite time consuming assessing what ignition advance angle is actually best for the engine on a particular circuit. Consequently it is quite easy for a tuner to make the wrong choice. With this in mind when Honda produced their 1998 model NSR500 V-twin production racer they equipped it with a two-phase detonation counter. At the conclusion of a practice lap the digital display shows the number of times each cylinder detonated in the range up to half throttle, and also the number of times each cylinder detonated between half and full throttle. Obviously such an arrangement enables the tuner to arrive at the correct spark timing angle much more quickly.

To do any timing changes on the less sophisticated ignitions you are going to

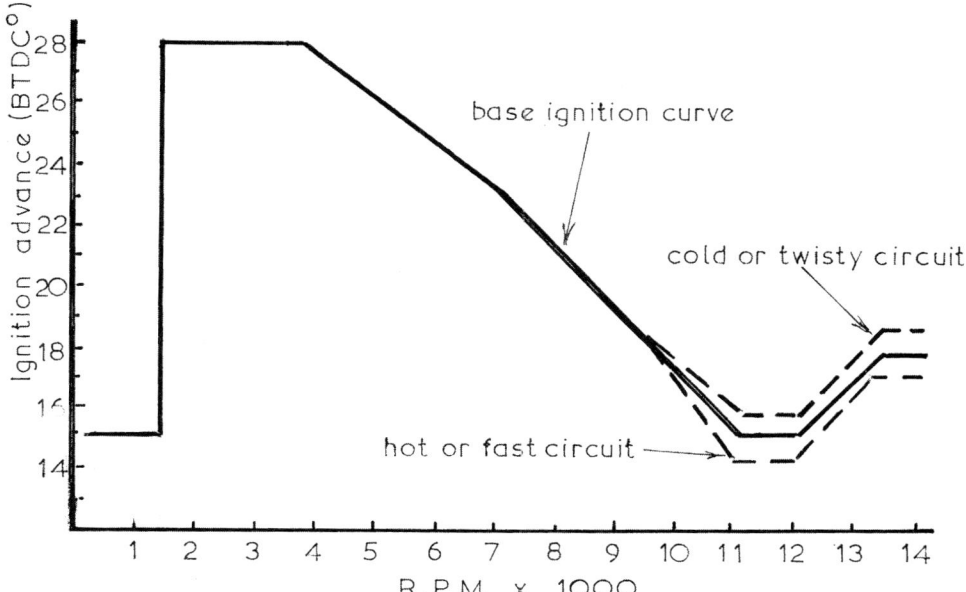

Figure 6.5 Modern digital ignition may allow the selection of an advance curve to best suit atmospheric conditions, circuit layout and riding style.

need a timing dial gauge. It is one of the most important tools in a two-stroke tuner's kit, so purchase a good one, after all they are only the price of a couple of new pistons. Some enthusiasts think that they can adjust the timing using an automotive strobe timing light, but this is not so. A strobe light will only indicate if the electronic advance/retard mechanism (in the case of CDI systems) is working correctly. The timing marks will always line up at lower engine speeds if the electronic components are functioning correctly, regardless of how far the ignition timing is under or over-advanced.

Some manufacturers quote their timing figures in degrees and suggest the use of a protractor or degree wheel to adjust the ignition to the correct advance. This is a very slow way. I feel that it is far better to convert their figures to mm before TDC using this formula:

$$T = L + R \times (1 - \cos A) - \sqrt{L^2 - (R \times \sin A)^2}$$

where T = timing in mm, A = timing in degrees, L = con-rod length in mm (usually stroke multiplied by 2), and R = engine stroke in mm divided by 2.

For example, McCulloch quote the timing for their 100cc go-kart engine as 26° before TDC. In mm before TDC this would be as follows:

A = 26°, L = 63.7mm, and R = 20.765mm.

$$T = L + R \times (1 - \cos A) - \sqrt{L^2 - (R \times \sin A)^2}$$

$$= 63.7 + 20.765 \times (1 - .8988) - \sqrt{63.7^2 - (20.765 \times .4384)^2}$$
$$= 63.7 + 2.1 - 63.05$$
$$= 2.75 \text{mm}$$

To adjust the timing on CDI systems using a timing dial gauge is quite easy, but you can waste a lot of time setting the dial gauge up if you go about it the wrong way. Assuming that the timing is to be adjusted to 2.5mm before TDC, this is the procedure to use. Insert a pencil in the spark plug hole and turn the crankshaft until the pencil rises to the highest point. In this position the piston will be approximately at TDC. Screw the timing gauge fixture into the plug hole and, after zeroing the gauge, insert it into the mounting fixture. Push the gauge down until it reads 3.0mm (ie about 0.5mm more than the timing figure) and lock it into position. Now gently rock the crankshaft backwards and forwards to find true TDC. When you have found TDC, hold the crankshaft in this position and turn the face of the gauge around until the zero mark aligns with the pointer. With that done, again rock the crank to ensure that the pointer does actually indicate zero when the piston is at TDC. When you are sure that the dial gauge is reading zero at TDC, rotate the crankshaft to align the timing marks, at the same time noting how many mm the pointer moves through. If the pointer has made two revolutions of the dial and is now indicating 0.5, then the timing is correct.

Using this method you must be very careful to note the movement of the pointer, as most timing gauges are graduated 0–50–0 not 0–100. Therefore 0.3mm, for example, would be either to the left or right of zero, the same as 0.7mm. Because of this drawback some prefer to set the dial gauge up this way: rotate the crankshaft to align the timing marks; insert the dial gauge into the mounting fixture and lock it into

The ignition timing is read off the dial gauge with the ignition timing marks lined up. Unfortunately the timing marks are not always easy to see, which leads to timing errors.

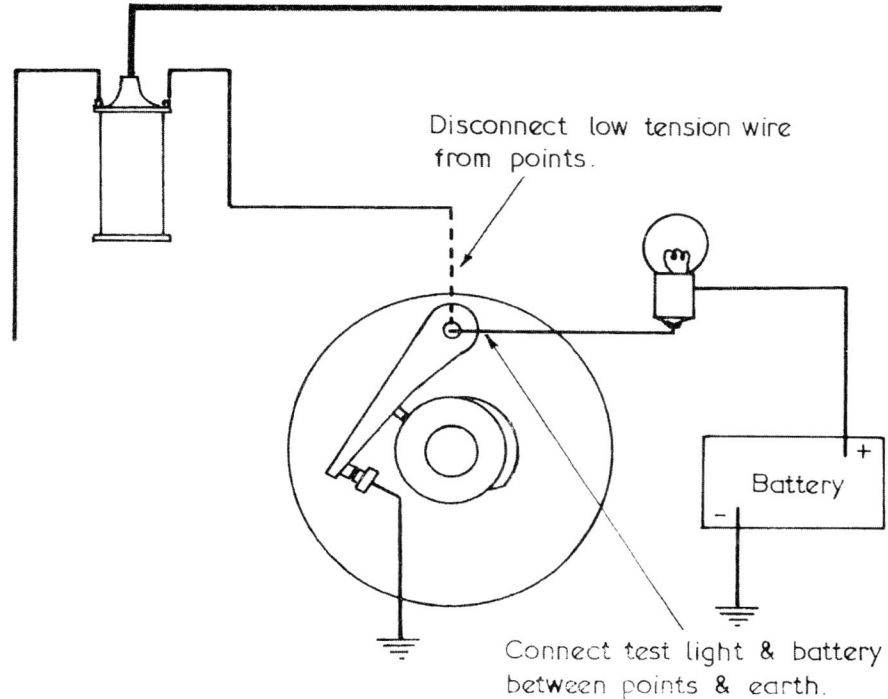

Figure 6.6 Using a test light to check when the ignition points just break open.

position; zero the gauge, being very careful to note that the timing marks remain aligned; rock the crank backwards and forwards, then ensure the timing marks align when the pointer indicates zero; rotate the crank to TDC and read the timing advance straight off the dial. It sounds easy, but you will find that it is quite difficult to check the timing this way. The problem is that the crank usually moves when you attempt to zero the dial, throwing the timing marks out of alignment.

With a points-type ignition system (either magneto or battery and coil) you will find it even more difficult to achieve accurate timing using the latter method. Points-type ignition is a little different to adjust than CDI, as you don't have any useful timing marks. Instead a continuity test light (or a buzz box) is connected across the points (Figure 6.6). When the light goes out, indicating that the points have just opened, the dial gauge reading will show whether the timing is correct. If the timing is incorrect then the points will have to be adjusted. Some manufacturers mount the points on a movable base plate, which can be rotated to achieve the desired ignition advance. Some, however, do not have this facility, so what you must do is increase or decrease the points gap to change the timing.

Perhaps the greatest obstacle to achieving accurate ignition timing is the inadequate timing marks which many manufacturers provide. Generally, I have found it impossible to adjust the timing to an accuracy of better than ±0.15mm of the desired figure, using the naked eye and the manufacturer's standard timing marks. To make the timing marks easier to see, they should be painted black and the surrounding area left unpainted, or else the entire area should be painted black and the marks painted white or silver. Often, you will find that the rotor face stands proud of the stator by anything up to 5mm, which means that it will be very difficult, if not impossible, to adjust the timing accurately because of parallax error, even if the timing marks are clearly visible. In this situation you have two options. Either you can make a fixture with a sight slot, like a gun sight, to sight through on to the timing marks, or you can pull the rotor off the end of the crank and very accurately extend the timing mark on to the rotor's periphery, using a three corner file or a scriber. Anything which you can do to improve the accuracy and repeatability of ignition timing adjustments will lessen the possibility of detonation and allow the engine to perform at its best.

If your machine has a points-type ignition system, don't attempt to adjust the timing using the stock timing marks. Instead, always set the advance with a dial gauge and a light connected across the points. Many of the piston failures experienced by

Careful tuners spend a lot of time peering into the spark plug magnifier, always looking for the tell-tale signs of detonation.

engines like the early Yamaha RD twins can be traced back to inaccurate ignition timing as a result of the tuner relying on the stock timing marks.

Detonation, and to a lesser extent pre-ignition, both damage many two-stroke engines. Detonation occurs when a portion of the fuel/air mixture, usually the 'end gases', begin to burn spontaneously after normal ignition takes place. The flame front created by this condition eventually collides with the flame initiated by the spark plug. This causes a rapid and violent burning of any remaining fuel (almost an explosion) which hammers the engine's internal components with such force that the cylinder wall and piston crown actually vibrate. This vibration makes the pinging sound which an alert ear can pick up. Pre-ignition is ignition of the fuel/air charge by a 'hot spot' before the spark plug fires to initiate normal combustion. Typically, this leads to a loss in performance much like excessive ignition advance in its early stages. If allowed to continue it can destroy an engine.

When engine damage results from either type of abnormal combustion, the culprit can usually be identified after an examination of the piston and spark plug. Pre-ignition damage is caused by the extreme combustion temperature which results, melting the piston crown and also, possibly, the ring lands. If a hole is present in the piston, it will appear to have been burned through with a welding torch. The metal around the hole will be fused and have a melted appearance. The spark plug may have the centre electrode melted away and, in extreme cases, the insulator nose and earth electrode will also be fused.

A piston damaged by detonation will show signs of pitting on the crown. The edge will be gray and eroded, as if sandblasted. In the very early stages, gray ash-like

These combustion chamber deposits can become incandescent and pre-ignite the fuel/air charge.

deposits form on the exhaust-side edge of the piston crown. In extreme examples, the piston will be holed. The hole will appear to have been punched through, with radial cracks and a depressed area around the hole. A spark plug subjected to fairly severe detonation will usually show signs of cracking at the insulator nose. Engines with plated aluminium cylinders will exhibit a sandblasted effect around the top lip of the bore.

Pre-ignition can frequently be traced to deposits in the combustion chamber or on the piston crown becoming incandescent. Since these deposits do not conduct heat well, very high temperatures can be reached within such accumulations. Spark plug heat ranges can also affect pre-ignition. If the electrodes retain too much heat from previous combustion cycles, they will glow and pre-ignite the fuel.

The conditions which lead to detonation are high fuel/air mixture density, high compression ratio, high inlet charge temperature, best power fuel/air ratio (ie 1:12.5), and excessive spark advance. A piston crown or combustion chamber overheated by pre-ignition can initiate detonation by excessively heating the 'end gases'. Go-kart engines with fixed gearing may also suffer from detonation when pulling out of low speed corners, if geared too high. Except in the latter two cases, detonation is eliminated by reducing the ignition advance and possibly by jetting richer (Tables 6.2 and 6.3).

Table 6.2 Effect of ignition advance on combustion temperature

Ignition timing (mm)	Spark plug electrode temp. (°C)
2.0	853
2.25	876
2.5	908
2.75	962

Table 6.3 Effect of fuel flow on combustion temperature

Fuel flow (litre/hr)	Spark plug electrode temp. (°C)
3.0	904
3.25	880
3.5	857
3.75	832
4.0	800
4.25	766

Due to increased combustion temperatures in a modified engine, consideration must be given to finding a spark plug with the correct heat range. A hot plug transfers combustion heat slowly and is used to avoid fouling in engines with relatively low combustion temperatures. A cold plug, on the other hand, transfers heat rapidly from the firing end. It is used to avoid overheating where temperatures are high, as in a racing engine (Table 6.4).

The length of the insulator nose and the composition of the electrode alloy are the primary factors in establishing the heat rating of a particular plug. Hot plugs have long insulator noses, and hence a long heat transfer path. Cold plugs have shorter nose

Ignition

Table 6.4 Effect of spark plug heat range on plug temperature

Plug type	Under-plug temperature (°C)
Champion L4G	234
Champion L3G	222
Champion L2G	213

lengths to transfer heat more rapidly from the insulator tip to the cylinder head, via the metal spark plug body (Figure 6.7).

Generally, two-stroke engines do not require a plug more than one or two grades colder than standard, even when extensively modified for very high power outputs. Providing the engine is in good condition, and the carburettor is correctly tuned, reading the nose of the plug will indicate if one with the correct heat range has been chosen. So that you do not end up with engine damage it is advisable to begin testing with a plug which is too cold, or else test the machine at moderate load and speed, and then check the plug before you engage in any full power running.

For the plug reading to be accurate, it will be necessary to run the engine at full throttle and maximum speed and then cut the engine dead. If you allow the engine to keep running as you bring the bike to a stop, the plug reading will be meaningless.

The signs to look for when reading a plug are indicated in Table 6.5. You will note that it is not just the colour of the insulator nose in which we are interested. The entire firing end of the plug exposed to the combustion flame must be examined and read.

Of course, spark plug heat range must be tailored to each race circuit. Tracks with a long, fast straight may require a plug one grade colder. Conversely, a tight, wet track may require a plug one grade hotter than normal.

Once you have determined the correct plug heat range, don't swap over to

Figure 6.7 The length of the heat path has a large bearing on spark plug heat range. The longer the heat path the hotter the plug.

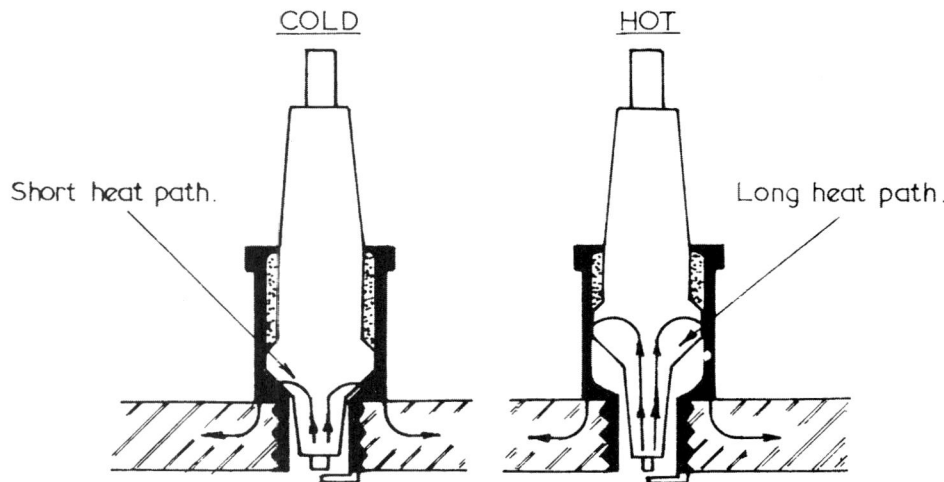

Table 6.5 Spark plug reading for heat range and other conditions

Spark plug condition	Indications
Normal – correct heat range	Insulator nose white or very light tan to rust brown. Little or no cement boil where the centre electrode protrudes through the insulator nose. The electrodes are not discoloured or eroded.
Too cold – use hotter plug	Insulator nose dark grey or black. Steel plug shell end covered with tar-like deposit.
Too hot – use colder plug	Insulator nose chalky white or may have satin sheen. Excessive cement boil where centre electrode protrudes through the insulator nose. Cement may be milk white or meringue-like. Centre electrode may 'blue' and be rounded off at the edges. Earth electrode may be badly eroded or have a molten appearance.
Pre-ignition – use a colder plug and remove piston and combustion chamber deposits	Insulator nose blistered or fused. Centre electrode and side electrode burned or melted away.
Detonation – retard ignition and richen mixture	Fractured insulator nose in sustained or extreme cases. Insulator nose covered in tiny pepper specks or even tiny beads of aluminium leaving the piston. Excessive cement boil where centre electrode protrudes through insulator nose. Specks on plug shell end.

another brand with an 'equivalent' heat range. Heat range conversion charts should be used as a guide only, when you swap from one plug brand to another, as individual plug manufacturers use different methods of determining the heat range of their plugs. If you cross-reference the conversion charts from all the plug manufacturers you will find that they disagree with each other, due to varying test procedures.

As well as the heat range, the gap style of the plug must also be considered to obtain the best performance, and in a few instances to avoid engine damage (Figure 6.8). The spark plug which I recommend for all two-stroke applications, with a few exceptions, is the fine wire type Champion Gold Palladium. This type of plug has a very wide heat range to resist both fouling and pre-ignition. It can be supplied with the standard Gold Palladium centre electrode or, for special applications, with a platinum centre electrode (Table 6.6). The small diameter centre electrode requires less voltage to fire than a regular electrode. This feature allows easier starting of all engines employing magneto type ignitions (either points or CDI) as the voltage available at lower cranking speeds is diminished. The insulator nose is a special 'open' design which allows more clearance within the firing end of the plug for better scavenging of deposits.

Ignition

Table 6.6 Champion spark plug heat range chart
(courtesy of Champion Spark Plug Co.)

14mm thread – ³⁄₄ in reach: N series

	Regular	Gold Palladium	Gold Palladium	Fine Wire	Retracted
Hot ↑	N4	N4G			
	N3	N3G	*N87G	N87	
					N62R
	N2	N2G	*N86G	N86	
	N60				
					N60R
		N59G	*N84G	N84	
	N1				
	N57	N57G	*N82G	N82	N57R
		N55G	*N80G	N80	
					N54R
Cold					+N52R

* Special plug for high compression engines
\+ Silver electrode

14mm thread – ¹⁄₂ in. reach: L series

	Regular	Gold Palladium	Retracted	Retracted
Hot ↑	*L82			
		*L6G		
	L4J			
	*L78	*L4G	L62R	
	*L77J	*L3G	L60R	
		*L2G	L57R	+L87R
		*L55G	L54R	+L84R
Cold				+L82R

* .472 in. reach.
\+ Special methanol plug.

14mm thread – ³⁄₈ in. reach: J series

	Regular	Gold Palladium	Gold Palladium	Retracted
Hot ↑	J5			
		J64G	*UJ7G	
	J4J			J62R
	J2J	J60G		J60R
	J79			
				+J57R
Cold				+J54R

* Plug has auxiliary gap.
\+ Not suitable for methanol – use L series methanol plug and N677 gasket.

Two-stroke performance tuning

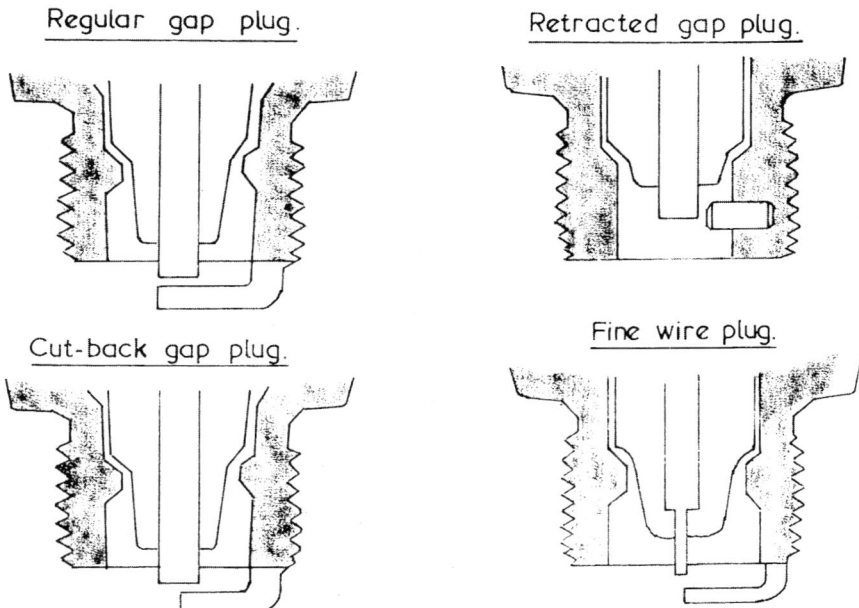

Figure 6.8 Spark plug gap styles suitable for two-stroke engines.

The conventional regular gap plug is my next choice after the fine wire type. It does not have such a good heat range and anti-fouling properties as the Gold Palladium, but it is cheaper. However, it has a heat range far superior to the retracted gap style. Really, the retracted gap plug should only be used when absolutely necessary. This type of plug has little resistance to fouling and it generates a poor combustion flame front, due to the way in which the plug masks the spark within its nose in a pocket of stagnant air. Combustion efficiency and speed depends to a large extent on turbulence within the combustion chamber causing the compressed fuel/air charge to rush through the plug electrode gap and propagate the combustion flame throughout the combustion chamber. When the spark generates a combustion flame in an area of relative calm, as in the end of a retracted gap plug, some time elapses before the flame radiates out into the turbulence of the combustion chamber. The fact that the insulator nose is in an area of such calm also means that fuel and ash deposits collect very easily to foul the plug.

A variation of the regular gap plug is the cut-back gap. This type has a shorter earth electrode which extends midway across the centre electrode. The main benefit of this design is that it requires less voltage to fire at high rpm than the regular gap plug. Champion plugs of this type have a 'J' suffix (eg: L4J, L77J). Of course, regular gap plugs can be modified by filing the earth electrode back when cut-back plugs are not available in the heat range required. This modification will, however, make the plug's heat range just a little cooler.

In some heat ranges the only spark plugs available have an auxiliary gap ('booster gap') to help resist low speed fouling. Since the booster gap increases the voltage requirement to fire the plug quite considerably, they can cause problems in competition engines. A booster gap plug can be identified visually by a small hole in

Ignition

the stud of the plug which ventilates the interior booster gap. In the case of Champion plugs, they can be recognised easily by a 'U' prefix (eg: UL81J, UJ7G). It is recommended that booster gap plugs be 'pinned' when installed in high-speed engines, by inserting a straightened paper clip down the vent hole. When you have pushed the wire in as far as it will go, cut it off level with the top of the plug terminal.

Another point worthy of consideration is the reach of the plug. A plug which is too short and does not extend the full threaded length of the spark plug boss in the head will reduce performance by masking the ignition flame. Additionally, it can invite a hot spot in the form of carbon building up in the unused portion of thread. A plug that is too long will have threads exposed in the combustion chamber. The threads fill up with carbon which damages or strips the threads in the head when the plug is removed. Also, the exposed threads, or the carbon deposited in them, may become a hot spot precipitating pre-ignition.

On any engine the spark plug reach should be checked, all the more so if the combustion chamber has been modified or if a temperature gauge thermocouple washer is fitted under the plug. In some instances a change to another plug reach may be in order, but in most cases the use of a single solid copper gasket will ensure the proper depth fit. The range of Champion gaskets for 14mm plugs is listed in Table 6.7. It should be noted that it is quite in order to use a solid gasket and the standard folded gasket together. This is necessary, for example, when ³/₄in reach Gold Palladium plugs are used in applications where ⁵/₈in reach plugs are normally fitted.

The width of the spark plug gap for best performance depends primarily on the compression pressure of the fuel/air charge, the engine rpm, the spark plug gap style and the high speed voltage output of the coil. Increasing the first two factors without

The standard reach spark plug protrudes into this combustion chamber now that it has been machined to a new profile. To correct the problem a thicker spark plug gasket is required.

Table 6.7 Champion gaskets for 14mm spark plugs

Gasket thickness (in)	Part No.
0.57	N675
0.080	N673X1
0.096	N673X2
0.135	N677
0.070/0.052	N678

Note – Part No. N678 is a thread-on gasket, the others are solid.

an increase in the latter calls for a decrease in the gap width. Therefore, it is fairly safe to say that all radically-modified engines will require a gap narrower than that recommended by the manufacturer.

Manufacturers generally stipulate a relatively wide gap (0.6 to 0.7mm) as this improves performance at lower rpm and reduces the risk of the gap being bridged by whiskers of carbon, or beads of lead, oil or petrol. As there is much less turbulence in the combustion chamber at low engine speeds, it is very easy for a blob of carbon or fuel to settle between the plug electrodes, shorting it out. With a wider gap the odds are better because the speck of carbon or the bead of fuel may not be large enough to bridge the gap. Later, when the engine is given a high-speed burst, the increased turbulence will 'blow' the electrodes clean. Also, because the spark generated in a wide gap is larger than that in a narrow gap, a more sizeable initial combustion flame is produced. This improves flame propagation through the fuel charge and allows for a more complete burn of the compressed mixture. Hence hp at lower revs goes up.

As the engine speed and compression pressures increase, the coil is not able to supply electrical energy of sufficient voltage to jump a wide spark gap and keep the air between the electrodes ionised for a period long enough to initiate combustion. What happens is that the coil has enough energy reserve to electrically bridge the spark plug electrodes but, before the spark generated can get a combustion flame started, turbulence within the combustion chamber will actually blow out the spark. This was a big problem with the early CDI systems, as the spark was produced for only a very short duration. Current CDIs have a shunt incorporated to lengthen spark duration, allowing marginally wider plug gaps. When a narrow gap is used, the magnetic field within the gap is much more intense, as it is confined to a much smaller space. Hence the spark 'holds together' for long enough to effect ignition, in spite of receiving severe buffeting from the turbulent gases within the combustion chamber.

From experience I would recommend that any competition engine with an operating speed in excess of 9,000rpm would use a plug gap of 0.5mm if fitted with a fine wire plug, fired by a CDI system. With a coil and battery, or magneto system, the gap may have to be reduced to 0.4mm. When retracted gap plugs are used, the gap will be 0.4 to 0.45mm with CDI and 0.35 to 0.4mm for other systems.

Engines operating at speeds of around 7,500 to 8,500rpm will require a gap of 0.55 to 0.6mm with a fine wire plug and CDI. If retracted gap plugs are fitted, the gap will be 0.4 to 0.45mm.

If the ignition system has been modified, or if a special ignition has been fitted, check to see that coil polarity is correct before you try experimenting to determine what plug gap gives the best performance in your engine. A coil with reversed polarity loses the equivalent of 40 per cent energy as the spark has to jump from what would normally be the earth electrode (the side electrode) to the centre electrode. Because the side electrode is many hundreds of degrees cooler than the centre electrode, there is much more restrained electron activity on the metal surface. This considerably increases the voltage required to cause the electrons to leave one electrode and jump to the other, thus ionising the gap and creating a spark. Cold engines are more difficult to get started for this very reason. The plug electrodes are cold, therefore a very high voltage is necessary to tear the electrons from one surface and have them jump the gap to the other electrode.

With a coil and battery type ignition system the polarity is correct when the wire from the coil to the points is connected to the coil terminal with the same polarity as the earth terminal of the battery, ie if the negative (–) battery terminal is earthed, then the wire running between the coil and the points should be connected to the negative (–) coil terminal. With other ignition systems the polarity is seldom marked, so it is a matter of following the manufacturer's wiring diagram precisely, otherwise the ignition may still function but the polarity could be wrong. Also take care that you do not fit mis-matched components, using a rotor/stator assembly from one bike, and a coil/electronic control unit from another. Even if the bikes are basically identical, but one is a year or two older than the other, it is quite possible to run into trouble.

A dished spark plug side electrode indicates incorrect polarity (Figure 6.9). The dish is caused by metal leaving the electrode each time a spark jumps across to the centre electrode. Normally, this would only be visible in road bikes where the plug has a service life of 3,000 to 5,000 miles.

The life of a spark plug in a two-stroke racing engine is not as short as many would suppose. Many have the idea that a new plug is required for each race, but this is just not so. With proper care, a plug should last at least 300 miles, and up to 500 miles. An exception would be in the case of engines using nitro, or if the engine blows, coating the insulator with metallic deposits.

A road machine should have the plugs filed, gapped and tested every 1,800–2,000 miles, and a race machine after each meeting. Bend the earth electrode back far enough to permit filing of the sparking surfaces shown in Figure 6.10. A points file should be used to file a flat surface with sharp edges on both the centre and

Figure 6.9 A dished earth electrode indicates incorrect coil polarity.

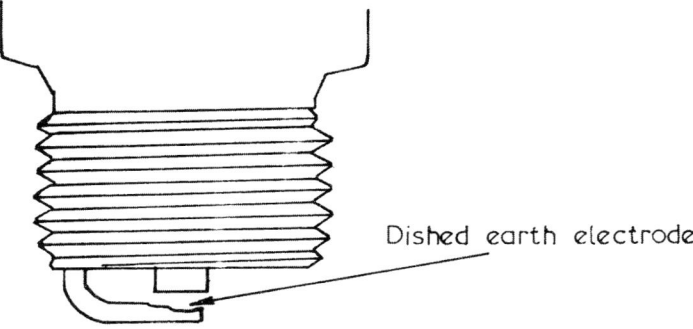

Two-stroke performance tuning

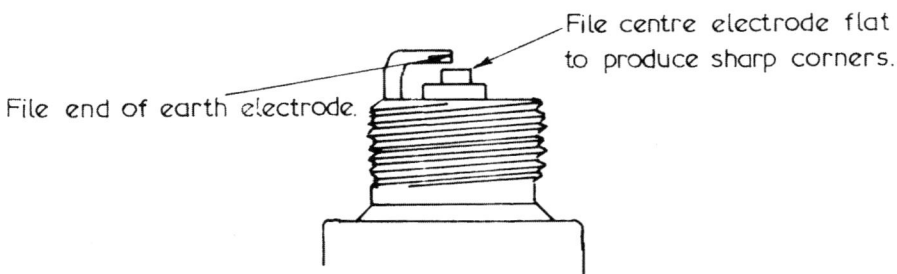

Figure 6.10 Conventional gap spark plugs are serviced by filing the centre and earth electrodes.

side electrode. This lowers the voltage required to fire the plug, firstly because electricity prefers to jump across sharp edges, and secondly because the electrical conductivity of the electrodes is improved. Combustion heat and pressure tends to break up and oxidise electrode firing surfaces, increasing the electrical resistance. Filing removes this 'dead' material and exposes new, highly-conductive metal.

Retracted gap plugs, naturally enough, cannot be filed. Also the centre electrode of fine wire plugs should not be filed, otherwise it will be damaged.

Spark plugs should never be cleaned with a wire brush, as metallic deposits will impregnate the insulator and short out the plug. I also do not recommend cleaning in an abrasive plug cleaner, as some abrasive material always seems to become wedged between the insulator and plug shell. If this cannot be probed out with a scriber, it will drop into the cylinder and possibly cause damage. However, if you choose to have your plugs abrasive blasted, be sure to remove all abrasive grit inside the plug nose and from the threads and gasket.

Personally, I prefer to leave plugs uncleaned. If they are fuel or oil-fouled, I clean them with a toothbrush and ether. Trichloroethylene or chlorothene are also excellent. Be sure to blow the insulator dry before refitting the plug. If the insulation is breaking down due to leaded fuel deposits, or other metallic deposits, I throw the plug away. Do not use carbon tetrachloride to clean plugs as this will leave a conductive deposit. Carbon tetrachloride will remove oil and fuel, but it leaves behind the makings of a fine carbon deposit that could short the plug.

The spark plug lead provides the high voltage electrical connection between the coil and plug. It also forms an effective insulating barrier to prevent the ignition current tracking to earth. If the insulation is damaged by coming in contact with hot metal, or by abrasion, a short circuit could result. Therefore, plug leads must be carefully routed to avoid such damage.

Apart from the plug lead, the spark plug cap can also be a source of high voltage leakage, or flashover. The cap must be free of dust, moisture, cracks and carbon tracks, both inside and out, to ensure that full voltage is reaching the plug. If it's raining or wet, the spark plug cap will be wet on the outside, but if it is of good design, like the KLG and some Japanese caps, it should remain free of moisture inside.

While a high voltage leak is the most common cause of ignition failure in wet weather, the low voltage system can also give trouble if not correctly waterproofed. Regardless of whether your bike employs a points or CDI system, the crankcase cover

Ignition

should be carefully sealed with Silastic, and don't forget to seal around both the inside and the outside of the rubber grommet which protects the wires entering through the cover. Condensation within the sealed cover is usually not a problem, but it is a good idea to give everything a light spray with WD-40 to prevent any trouble from moisture in the air. Be careful that not too much WD-40 is applied, otherwise it will run over the surfaces to be sealed and prevent the Silastic from adhering.

If, after this, you still have a problem with water entering the ignition cover, a vent hose will have to be fitted to run from the cover to high up under the seat or the fuel tank. What can happen is that the engine heats and expands the air within the ignition cover, building up pressure which ruptures the Silastic seal. Then when you ride through a water hole the sudden drop in temperature cools the air, causing it to contract and suck in water.

The kill button can also give trouble, so be sure to fit a good, sealed Japanese button. Then seal it with Silastic to further enhance its water resistance.

Chapter 7

The bottom end

The bottom end is certainly the least glamorous part of the two-stroke engine, and if you are like me it is also the part you would prefer to forget about until something actually goes wrong. Because the two-stroke is so easy to dismantle, the bottom end seems difficult to get at by comparison.

Fortunately the modern day engine has a crankshaft-rod-bearing assembly that in normal service is very reliable and requires little attention. But this is not to say there is nothing that can be done to improve the crankshaft assembly. Your careful attention in this area will not only pick up power and reduce fatigue induced by vibration, it will also decrease the number of crank rebuilds required and lower the cost of being competitive.

Most two-stroke crankshafts are a pressed together affair. During assembly at the factory, or from usage, the crank can get out of alignment. This sets up vibration in the engine which soaks up power, wrecks the bearings and fatigues you. The only way to overcome the problem is to blueprint the crank assembly. If you have a single cylinder engine this can wait until the bearings or crankpin are due for replacement. However, if your machine is a twin cylinder road racer or a street sports bike modified for road racing, I would encourage you to set up properly even brand new crankshafts. When you do this you can be assured of 700 miles trouble free from RD250 LC and RZ350 crankshafts.

If you don't have a press, dial gauge and centres, you should take your crankshaft to a reputable firm to have the work done. However, don't assume every motorcycle or engineering shop will do the crankshaft work to the accuracy required. Before you hand your crank over, have a talk with the shop foreman, tell him what you require and why you insist on accuracy. If he wants your job, he will probably show you other crankshafts he has done and prove their accuracy.

Before you split the crankcase and remove the crank there are a few clearance checks which must be made to avoid problems when the time comes to reassemble everything. The first thing we need to know is the crankshaft end-play. To determine

The bottom end

this attach a dial indicator to the engine so that the plunger directly contacts the end of the crankshaft. Then, with the crank pushed as far as possible to one side of the crankcase zero the dial gauge. Next, push the crankshaft back toward the gauge, noting the dial reading. The correct end float is 0.3–0.4mm. If the end-play is outside this range we may be able to correct it when we rebuild the crank.

With this in mind the next measurement is to determine the big end bearing cage side clearance. To find this, insert a feeler blade between the crankwheel and big end thrust washer, and repeat until you find a feeler that is a nice slide fit in the gap. The minimum we want is 0.5mm. Up to 0.7mm will increase big end cooling and lubrication, but I prefer to keep between 0.5 and 0.6mm with new thrust washers.

Having noted both the crank end-float and the big end side clearance you next have to calculate if it is possible to reassemble the crank to a different width to achieve the ideal big end side clearance and still get acceptable crank end-play. For example if big end clearance is only 0.4mm with worn thrust washers it probably means the clearance with new washers will come down to 0.2mm if the same width across the crankwheels is maintained. We really want a minimum of 0.5mm with new thrusts, so what will the end-float be reduced to with the crank reassembled wider? Well if the crank end-play is now 0.5mm spreading the crank 0.3mm to give the necessary big end side clearance will reduce crank end-float down to 0.2mm, which is the minimum you want to run.

On the other hand if assembling the crank to give the necessary big end side clearance reduces crank end-float below 0.2mm the fix is more difficult. We can't run with less than 0.2mm end-play and for good big end lubrication and cooling we need 0.5mm side clearance. The only options if the crankcase is split vertically is to either

Measuring big end cage side clearance prior to splitting the crankcase. With only 0.019in clearance on worn thrust washers this big end is a touch too tight.

Two-stroke performance tuning

use a thicker gasket between the two case halves, or else machine a main bearing housing deeper to allow one main bearing to seat deeper. With horizontally split cases only the latter option is open to increase crank end-play.

With these two clearance checks made and recorded the crank can be removed from the crankcase, but before being dismantled measure the width across the crankwheels. Record this measurement as it determines the big end side clearance and the crank end-float when the crank is rebuilt.

The next step is to press the crankpin out and separate the crankwheels. Then have the wheels magnaflux crack tested. Next, check each crankwheel for concentricity. A crankwheel is concentric when the axle is exactly in the centre of the flywheel. Generally, the shafts are not exactly in the centre of the flywheels, which produces an imbalance and vibration. For example, the radius from the shaft's centre to the top of the flywheel might be 2.498in. The radius to the bottom of the flywheel might be 2.502in, indicating that the shaft is 0.002in off centre. The dial gauge would show the runout to be 0.004in. What we want is not more than 0.001in runout, so the crankwheel will require very light machining in the lathe to bring the radius of the flywheel to within 0.0005in of centre.

Another important inspection which has to be done on used crankwheels before correcting their runout is to check the condition of the axles where they ride in the main bearings. If the axles are worn this will allow the crank to flop about which increases big end loading and deflection. Also the crankcase seals may not be able to seal effectively which can drop engine performance or even allow the engine

This con-rod fork, which is made of 50 x 6mm mild steel, locks the crankshaft to allow the ignition rotor and primary gear fixing nuts to be loosened or tensioned during crankcase disassembly or assembly.

to suck sufficient air to cause an engine destroying lean-out. If the axles are worn they can be hard chromed and then machined. However to avoid embrittlement and subsequent cracking ensure that the hard chroming is carried out to strict aircraft standards.

After the crankwheels have been trued it is important to select a crankpin of the correct diameter to ensure an interference fit between the pin and crankpin holes of 0.002–0.003in per inch of diameter. If the pin is too loose, the crankshaft will not stay in alignment.

When a crankpin of suitable diameter is found slip a new big end bearing on the pin and measure the bearing diameter across the rollers. Next, measure the big end eye diameter in the con-rod. The difference of these measurements is the bearing radial clearance. For most applications a clearance of 0.0012 to 0.0015in (0.031–0.038mm) is about what is required. If the clearance is more than this it probably means that the big end eye is worn excessively, so the con-rod should be replaced.

Figure 7.1 Critical crankshaft measurements.

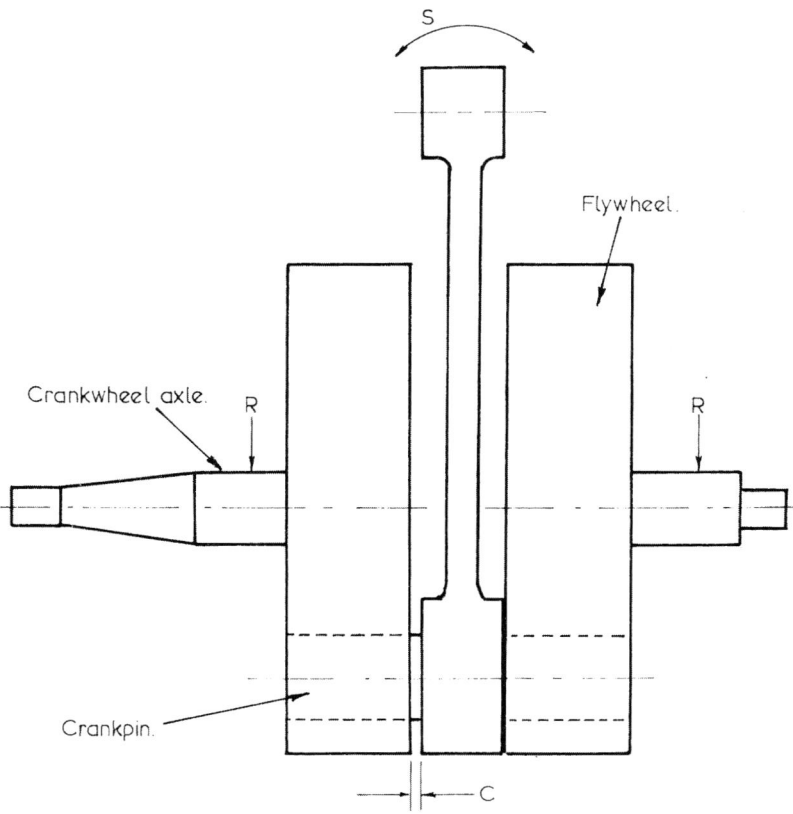

C - big-end side clearance (0.5mm to 0.6mm)
R - crankshaft runout (0.04mm maximum)
S - little-end side shake (0.8mm to 1.3mm)

Two-stroke performance tuning

The owner failed to heed the warning of excessive little end side shake with the result that the big end failed, destroying a brand new piston, as well as a serviceable rod and pin. Those small pieces of metal are what was left of the big end bearing.

With an assembled crank we can also get an indication of wear on the pin, bearing and con-rod by measuring the small-end side shake (Figure 7.1). The amount the little-end moves from side to side indicates the big end radial clearance. With new parts the side shake will be around 0.030–0.040in. The service limit is about 0.065in, but once the side shake increases past about 0.055in it is wise to dismantle the crank and at a minimum, replace the big end bearing.

Some competitors ignore increasing side shake hoping to get a few more miles out of the crank before a rebuild. This is most unwise as it could be that the pin and con-rod are still serviceable and only the bearing requires replacement. Left to run a few more races the bearing will fail and massively increase the cost of a rebuild by wrecking the con-rod, the pin, the piston and ring, the bore and the squish band. On top of the engine damage you may be spat off on to your head when the engine locks up at high speed.

For very high speed or endurance engines we need to be much more precise to achieve the ultimate in big end bearing reliability. To achieve this a lot of painstaking work is required. Basically, the big end assembly will give the most reliable service if the bearing rollers do not skid on the pin or in the con rod eye. This ideal situation can only be achieved by working the big end assembly to several very close tolerances.

Looking at Tables 7.1 and 7.2 you will note the dimensions in which we are interested. Naturally, we have to select components which not only fall into either of the two selective fit categories dimensionally but which also exhibit true parallelism. If the pin or big end eye, or any of the individual rollers, are not parallel, the rollers will still skid no matter how carefully we match these parts for fit.

Table 7.1 Typical big end bearing radial clearance

Crankpin diameter (mm)	Radial clearance (mm) Minimum	Maximum
18	0.023	0.035
20	0.025	0.037
22	0.028	0.040
25	0.031	0.043
27	0.034	0.046
30	0.038	0.050

Note – the above clearances are for high speed racing engines. Low speed road and play bike engines could use clearances 25 per cent less.

Table 7.2 Typical big end assembly tolerances

	Crankpin	Big end eye	Bearing rollers
Nominal dimension (mm)	20	26	3
Tolerance	−0.006 / −0.010	+0.010 / +0.020	−0.002 / −0.006
Selective fit A	20 $^{-0.008}_{-0.010}$	26 $^{+0.010}_{+0.015}$	3 $^{-0.004}_{-0.006}$
Selective fit B	20 $^{-0.006}_{-0.008}$	26 $^{+0.015}_{+0.020}$	3 $^{-0.002}_{-0.004}$

Also, if we are to avoid skidding rollers, we must ensure that the connecting rod has been machined true. To determine this you will have to make a pair of dummy pins about 100mm long to fit the little end and big end eyes. Measuring between both ends of the dummy pins will determine if the rod is bent or has the eyes machined out of parallel (Figure 7.2). Next check that the eyes have been machined in the same plane (ie not twisted). To do this set the big end of the rod up with a dummy pin fitted, on a pair of parallel V-blocks. Then, with a dial gauge, measure to see that both ends of the pin fitted in the little end are the same dimension from the surface plate.

In multi-cylinder engines crankshaft balance must be maintained to avoid vibration damage to the crankshaft and bearings. This means that the weight of each big end assembly must be equal, and the weight of each little end assembly must be equal. The big end assembly is made up of the crankpin, big end bearing and thrust washers, and the rod big end. The little end assembly comprises the rod little end, the little end bearing and thrust washers, and piston pin. Unfortunately, few tuners have the equipment to do this balancing themselves, so this usually means that all these components must be sent to some automotive firm for balancing. If this is true in your case, be sure to pack each con rod assembly in a separate plastic bag and instruct the firm doing the balancing that under no circumstances are parts to be swapped from

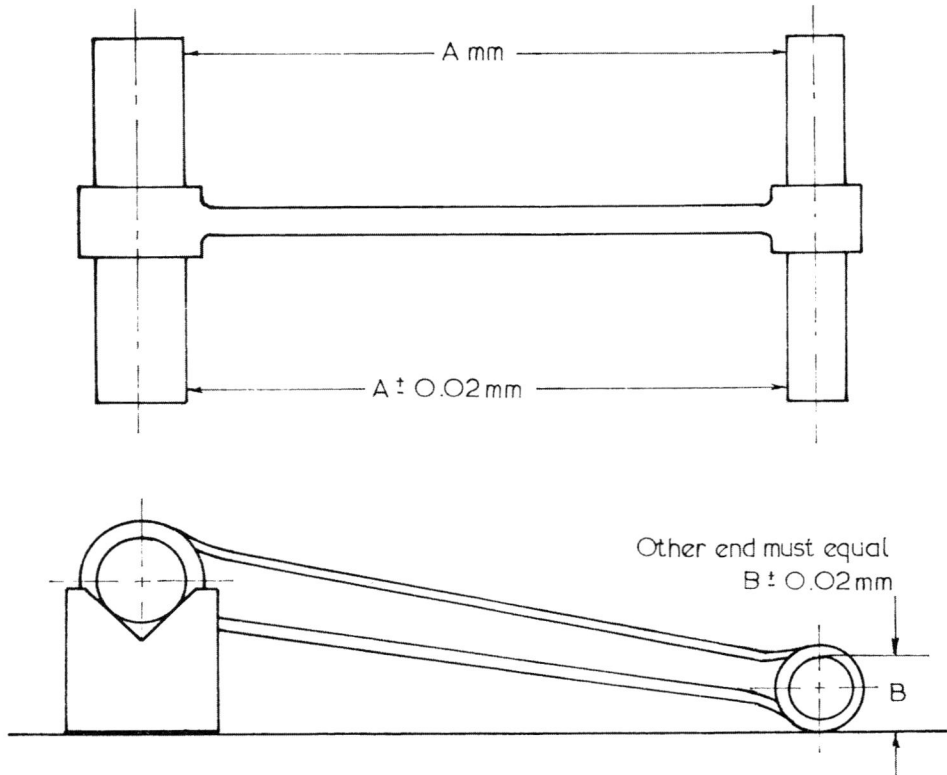

Figure 7.2 Checking connecting rod trueness.

one assembly to another, otherwise all the time spent on obtaining proper big end tolerances will have been wasted.

The pistons, of course, will also have to be balanced, using either an accurate pair of laboratory scales or a simple beam balance. When the lightest piston is found, remove metal from inside the piston skirt and around the pin bosses to reduce the weight of the other pistons to within 1 gram of the lightest piston.

Whenever the big end or main bearings are replaced, don't just use any bearing which will fit. The loads experienced by the bearings in two-stroke engines demand the use of high quality parts if reliability is to be maintained. Therefore only those bearings equivalent to, or superior to the original components, should be utilised.

If you wish to use bearings better than those fitted as standard, you may be able to obtain a suitable replacement from the German INA bearing company. Their two-stroke bearings are the best available. Try to get main bearings with fatigue resistant plastic or fibre cages rather than riveted steel cages which seem prone to cracking up.

Main bearings with plastic cages demand plenty of lubrication to enable cool running. If the bearing overheats the plastic cage will distort or melt, causing bearing failure. To improve lubrication you may have to drill an oil feed hole in the crankcase to each bearing, similar to that illustrated in Figure 7.3. The hole should be about $5/32$in diameter, drilled from the transfer slot or the barrel spigot recess to the main bearing

The bottom end

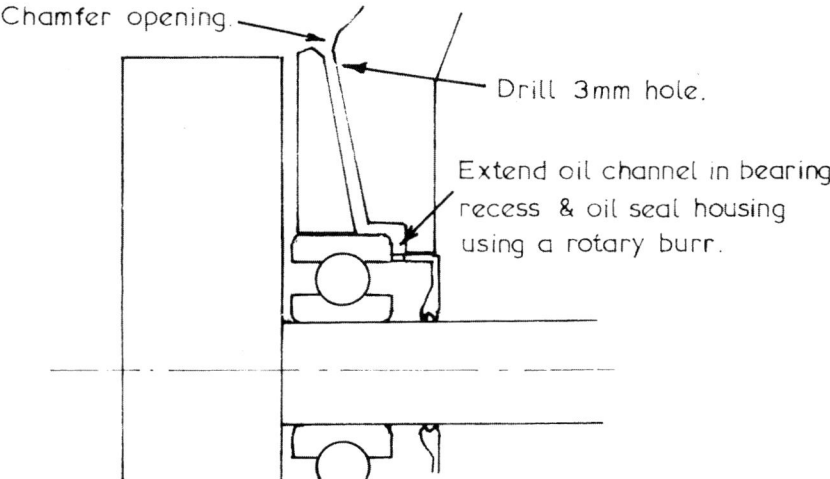

Figure 7.3 Main bearing oil feed hole and channel.

housing. The oil feed channel must be extended down around the main bearing recess using a rotary burr. Take care that the case seals do not block off this channel. Additionally check that the shoulder of the main bearing does not block the channel. Some bearings have a very wide round shoulder which allows good oil flow around the edge of the bearing. However some main bearings have very narrow square shoulders which will almost completely block the channel in the corner of the bearing recess.

The big end bearing must be as light as possible, otherwise the inertia generated by the swinging of the con rod as it passes top and bottom dead centre will cause the rollers to skid and over-heat the bearing and rod big end. A lightweight bearing can be accelerated and decelerated quickly, but a heavy bearing will continue to rotate at a more constant speed, rather than staying synchronised with the relative rotational speed of the crankpin.

Most people are surprised to know just how much influence the angular swing of the con rod has on the rotational speed of the big end bearing. Normally, the bearing should rotate at half the relative crankshaft rpm. On the surface it would appear that a big end bearing in a motor spinning at 11,000rpm would be rotating at 5,500rpm. However, when you look at Figure 7.4 you can see that this is not so. At TDC the angular swing of the rod is in the opposite direction to rotation, but in the same direction at BDC. With a 2 to 1 rod length to stroke ratio (eg: engine stroke 54mm; rod length centre to centre 108mm) the instantaneous rotational speed of the rod in relation to the crankpin is 25 per cent greater or less than the crank speed. Thus, at an engine speed of 11,000rpm the relative rotational speed will be 13,750rpm at TDC, and 8,250rpm at BDC. Remembering that the bearing rotates at half these speeds, we can see that its revolution rate must drop from 6,875rpm to 4,125rpm, and increase back again to 6,875rpm twice per crankshaft revolution. If the bearing has enough weight, it will resist this rapid oscillation, forcing the rollers to skid.

Most modern two-strokes have steel big end bearing cages plated with tin or

This phenolic cage class 4 main bearing (left) failed due to inadequate lubrication. Oil flow was restricted by the bearing's narrow shoulders almost closing off the oil passage in the crankcase. Comparing the two bearings (right) it can be seen that the bearing on the right has more square shoulders (it is 11.2mm across the flat) which almost completely blocked the crankcase oil feed passage. The other bearing measures 10mm across the flat.

copper, to provide a low friction surface. These can be beneficially replaced by very light INA bearings with a special lightweight silver plated cage. Such a move could raise the red line speed of a street engine modified for road racing by 2000rpm. The Yamaha RZ350 is very popular for road racing, but its standard bearings are not up to the task. The simple solution is to substitute big end bearings out of Yamaha's TZ250 road racer. With these bearings, the RZ350 will run reliably for hours at 10,000rpm.

From time to time a few tuners get hooked on the fad to lighten the crankshaft. They feel the flywheels should be machined to a 'T' or 'V' shape to reduce their weight and increase engine acceleration. Acceleration will increase, but you will have to change gears so much more that the machine will be slower around the track.

This is not to say all crankshaft lightening is a 'no no', as a very small number of engines will benefit from a moderate reduction in rotating mass. Generally, we can forget about the majority of Japanese engines, as these already have very light flywheels. The exceptions would be some single cylinder 250cc motocross engines and also the Yamaha RZ350/RD400 when these are modified for road racing. The amount of metal removed is quite small, usually not more than 6oz (170gm) from the inside of each flywheel. This reduces the weight of the RD400 crank, for example, by 1.5lb (680gm).

Generally this crank lightening is done in the interests of crank reliability. Depending on crankwheel and big end pin stiffness all two-stokes will eventually reach a critical rpm where the flywheels begin vibrating, with the crankwheels

The bottom end

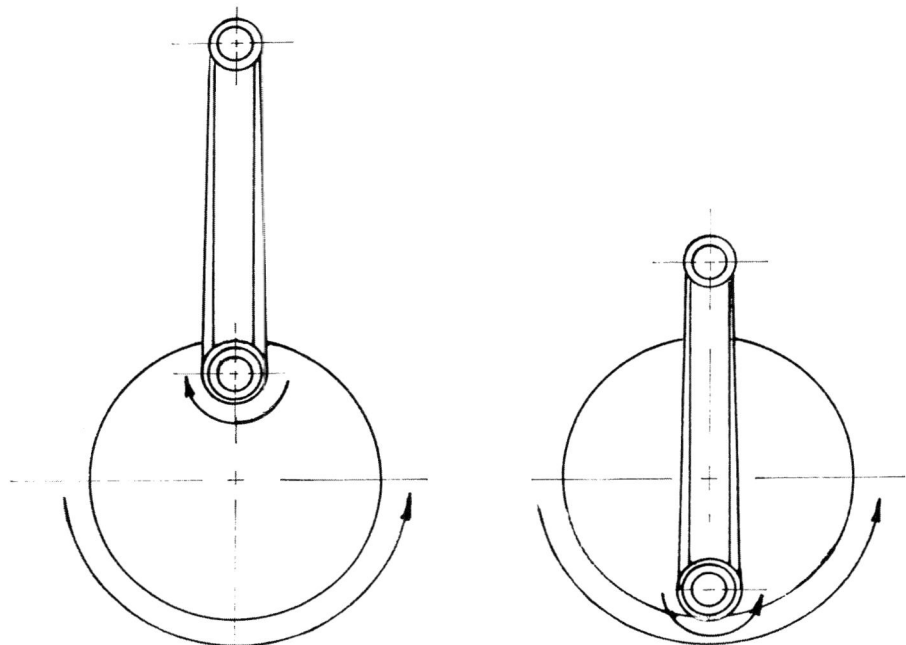

Figure 7.4 The big end bearing rotational speed oscillates due to the angular swing of the con rod.

flapping toward and then away from each other. The heavier the flywheels and the ignition rotor attached to them, and the more easily the crankpin will bend, then the lower the engine speed when this dangerous crankwheel-cracking vibration occurs.

Some people accept this situation and either limit the potential of the engine by

Figure 7.5 Dishing the centre of the crankwheels raises the crank's critical speed while maintaining maximum flywheel effect.

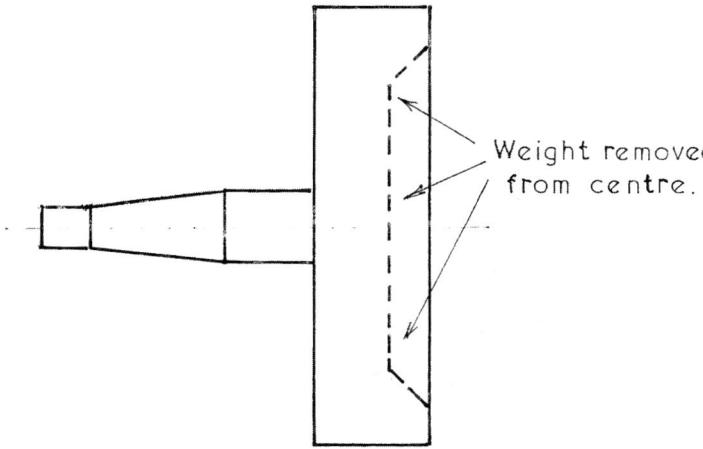

Two-stroke performance tuning

running it lower than this critical rpm or else regularly replace wrecked cranks and at times also wrecked crankcases. However there are two lines of attack which will improve both crank reliability and also allow higher engine speeds. First off the stock ignition rotor on modified street bikes should be replaced with something lighter. Then lighten the crankwheels in the manner shown in Figure 7.5. Dishing the centre like this reduces the mass of the crankwheels and raises the crank's critical engine speed, while the full width band of metal left at the outer edge gives maximum flywheel inertia effect.

Before the crank is reassembled, the rods (even if new) must be crack tested. If the engine has a history of rod failures, the rods should be polished along the beams and then shot peened. Also polish and radius any corners formed by the oiling slots in the big end (Figure 7.6).

The tough skin formed on the con rod by forging gives it much of its strength and fatigue resistance. Therefore the rod should never be polished unless you intend to follow up with shot peening to create another work-toughened skin. I consider it a waste of time polishing the entire rod.

If you have a look at a rod you will see, along its edges, a rough band where

Figure 7.6 Con rod strength is improved by polishing and shot peening in critical areas.

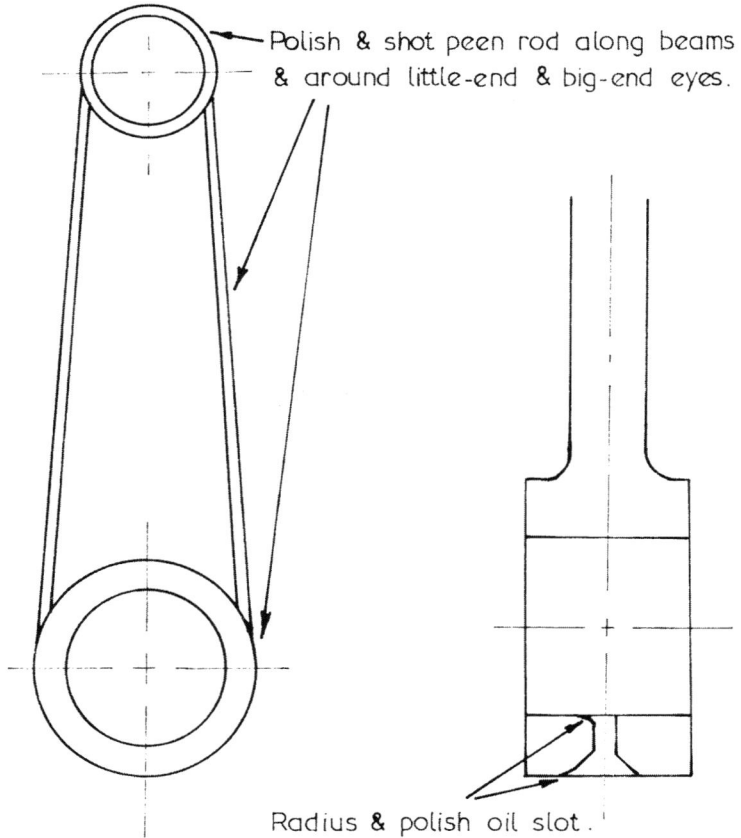

metal appears to have been sawn away. That is where the excess metal called flash was squeezed out from between the forging dies when the rod was being made. Later, most of the flash is trimmed off, but a bead is left, as you can see. Of course, there is no hard skin along this ridge, in fact its roughness is a stress raiser, so the ridge should be removed with a sanding belt. Give the entire beam a polish with fine emery cloth and then follow up with buffing and shot peening.

After the rods have been prepared and the crankpins and bearings matched, the crankwheels can be reassembled. Scrupulous cleanliness is essential, and care should be taken to ensure the wheels are started on the pin as accurately as possible. Use a straight edge across the wheels to check this. If the pin is shouldered, the wheels are pressed hard home, but if a straight pin is used, the necessary rod side clearance must be maintained by inserting two feeler strips of the appropriate thickness, usually 0.25–0.3mm, on either side of the big end and pressing until the strips are just ensnared.

The alignment should be checked between centres, using a dial gauge in contact with the bearing seat of a crankwheel axle. Any runout is eliminated by holding one wheel and striking the other with a copper or lead hammer. Runout must be kept to a maximum of 0.0015in, but 0.001in is preferable (Figure 7.1).

Some people weld cranks to keep them in alignment but I do not agree with this practice. Welding hardens steel and makes it prone to fatigue fractures. Therefore, I use Loctite on the fit between the crankpin and crankwheels. Apply a small amount of Loctite to the pin, and a larger amount in the crankwheel holes, before pressing the shaft together. Take care that you do not allow Loctite into the bearing.

Before the crankshaft is refitted, the crankcase will require some reworking to reduce friction in the engine and increase piston and ring life. At all times the piston should be perpendicular to the crankshaft, but production tolerances being what they are, this is seldom the case. For the piston to be 90° to the crankshaft the con rod must be straight, the cylinder bore axis must be at 90° to the base of the barrel, and the top

Some engines use a hollow crankpin and utilise expander plugs, which are driven into the ends of the crankpin after the crank is trued, to lock the crankpin in position in the crankwheels.

of the crankcase must be parallel with the crankshaft centre.

To check the parallelism of the crankcase first bolt the two halves together loosely. Then fit the cylinder and tension the cylinder retaining bolts evenly to align the case halves. Next, tension the crankcase bolts. With this done, the cylinder can be removed and the cases measured for trueness. The simplest way to effect this with accuracy is to fit a mandrel in the main bearings and take a measurement from the mandrel to a straight edge laid across the top of the crankcase (Figure 7.7). The dimension between the mandrel and straight edge must not differ by more than 0.001in from one side of the crankcase to the other. Usually, it will be found necessary to machine the top of the cases to bring them into line with the crank.

When the crankcase has been trued, the crank, together with new main bearings and seals, can be fitted. Note that the seals are 'handed' and they must be fitted with the directional arrows going the same way as the direction of crankshaft rotation. Take the time to lubricate the seals and bearings before fitting the crankshaft. A dry start will quickly wreck any engine.

Engines with horizontally-split crankcases can experience problems with the main bearings attempting to spin in their housings. The early one-piece crank TZ250 Yamaha is particularly prone to this. The outer races do have little pips on them that fit into a small cavity where the case halves join, but this hasn't stopped the trouble. About the best move is to go over all the holes in the case faces and chamfer them. Studs tend to pull metal up around their threads and this can stop the cases mating

Figure 7.7 The crankcase halves should be checked for parallelism and then trued to avoid power loss to unnecessary friction.

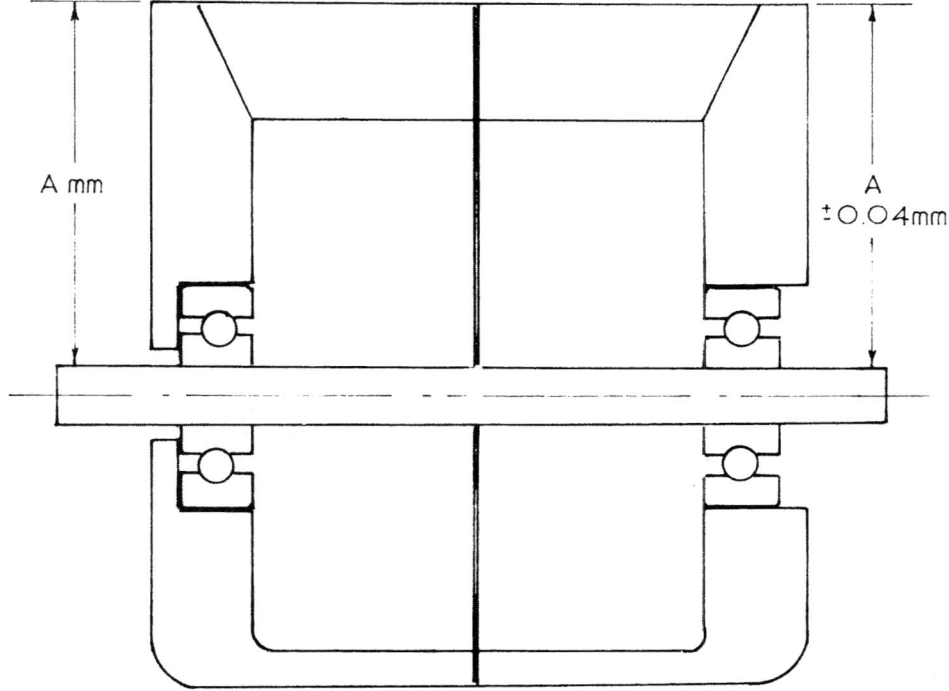

Crankcase seals may be directional like the lower seals or non-directional like the seal at the top. Because the directional arrows are frequently very small all case seals should be examined with a 4X magnifier to ensure that they are fitted in the correct side of the crankcase.

tightly. When the shaft is fitted apply some Loctite to retain the bearings in the case.

Possibly the part of a two-stroke that takes unequalled abuse and gives the tuner the most trouble is the piston. Fortunately piston technology is constantly moving ahead, and piston related unreliability can, to a large degree, be eliminated by regular piston replacement and correct installation of the part in the first instance.

The largest improvement to be made to pistons came when the means were discovered for adding large quantities of silicon to the aluminium alloy. This has reduced the piston expansion rate drastically, minimising the incidence of seizure. Silicon also imparts more strength to aluminium at high temperatures and increases wear resistance.

Quality pistons for competition use generally contain around 18–22 per cent silicon. Unfortunately, there are pistons being sold that do not contain very much silicon at all, even though the manufacturers claim they are racing pistons. This occurs because high silicon content pistons are difficult to manufacture and expensive to machine. Consequently, I stick with the manufacturer's original pistons if the engine was designed for road racing, enduro, motocross, etc., in the first place. I have usually found original pistons to be of good quality. This particularly applies to Japanese pistons; they seem to be able to produce a very good product.

There is one area for concern with standard replacement pistons which must always be checked; a percentage are cracked from new. The best insurance against this is to have all your new pistons Zy-Glo crack tested. If you can't find an engineering shop with a Zy-Glo test kit, check out aircraft repair workshops in your area.

Another problem with standard replacement pistons is that some do not have any circlip extractor slots. This means that only tail-type wire circlips can be used and unfortunately this type of circlip wrecks engines. The constant rubbing of the gudgeon

Two-stroke performance tuning

Circlips with tails (left) should be replaced frequently. One single-tail circlip in this group dropped its tail, destroying the piston and the barrel. This close up view (right) shows how the piston pin has worn right into the tails of the circlips.

pin against the circlip wears through the tail, allowing it to drop into the cylinder, scoring the bore and possibly seizing the motor. If tail-type circlips are replaced regularly, say after every second race meeting, this kind of damage can be avoided.

A better solution is to machine extractor slots into the piston so that tailless circlips (or tail-type circlips with the tail cut off) can be fitted (Figure 7.8). The slot need only be 1/8in wide to allow a small electrical screwdriver or the point of a scriber to fit under the circlip so that it can be flicked out. It should be cut in the position

Figure 7.8 A circlip extractor slot facilitates the extraction of tailless circlips.

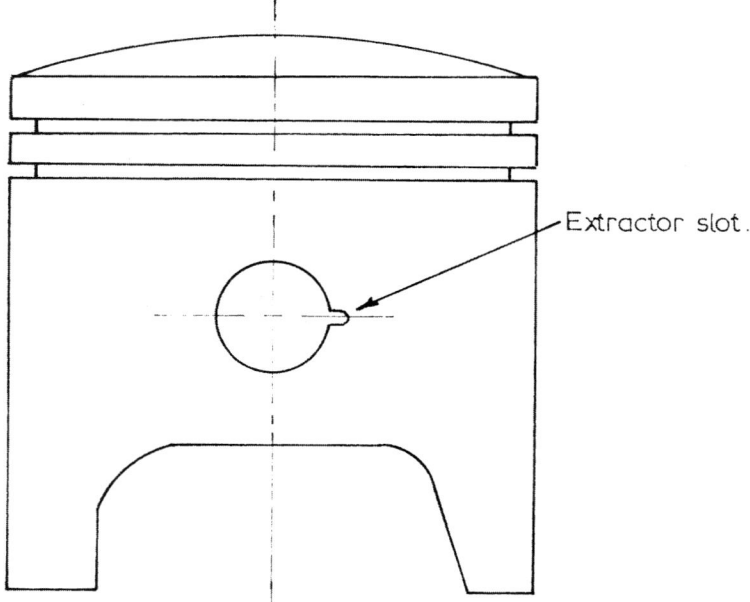

shown, using a small round key file or a $1/8$in diameter mounted grinding tip. Do not use a hacksaw blade or three cornered file to make the extractor slot, as the abrupt corner will form a stress point and eventually cause the piston to crack.

We tend to think of pistons as being round, but actually the skirt is cam ground an oval shape. The piston also tapers from top to bottom (Figure 7.9). Both ovality and taper are necessary to prevent seizure. The top of the piston gets twice as hot as the bottom of the skirt, therefore it expands more and, due to the extra material around the pin bosses, more heat is directed to this area, elongating the piston across the piston pin axis. To compensate for this, the piston is also ground oval. Therefore you must be careful to measure piston clearance only on the thrust faces, and at the bottom of the skirt.

Before the piston is fitted, there are several clearance checks to be made. The first of these is the fit of the piston pin. It should be an easy slide fit, slipping through the piston under its own weight. A tight pin is to be avoided as this will overload the sides of the piston when the engine is running. At high engine speeds the crankshaft tends to whip and, if the pin is tight, this load will be transferred to the side of the piston, possibly distorting it and causing seizure. A tight pin may also change the expansion characteristics of the piston and cause scuffing or seizure on the piston thrust faces. When the piston is unable to expand correctly along the pin axis it will expand at 90° to the pin and force the thrust faces of the piston hard against the cylinder wall.

When you are handling pistons take care not to drop them, as this can distort the piston skirt and lead to engine seizure. Also keep in mind that you must never bang a piston pin out using a hammer and drift. Beside the risk of bending the rod if it is not properly supported, you may easily push the piston out of shape. If the pin will not push or tap out easily, heat the piston in boiling water or oil and then gently tap the pin out. Often you will find that the pin will not budge because the piston pin holes are slightly closed over by metal frazes. After the ends of the pin holes are cleaned out with a sharp knife or bearing scraper, the pin will push out effortlessly.

Figure 7.9 Pistons are ground both tapered and oval to reduce friction and seizing.

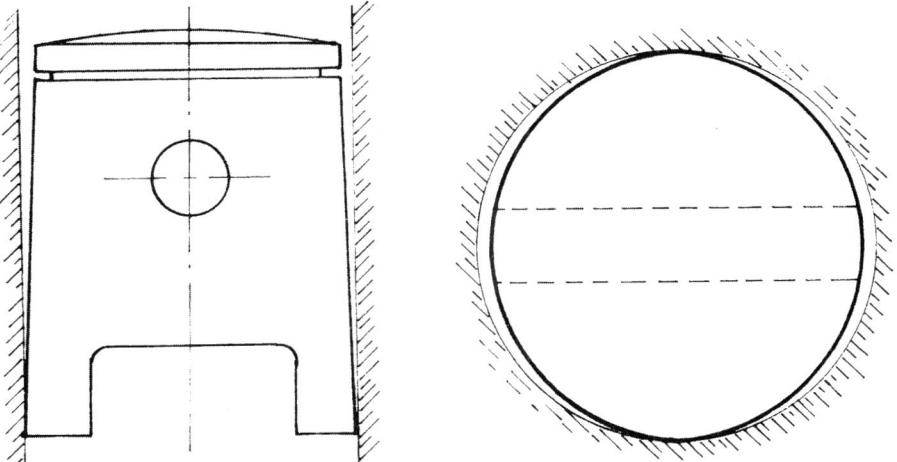

The clearance of the piston in the cylinder is most important. A piston without enough clearance will seize at worst; at best the engine will be down on power due to increased friction. A piston with too much clearance wears quickly, can't maintain a good ring seal, and overheats because heat transfer to the cylinder wall is reduced.

Just how much clearance the piston should have, varies from engine to engine. Dissimilar piston and cylinder materials expand at different rates. Large diameter pistons expand more than smaller pistons. Most manufacturers using cast iron linered aluminium barrels specify a minimum clearance of 0.002–0.0025in for cylinders up to 56mm, 0.0025in–0.003in up to 72mm, and 0.0028–0.0033in up to 85mm. Engines with plated aluminium cylinders typically run at clearances up to 0.001in tighter.

Some engines do not have enough little end side clearance. If the clearance is tighter than 0.25–0.3mm the top of the con rod tends to overheat. This is indicated by blueing of the little end or by the presence of burnt oil in the eye of the rod. Engines fitted with thrust washers are easily cured by lapping the washers on 180 grit paper. If the engine doesn't use washers, either the rod or piston will have to be machined to increase the side clearance.

Because the two-stroke piston has to function with just the scantest amount of lubrication, some thought must be given to modifying the piston to encourage more oil up the cylinder walls. With the piston at TDC, only about 50 per cent of the cylinder is directly bathed with oil mist. Therefore we have to rely on the piston and rings collecting oil on the down stroke, and distributing it at the top of the cylinder on the compression stroke.

Unfortunately, this does not work out too well in practice, as the square edge of the piston skirt tends to scrape most of the oil off the cylinder wall as the piston descends (Figure 7.10). The way around the problem is to put a nice chamfer on the piston to extend about 2mm up the skirt. I do not advise a larger chamfer as many engines have a piston skirt barely long enough to cover and seal off the exhaust port

Figure 7.10 Chamfer the piston skirt to improve piston lubrication.

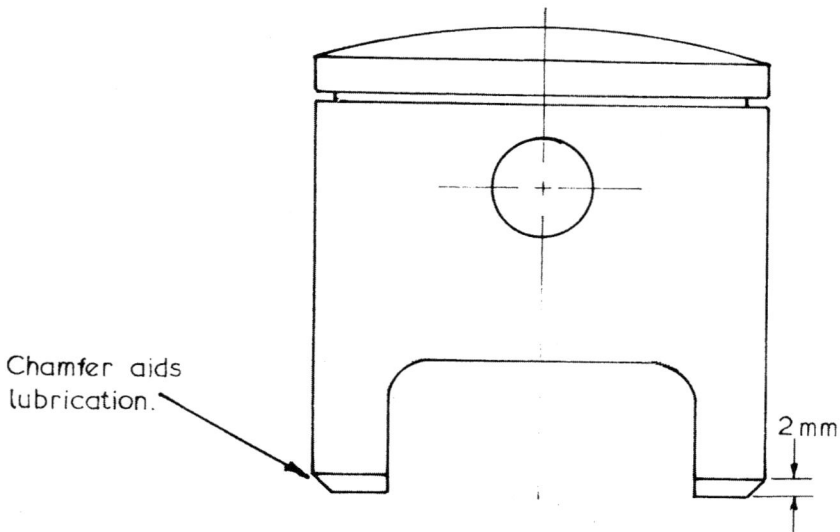

The bottom end

When thrust washers are fitted at the little end it is essential to ensure that there is sufficient clearance between the washers and the piston. The minimum is 0.010in. These washers have been sanded on 180 grit paper and the clearance is now 0.011in.

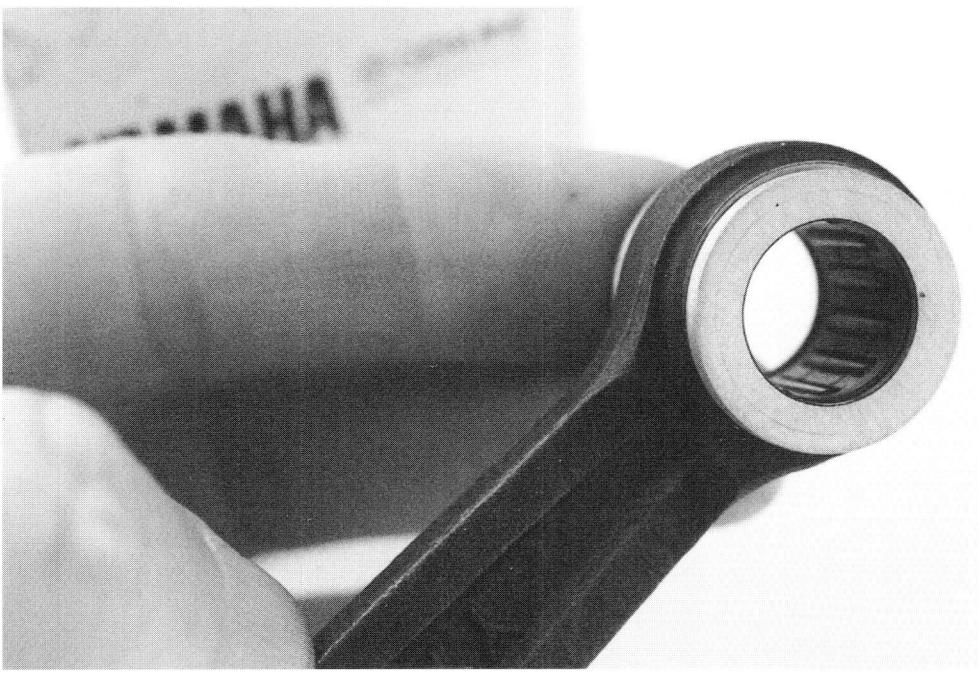

when the piston is at TDC. A larger chamfer will encourage exhaust leakage into the crankcase during the induction cycle.

When the piston skirt is chamfered, more oil will remain on the cylinder wall to lubricate the rings and the top of the piston, then on the compression stroke the rings will carry some lubricant to the upper cylinder area. Of course, oil does reach the top of the cylinder when the transfer ports are exposed to admit the fuel/air charge, but very little settles on the cylinder wall directly above the exhaust port.

Additional to improving lubrication, and hence engine life, an oil coating at the top of the cylinder and on the piston rings increases power by improving the seal between the rings and the bore wall. A decrease in compression leakage results in more power.

Leakage is a bigger problem at lower rpm, simply because there is more time for the gases to find their way through ring gaps and around the edges of the rings. This is why you will find that street bikes have two rings and high rpm road racers only one ring.

Many people mistakenly believe it is the ring's inherent radial tension that holds it against the bore wall to effect a seal, but this is not the case. Radial tension does help but it is gas pressure behind the back of the ring that forces the ring face against the cylinder wall (Figure 7.11).

There is a detrimental phenomenon that can occur in high rpm engines, called ring float or ring flutter. As the piston approaches TDC it is slowed by the con rod, but the rings try to keep on moving. If they have enough weight they will leave contact with the lower side of the piston groove. When this happens, the ring seals off the gas pressure in the combustion chamber, preventing the gas getting behind the back of the ring to push it against the bore. Any gas pressure that may have been behind the ring quickly leaks into the crankcase, and combustion pressure forces the ring to collapse inward, causing it to break contact with the cylinder wall. This allows the combustion gases to blow by into the crankcase (Figure 7.11).

Radial tension in the ring is unable to prevent this type of blow-by caused by ring flutter. However, a certain degree of radial tension is necessary for good sealing,

Figure 7.11 Gas pressure behind the piston ring forces the ring to seal against the cylinder wall.

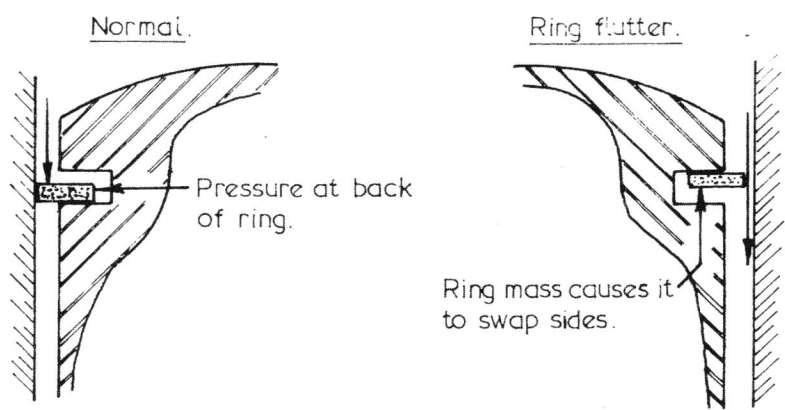

otherwise the pressure at the back of the ring would only equal the pressure trying to force the ring off the bore wall. This would allow blow-by, and it is this type we see occurring when the rings are badly worn and have lost their tension.

Ring flutter can also wreck engines due to an increase in piston temperature. When the ring loses contact with the cylinder it is unable to conduct heat away from the piston crown. This may lead to severe detonation and melted pistons.

In a racing engine it may be found that the standard rings are too wide for operation at the engine speed at which we desire to run. This problem usually arises only when a street engine or motocross engine is modified for road racing. At times, the manufacturer makes a racing engine with the same size piston as the stock street motor. If this is the case it is a simple matter to substitute the racing piston in the modified motor. Some manufacturers have changed to thinner rings in later motors of basically the same design, and the later model piston and rings can be used.

Besides raising the speed at which flutter occurs, thin rings also reduce power loss due to friction. This is really only significant when engine speed is in excess of 9,000rpm. The same can be said for single ring designs. Below about 8,000rpm a two-ring setup gives marginally more power, but above 9,500rpm the situation reverses.

In this day of advanced metal technology, ring breakage is rare, and can usually be attributed to one of the following causes: excessive piston to cylinder clearance allowing the piston to rock and twist the rings as it passes TDC; worn piston ring grooves that let the rings jump about; excessive bore taper causing radial ring flutter; exhaust port widened excessively or ground an incorrect shape; sharp edges left in exhaust and/or transfer ports; insufficient ring gap; ring grooves not properly cleaned before fitting new rings; ring jamming in the groove and snagging the exhaust port due to excessive piston crown temperature or insufficient lubrication.

Generally, a two-stroke engine should be set up with groove clearances between 0.04 and 0.1mm. Tighter than 0.04mm clearance will cause the ring to stick in the piston groove as carbon and varnish builds up. Before new rings are fitted, the piston grooves must be carefully cleaned to remove all traces of carbon. Then when the rings are fitted, measure the side clearance and ensure they are not jamming in the grooves. As a final check hold a straight edge along the side of the piston to confirm that the rings are seating correctly in their grooves (Figure 7.12). If the grooves have not been thoroughly cleaned, the rings will be flush or possibly even project past the ring lands.

When new rings are fitted they should be checked for end gap. Usually the ring gap is 0.1 to 0.12mm per inch of bore. This means an engine with a 54mm bore would need a gap of 0.2 to 0.25mm. Engines with 0.63mm rings require a much wider gap than this, as the rings are too narrow to lap over the ring locating pin. In this instance the required gap must be increased by the diameter of the locating pin. Therefore, an engine with a 54mm bore and a pin 1.2mm in diameter will need a ring gap of 0.2 to 0.25mm plus 1.2mm = 1.4 to 1.45mm.

The end gap of each ring should be measured with the ring fitted squarely into the top of the cylinder bore. To ensure an accurate measurement, all traces of carbon must be removed from the bore, using a scraper. Then fit the ring in the unworn part near the top of the cylinder. If the gap is insufficient, carefully file the ring ends, using a wet oilstone or a very fine file.

If an engine is assembled with ring gaps that are too narrow, damage may easily result. When heated, the rings will expand and cause the ends to butt together. This

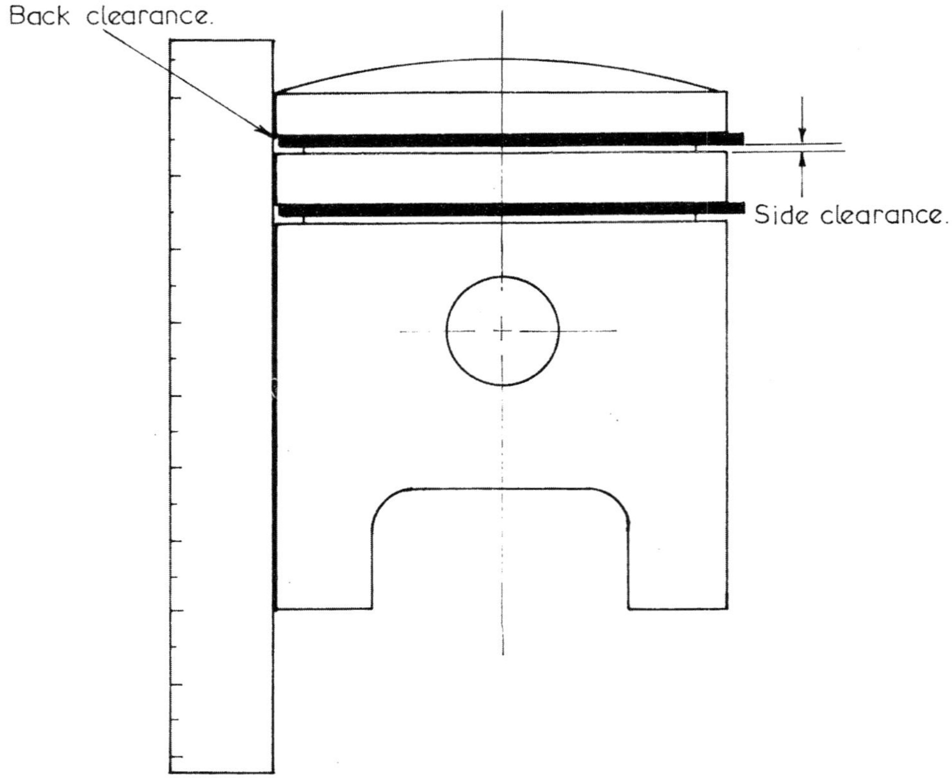

Two-stroke performance tuning

Figure 7.12 During engine assembly check the piston ring clearance in the piston groove.

Figure 7.13 Dressing the ends of the ring helps stop the sharp ends scoring the bore or loosening the ring locating peg.

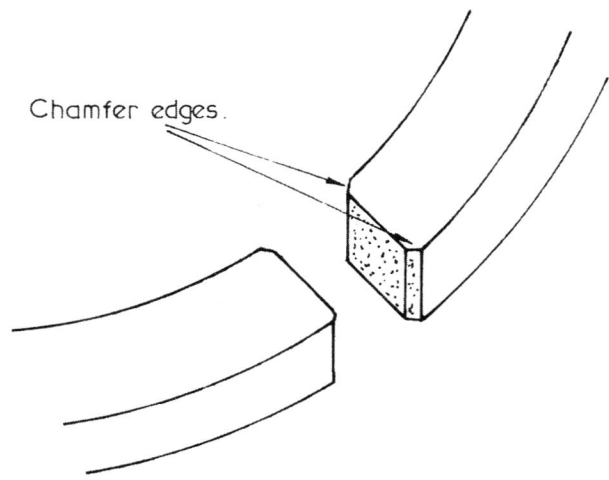

The bottom end

results in ring breakage or, if the pressure is not high enough to cause the rings to break, the cylinder wall will be scuffed.

It is always a sensible practice to dress the ring ends with an oilstone. Chamfering the outside edges reduces the chances of the sharp ends scuffing the bore. It is also a good idea to stone the inside corners of the ring ends too. The slight radius formed helps to stop any tendency the ring may have to pull the ring locating peg out of the piston (Figure 7.13).

When rings are being fitted, care should be taken to avoid fitting them the wrong way up and to prevent damage by incorrect fitting technique. Piston rings can be twisted permanently if they are fitted in the groove at one end and then gradually screwed around until the entire ring is in place. Instead, they must be expanded sufficiently to fit over the piston and then allowed to drop into the groove. Special expander tools are available for this purpose but I prefer to use two 0.4mm feeler blades held between the ring and piston. The blades provide a bearing surface and stop the ring digging into the piston.

It is obvious that type 'A' rings in Figure 7.14 can be fitted only one way up. If the rings are upside down, the ends will foul on the locating pin, preventing the ring from seating in the groove.

On the other hand, type 'B' rings can be fitted incorrectly, so it is necessary to know something of the theory behind the various ring sectional shapes so that you may determine which way up the rings should be fitted.

Figure 7.14 Two methods of pinning the piston rings.

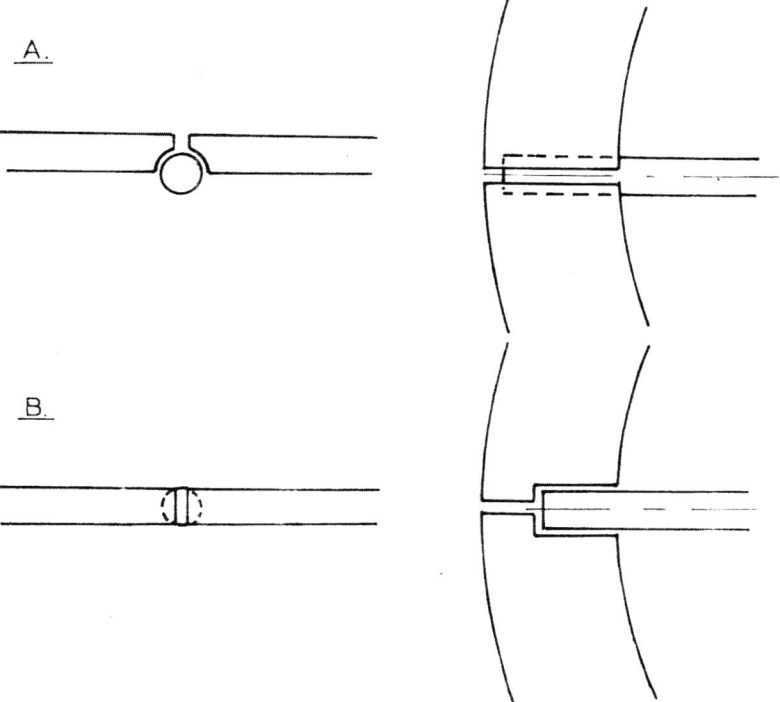

Two-stroke performance tuning

The first ring, shown in Figure 7.15, is the type most commonly used in racing engines. It is called a rectangular section ring, for obvious reasons. The edges are usually chamfered equally so this type of ring can be fitted either side up.

Thinner rectangular section rings (ie 0.63–1.0mm) may be ground with a barrel face to reduce friction and improve the gas seal. Again, this ring may be installed any way up.

The keystone ring fitted in many motocross and road bikes is ground in the shape of a keystone. The idea behind this design is to reduce the incidence of ring sticking. As the ring moves in the groove it is supposed to scrape the groove clean. The cutaway section also provides a large space for the combustion gases to find their way into, to push the ring out against the bore wall. When fitting the keystone type ring, the chamfer should be uppermost. (Note – some keystone rings also have a slight taper on the lower edge of about 7°.)

A few rectangular section rings have an exaggerated chamfer on one inner edge. This type is called an unbalanced section or torsional twist ring. The chamfer causes a slight dish in the ring face so that the lower edge makes high pressure contact with the cylinder wall. Some rings of this design have a step cut into the inner edge instead of the large chamfer, but, whatever method is used to create a high pressure area on

Figure 7.15 Common types of two-stroke piston rings.

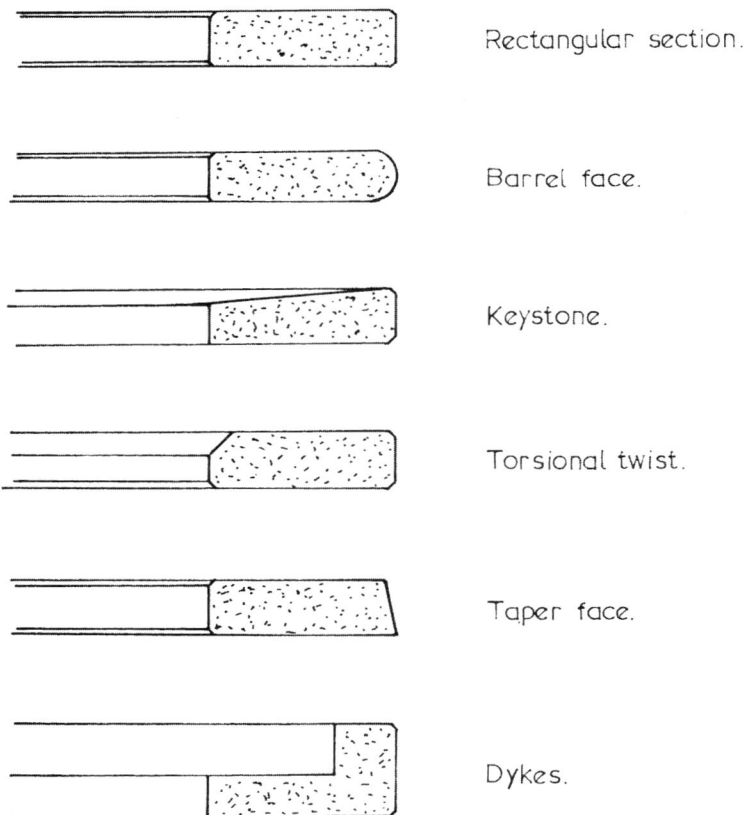

the ring face, the ring must be fitted on the piston with the large chamfer or step to the top.

Taper face rings work in the same way as the previously mentioned rings, but in this instance the ring is made with a tapered face to exert a high pressure on the cylinder wall. These rings are always marked TOP, to identify the side that should be uppermost.

The other type of ring used in two-stroke engines is the 'L' section ring designed by Paul de K.Dykes. Obviously this type cannot be incorrectly fitted but, nonetheless, it is good to know why it is made in such a shape.

Dykes was involved in research into the problem of ring flutter and he came up with this design, which for many years was used in both four- and two-stroke racing engines. As piston ring technology has advanced, the Dykes ring has lost favour. Today it is used in a small number of two-stroke racing engines, and it is just being reintroduced into drag racing engines.

The main difficulty with the Dykes ring, as far as the two-stroke tuner is concerned, is ring sticking. I have never experienced this trouble but many two-stroke mechanics will not use it for this reason. Perhaps if the engine is not dismantled reasonably frequently (ie every 300 miles) sticking could occur.

If you take a look at Figure 7.16 you can see how Dykes designed his ring to be resistant to flutter. Note that the piston is also designed to complement the ring by having considerably more clearance above the vertical leg of the ring, as compared with the normal groove clearance for the horizontal leg. Therefore, even if piston acceleration is high enough to cause the ring to swap sides in the groove, it can not lift high enough to close the gap above the vertical leg. This feature ensures that gas pressure can always be maintained behind the ring to force it against the cylinder wall and prevent blow-by.

To make life pleasant for the piston and rings, the cylinder must be faultless. There is only one type of bore which gives good performance. This is one that is

Figure 7.16 The Dykes type ring resists high speed flutter.

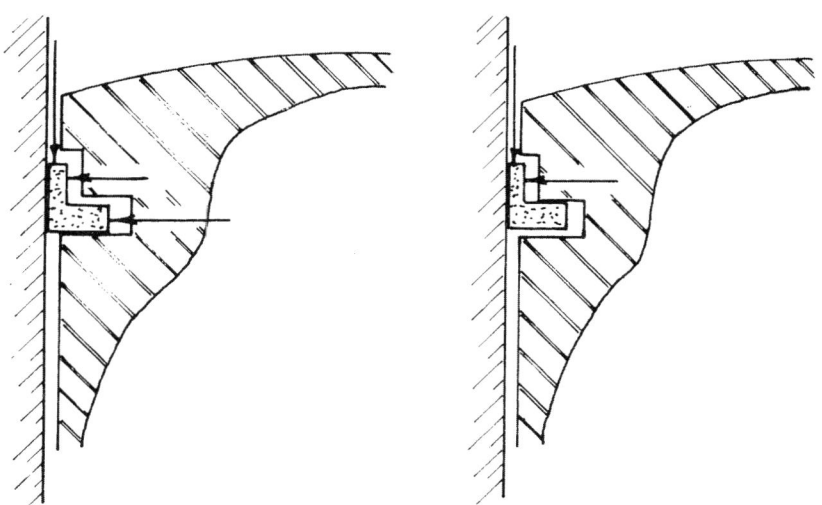

perfectly round, parallel (ie no taper from top to bottom), and square to the cylinder base. Providing you have taken time correctly setting up the crankshaft and crankcases, a cylinder trued to these requirements will enhance performance and reduce wear to the piston, rings and bore.

When taking the barrel to a machine shop for reboring, have the base checked for trueness. If the cylinder is not square to the base it will be necessary to mount the cylinder on a mandrel and then skim the base in a lathe. You may find that such machining will render your barrel illegal in certain classes of kart racing, where a minimum cylinder height is specified in the regulations.

After squaring, the cylinder can be rebored. Make certain the machinist understands that the barrel base must be mounted directly on the boring jig's parallel bars. The barrel must never be shimmed to bring the cylinder in line with the boring bar. The bore should be taken to within 0.1mm of the required size and then honed to give the necessary piston clearance.

The cross-hatch pattern the hone leaves on the cylinder walls is critical if the rings are to bed-in quickly and last for a long life. Personally, I prefer a 45° crosshatch with a finish of 10–12 micro inches. This type of finish makes it necessary to run the rings in, but they will wear well and not leak. A smoother finish does not retain enough oil and consequently allows a glaze to form on both the ring face and bore wall. Power is lost due to the poor ring seal, allowing gas leakage into the crankcase. A finish rougher than 12 micro inches will greatly reduce the amount of time required to bed the rings, but ring life is shortened. Keep in mind that glazing can, again, be a real problem, but not due to a lack of lubrication. A rough finish acts like a file on the rings, the extra friction increases their temperature and causes glazing to form.

A small carborundum stone is needed to smooth and shape the sharp port edges. Left undressed the sharp edges will damage the piston and ring.

The bottom end

After being honed, the top lip of the cylinder should be chamfered to remove the sharp edge produced by boring. A smooth cut half round file is ideal for the job, but take care that you do not let it slip and nick the bore wall. If the sharp lip is not removed you will soon destroy the engine by pre-ignition or detonation.

Each time the engine is rebored, or even just honed, it is essential that all port openings be carefully filed and then dressed, otherwise the sharp edge formed will damage the rings and piston. To the naked eye the openings may appear smooth enough, but try running your finger around the edges and you will find just how razor sharp they are.

This problem of razor-edged ports exists with some new engines too. I have found the TZ Yamahas particularly bad, so if you own one of these be sure to dress the port openings before you run a new barrel. The idea is to smooth off the edges carefully with a cigarette size oilstone. The stone should be either 180 grit silicon carbide or hard grade Arkansas stone. The hard chrome in the bore of the TZ is not easy to dress, consequently this job can take up an entire day.

Barrels with iron cylinders are much easier to work with, even though it is equally difficult to get at the port openings in cylinders with a bore of less than 60mm. Initially, I like to shape the edges of the port using either a high speed grinder or a ¼in chainsaw file. (Note – this step applies to cylinders with iron liners only, not those with any type of plating directly on the aluminium.) This work with the grinder is done not only to deburr the sharp port edges, but also to make life easier for the rings in another way. If we put a chamfer around the port opening similar to that illustrated in Figure 7.17, the piston ring will be eased back into its groove as it closes over the port, reducing the possibility of ring bumping and/or breaking. The chamfer will be about 1.5–2mm wide and taper to a maximum depth of 0.5mm around the exhaust port. Because the other ports are much smaller, the chamfer can be reduced to 1.0–1.5mm wide and 0.3mm deep. After shaping the port edges, smooth them off with an oilstone.

When all the machining work is completed, the barrel must be thoroughly washed with hot, soapy water. Be sure to get all traces of honing grit scrubbed out of

Figure 7.17 Chamfering the port edges avoids damage to the rings and piston.

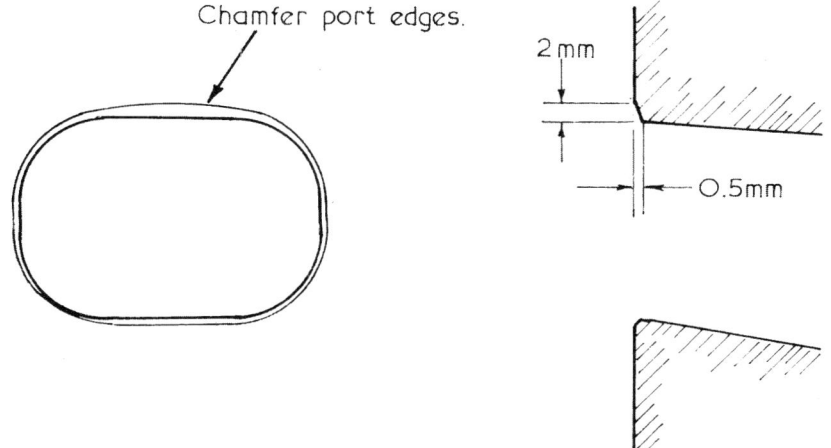

the cylinder, using a bristle scrubbing brush. Next spray the cylinder with a water dispersant such as WD-40 and blow it dry with compressed air. Apply another coat of WD-40 to the bore.

Today most racing engines are using barrels without any type of cylinder sleeve. Some people have suggested that factories are doing this to make their bikes lighter, since an iron sleeve weighs a kilogram or more. Actually, the real reason is associated with heat transfer.

Initially, manufacturers pressed the iron sleeve into the barrel, but the minute gap existing between the two formed an insulating barrier which seriously limited heat transfer to the cylinder's cooling fins or water jacket. This reduced the power potential of all two-stroke engines.

Later, the aluminium cylinder was cast around the iron sleeve and bonded to it. This resulted in improved heat transfer and a corresponding increase in performance. However, no matter how effectively the two materials are bonded, there is always less than perfect heat conduction from the cast iron sleeve to the barrel.

The next development involved the total elimination of the iron sleeve. Because the piston rings would quickly wear and score a plain aluminium cylinder, the bore is plated with porous hard chrome by a special process. The chrome plating is usually 0.08–0.1mm thick and offers a reasonably long service life in racing engines. At times the chrome has been known to flake, and it is easily damaged by dirt inducted into the motor. Yamaha have been using chromed bores on their TZ range of motors for many years now, and Honda went the chrome cylinder route when they introduced their new CR250R motocrosser in 1978.

The German Mahle firm has been working with a superior electro-chemical plating called Nikasil. This plating was originally developed for Mercedes when they were building experimental Wankel rotary engines. Then Porsche began using Nikasil plated cylinders in the 630hp air-cooled 917 model Le Mans racer. This engine later produced 1,100hp in turbocharged form for the Can-Am series. Today, Nikasil cylinders are in use on tens of thousands of chain saws and other industrial two-strokes throughout Europe. It has proved to be very successful in racing two-stroke engines also; the Morbidelli 125 and Rotax 125 and 250 production racer engines all exhibit excellent low wear characteristics for the cylinder, piston and piston ring. In more recent years Japanese manufacturers such as Honda have also begun using Nikasil cylinders on some of their engines.

The Nikasil coating is a nickel and silicon carbide matrix about 0.07mm thick. The nickel matrix is very hard, but it is comparatively ductile, whereas chrome is brittle. Dispersed through the nickel are particles of silicon carbide less than 4 microns in size. These extremely hard particles make up about 4 per cent of the coating and form a multitude of adhesion spots on which oil can collect. So beside providing a very long wearing surface for the piston and rings to bear against, the silicon carbide particles also contribute to long engine life by ensuring good cylinder lubrication.

Another type of cylinder 'plating' was revealed to us at the release of the Kawasaki KX125 and 250 motocross bikes. Their patented electrofusion process involves the exploding of wire inside the cylinder in order to plate the bore. After a hone, the coating is about 0.065mm thick. Fifteen separate explosions plate the cylinder, first with three layers of pure molybdenum, followed by six alternate coatings of high carbon steel and molybdenum and then six layers of high carbon

The bottom end

steel. When the cylinder is honed, the last three coats are removed.

The two types of wire are exploded in the centre of the cylinder by a 15,000 volt burst of electricity, which gasifies the wire. The gas expands out to the cylinder wall, burning up any oxygen in its path. This eliminates any risk of oxidation and ensures a good bond between the cylinder and the plating.

Originally, the electrofusion plating was applied only to the Kawasaki motocross engines, but now that the process has demonstrated its worth, Kawasaki are coating their road-going two-stroke engines as well.

When you have prepared the piston, rings and barrel as outlined, these parts may be fitted to the engine.

First, oil the little end bearing and the gudgeon pin, and then connect the piston to the con rod, taking care to install it the correct way round. After the gudgeon pin has been pushed in carefully fit the wire circlips. Always use new circlips and be sure to stuff a clean rag into the top of the crankcase. If you happen to drop a circlip, you don't want it falling down in there. Double check to ensure the circlips are properly seated in their grooves.

Before fitting the cylinder, give the piston and rings, and also the bore, a liberal coating of oil. Use straight oil and don't mix anything like STP with it, as piston rings are not designed to cut through goo like this.

An engine that has been properly prepared can be run-in in about 45–60 minutes. Start running at reasonably low speeds with a rich mixture and additional oil in the fuel. If you normally use a 20:1 mix, run-in on an 18:1 ratio. After about 15 minutes of operation at a fairly constant speed, try varying your speed and occasionally use up to about 3/4 throttle. Continue changing your speed for another 15 minutes, but take care not to accelerate too briskly. Prolonged operation of a two-stroke at part throttle followed by a burst of full throttle may damage the engine. This occurs because only a small amount of oil is present in the engine during constant light throttle running. Suddenly opening the throttle severely loads the engine but only a small quantity of oil is available to lubricate at slow speeds.

During the next 15 minutes gradually build up to racing speed, but do not use full throttle for more than 200 yards on the main straight. Accelerate hard out of the corners for the last 15 minute segment, easing off after each burst to allow the rings and piston time to cool before the next blast. The hard acceleration is necessary to bed the rings into the bore. Do not hold full throttle for more than a few seconds or the rings and bore could overheat and glaze.

Some engines begin to 'lock up' during the run-in period and then seize under actual race conditions. As a safeguard against this, I would recommend that you remove the barrel after running-in and examine the piston for any sign of scuffing. If you find any high spots on the piston, remove it from the engine and, using a smooth cut millsaw file, gently file the high spots off the piston. This does not sound very scientific, but I can assure you that it works.

Not many engines require this treatment but there are a few that always seize where the piston bears against the exhaust port bridge, and there are others which lock up just under the bottom ring land at two points about 30° around from the gudgeon pin. Increasing the piston clearance usually doesn't help with either of these problems (Figure 7.18).

When a street engine is radically modified for racing, you may find it necessary

Two-stroke performance tuning

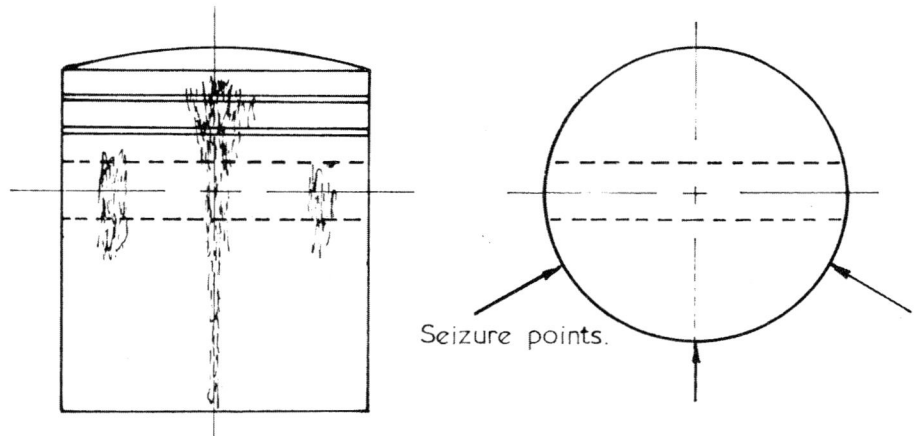

Figure 7.18 Possible seizure points on the exhaust side of the piston.

to machine what we call a 'clearance band' around the top of the piston (Figure 7.19). This modification should be required only when the standard pistons are retained. The actual band can be machined in a lathe. It should be about 0.06–0.1mm deep and extend from the top of the piston to a point approximately 3mm below the bottom ring groove. Pistons in a racing engine operate at temperatures considerably higher than those experienced in street engines. The additional heat is concentrated in the piston crown and this results in abnormal expansion. The clearance band makes allowance for the crown to expand more without risking a lock up.

Figure 7.19 Piston clearance band may be necessary on low silicon content street pistons when used for racing.

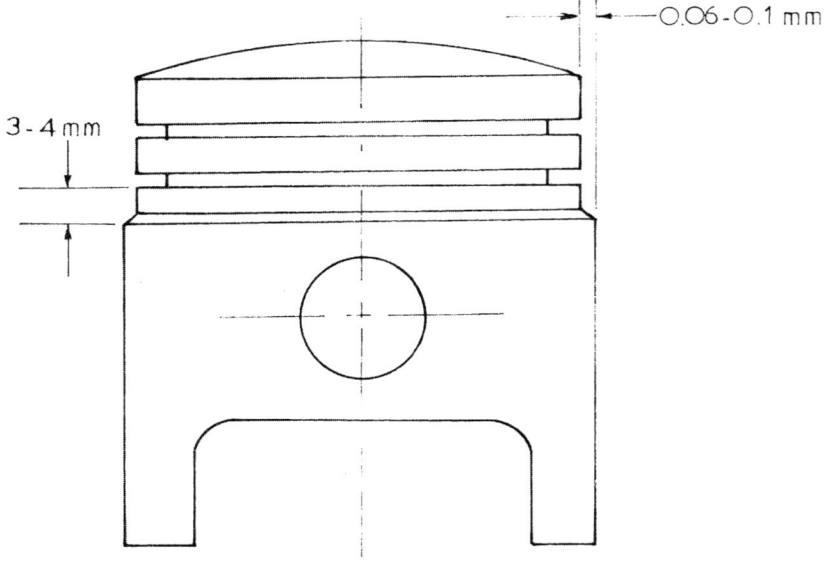

The bottom end

Only the ignition rotor and one crankwheel were salvaged from this engine which blew because the rider persisted in holding the throttle wide open each time the machine became airborne. The data logger recorded an engine speed of just over 23,000rpm! The message is, rpm limits do matter.

A bottom end assembled as outlined should prove relatively reliable in competition, providing you keep an eye on the tachometer red line. Deciding what the red line rpm should be can only be determined by actual experience with that particular make of engine.

You cannot arbitrarily say, for example, that all single cylinder 125s can be safely run at 12,000rpm. As a starting point, run the engine initially at a speed equivalent to a mean piston speed of 4,000 feet per minute and work up from there to a maximum of about 5,000 feet per minute.

The mean piston speed is calculated using the formula:

$$\text{Mean piston speed (ft/min)} = S \times 0.00654 \times \text{rpm}$$

$$S = \text{stroke in mm}$$

Chapter 8

Lubrication and cooling

The reliability of any engine is closely related to adequate lubrication and efficient cooling. Unfortunately, it is in these areas that the two-stroke engine is most vulnerable. It has to rely on just the scantiest supply of lubricating oil to resist piston seizure in a cylinder badly distorted by the steep temperature gradient existing between the hot exhaust side and the much cooler inlet side of the barrel.

The lubricating oil must be able to prevent metal to metal contact of moving engine parts and at the same time assist in conducting heat away from the piston crown to the cylinder wall. Additionally, it must form a seal between the piston rings and cylinder wall to contain the pressure of combustion effectively. If the oil film is too thin, blow-by will result, reducing the amount of energy available to power the piston down.

There are basically three types of oil: mineral oil derived from crude stock; vegetable oil from the castor bean plant; and synthetic oil, which is man-made or man modified and used straight or blended with mineral or vegetable oil.

In the past most two-stroke oils were mineral based, with a variety of additives blended in to improve them for certain functions. There are some good mineral oils available which will provide adequate lubrication and better wear resistance than many synthetic and some castor oils. All mineral oils will dirty the plug and leave carbon on the piston and in the combustion chamber and exhaust port. However, I no longer recommend them for use in anything but lightly stressed play bikes as even the best mineral oil is not able to provide the ultimate scuff resistance of the best synthetic and castor/synthetic oils.

For many years I specified Castrol R30 or R40 castor based oil for all of my competition engines. R30 was the oil of choice for water-cooled engines and R40 for air-cooled, while a similar product, Castrol M, was recommended for methanol-fuelled two-strokes. All three vegetable oils have incredible affinity for adhesion to metal surfaces and as such provide excellent anti-seize properties along with reduced piston, ring, bore and big end wear.

Many tuners do not like castor oil or blended castor/synthetic because of some problems associated with the use of an oil of this type. Some claim that castor gums up the rings and causes ring sticking but I have never found this to be a problem, even on engines that are required to run 500 miles between rebuilds. Generally ring sticking occurs only when the fuel-to-oil ratio is leaned right off to less than 25:1.

Personally I would classify castor oil as a 'dirty' oil in that it leaves behind a considerable build-up of deposits in the combustion chamber and exhaust port, and on the piston crown. Also, it leaves a varnish deposit on the sides of the piston. Consequently regular engine maintenance is necessary otherwise these deposits will reduce piston clearance and combustion chamber volume. Reduced clearance near the top of the piston could see the piston seizing or scuffing the bore. Reduced combustion chamber volume could see you disqualified in compression controlled kart classes, and in other types of competition the increase in compression ratio could put the engine into detonation.

The principal area of concern, and this is the main reason why oil companies try to discourage their use, is that castor-based oils are hygroscopic. This means that they will absorb moisture from the atmosphere. Therefore, once a container is opened, its entire contents should be used, or if oil is left over this should be poured into a smaller container so that no air space is left above the oil from which to absorb moisture. Remember, too, that castor oil will also absorb moisture after it has been mixed with fuel. Therefore, do not use fuel more than three days old, and don't forget to drain the fuel from the tank and carburettor bowl.

While we are on the subject of castor bean oil, don't think for a minute that all castor oils are as wear-resistant as Castrol R. This all depends on how well the manufacturer de-gums the basic castor stock and what additives are used. Some castors provide wear protection no better than average mineral and synthetic oils. Castrol have a long experience with castor oil and their R30/R40 formulation which includes some synthetic base stock results in a high quality budget price competition oil.

Today the majority of two-stroke tuners advocate the use of synthetic lubricant. There are several points in favour of synthetics, namely: less exhaust smoke, less incidence of plug fouling, and less build-up in the combustion chamber and exhaust port and on the crown and sides of the piston. Some also claim better wear protection and more power.

In general terms I agree with the former comments but I disagree with the two latter assertions. Yes, synthetics are very 'clean' oils but only a relative few will provide a higher degree of protection than the best castor oils. Consequently my prime recommendation for road race bikes and road race and spirit karts running on petrol is Castrol A747 castor/synthetic or Shell Advance Racing X2 synthetic. For enduro and motocross bikes used in competition I prefer Shell Advance Racing X2 synthetic. Castrol R30 or R40 will not provide quite the same level of protection as A747 or X2, but both are good budget lubricants giving excellent protection providing the engine is regularly 'decoked'. For high performance sports road bikes and also enduro and MX bikes not used in competition Mobil Racing 2T synthetic and Castrol TSS synthetic are excellent. Both of these oils have an SAE 20 rating so are suitable for oil injector systems (Castrol A747 and Shell X2 are SAE 50 grade) and being very clean burning engines can run for a long time before deposits cause problems with the exhaust power

Castor oil can cause heavy varnishing on the sides of the piston as well as pronounced carbon deposits in the exhaust port and combustion chamber and on the piston crown, when the engine is overhauled infrequently.

valve sticking or the piston becoming heavily varnished. Competition two-strokes running on methanol have to use either Castrol M castor or Shell Advance Racing M castor as both are specially formulated not to separate out or 'layer' in this fuel when left standing.

The manufacturers of synthetic oils claim their oils will give better power because the amount of oil in the fuel can be reduced (eg: Bel-Ray MC-1 is mixed 50:1 as compared with 20:1 for most mineral oils). But why should it ever be imagined that a smaller quantity of lubricant entering the engine will give a power increase? At the races it almost seems as if there is as much glory to be gained from running a fuel/oil ratio of 60:1 as there is in actually winning the race; by the pit bragging going on it would seem to be so!

My experience has shown that the more oil you pour into a two-stroke, the harder it runs. Just how much you should pour in depends on several factors, but it usually works out that the longer you hold the throttle wide open, the more oil you should use. This is due to the fact that the fuel/air ratio will be leaner at full throttle than at half and three-quarter. Therefore, with less fuel entering the engine at full throttle, proportionally less oil will be available for lubricating the piston at a time when it requires the most lubrication. Spelled out, it means that on a track with long

Lubricating and cooling

straights you will have to use more oil than on a tight twisty track.

Keep in mind that your engine only needs enough oil to lubricate one stroke at a time and then the excess is burnt up. If your bike is dribbling oil out of the exhaust then you are running too much oil for its needs, or for your riding speed. A faster rider on the same machine may need more oil, because he is holding full throttle for longer periods.

When you start experimenting with oil ratios, always use the engine manufacturer's recommendation as a reference point and work from there. If you go too rich, the spark plug will be coated with black soot and the exhaust pipe will be wet. If there is not enough oil, the plug could look white or grey, the pipe will be very dry, the piston crown will be white or light grey, possibly with 'death ash' forming under the crown. Any of these signs indicate that you are bordering on a seize up.

Generally, I would say that road racing engines using Castrol R30 will work best at a 16:1 to 20:1 fuel/oil ratio, depending on the nature of the circuit. Running Castrol A747 or Shell X2 synthetic will decrease this to 25:1 to 32:1. Desert racers require a similar mix ratio, but if plug fouling is a problem at 16:1 on R30, try 18:1 or 20:1. For enduro and motocross bikes running R30 castor, 20:1 or 22:1 is the best ratio, but if using Shell Advance Racing X2 synthetic which I prefer, I would mix at 35:1 for enduro and 30:1 for motocross and lean the oil down to a minimum of 40:1 for slow circuits. Fixed gear sprint karts running Castrol R40 should use 14:1 to 16:1 preferably, but on slower circuits 20:1 may provide sufficient oil. On Castrol A747 the mix ratio will be in the range of 20:1 to 25:1 depending on the circuit. High performance engines not operating in competition on synthetic Mobil Racing 2T or Castrol TTS will usually give good engine life on mix ratios in the range of 32:1 to 40:1.

All of the above fuel/oil ratios are for the specific oils to which I have referred. Other types of synthetic oil are an entirely different kettle of fish. Some may be suitable at the above mix ratios but many synthetics have to be used at the fuel/oil ratio that the oil manufacturer recommends. This is because the oil may have been loaded up with additives in an attempt to give the oil acceptable scuff resistance when mixed at 50:1. Blended at 25:1 there will be twice as much chemical and metallic additive and detergent being inducted into your engine and this could very easily cause carbon build-up and plug sooting serious enough to stop or even damage the engine.

It seems that the trend towards leaner and leaner oil ratios has resulted from the desire of two-stroke manufacturers to get rid of unsightly exhaust smoke and to eliminate plug fouling completely in two-stroke mower, outboard and chain saw engines. These engines are seldom serviced and the plug is probably only changed each time the rings are replaced. To cut down on spark plug deposits the manufacturers decided on less oil, and unfortunately, this idea has carried over into competition two-stroke circles.

Years ago I did some power tests at various oil mix ratios on a fully worked Suzuki RM125C motocross engine (Table 8.1). As you can see, reducing the oil content from 20:1 to 27:1 (I wasn't brave enough to lower it any further) resulted in a power loss of about 8 per cent at the top of the power range – a heavy price for the sake of a clean plug. On top of that the piston showed signs of scuffing bad enough to

Two-stroke performance tuning

deter me from testing at 32:1 which, according to a lot of tuners, is the best mix when using R40.

Table 8.1   Suzuki RM125C horsepower/oil tests

rpm	Test 1 (hp)	Test 2 (hp)	Test 3 (hp)
8,000	15.7	15.4	16.0
8,500	18.8	18.7	18.7
9,000	20.4	20.5	19.2
9,500	21.3	21.7	19.6
10,000	21.9	22.1	20.3
10,500	22.6	22.9	20.7
11,000	23.2	23.6	21.4
11,500	17.3	17.6	15.8

Test 1 – Castrol R40 mixed at 20:1 with Shell 115 MB racing fuel. Champion N-57G plug – no sign of carbon. Light coat of 'varnish' on sides of piston.

Test 2 – Castrol R40 mixed at 16:1 with Shell 115 MB racing fuel. Champion N-57G plug – slight trace of carbon on insulator, heavier deposits on plug shell and earth electrode. Less 'varnish' on piston than with 20:1 mix.

Test 3 – Castrol R40 mixed at 27:1 with Shell 115 MB racing fuel. Champion N-57G plug – very clean – cleaner than plugs from 20:1 and 16:1 tests. Heavy 'varnish' coating right around ring lands and down exhaust side of piston.

When the fuel/oil mix was raised to 16:1, power was marginally improved by about 2 per cent, which is almost too small to measure on the dyno. Interestingly, the piston was much cleaner and the rings showed no sign of gumming up. This indicated that with more lubricant the rings were transferring more heat out of the piston crown, and they were sealing better.

To keep a two-stroke alive you really have to go further than simply looking at fuel/oil mix ratios. What a lot of people fail to realise is that when you shut the throttle at high rpm you not only shut off air and fuel flow into the engine, you also cut the oil supply. Now, depending on how long it is before you open the throttle the engine could run dry and begin to seize. This means you have to run the bottom end rpm very rich, perhaps richer than desirable for best horsepower. However the excess fuel will do two things: first it will help cool the big end bearing and the piston, and second, it will carry a lot more oil in to provide a bit more lubrication.

With this in mind karters must be careful not to run lean on the slow speed mixture screw. There has to be proper balance, if the engine is too rich on the low speed screw and the high speed is backed off to get the top end running cleanly the engine will be too lean at the end of the straight. Conversely if the engine is too rich on the high speed screw and the low speed screw is backed off, the engine will run dry of oil when you lift off the throttle.

Desert racers face an even greater problem. The throttle may be held wide open for miles, then it will be closed off while the bike slows to a crawl. This slow down on sand isn't like pulling a road racer down to hairpin corner speed, it takes a good distance, so there is considerable risk that the engine will dry out and seize. There are

Lubricating and cooling

This big end overheated, destroying the bottom end when lubrication was cut off during deceleration. The oil was not at fault, rather the carburation was way too lean at the bottom of the rev range.

two things which will avert this problem. Obviously the jetting at all speeds has to be rich, with the engine almost blubbering. Additionally you need to fit a choke or tickler control on the handlebars. Then, before throttling back jab the control a few times with the thumb until the engine four strokes. This will ensure that a lot of fuel is drawn in to cool things off and there should be enough oil present to adequately lubricate everything.

Another way to tackle this problem on desert bikes is to run a Keihin Power Jet carburettor and wire up a solenoid override control, mounted on the handlebars. With such a setup the power jet fuel circuit is opened by manual control to dump more fuel into the engine prior to the throttle being shut off for slow sections to be negotiated.

Accuracy is of utmost importance when blending oil and fuel. It is of no use mixing one and a half beer cans of oil to each drum of fuel, you have to be precise. For measuring the oil you need either a laboratory measuring cylinder or a graduated beaker. Fill the measuring container with the required quantity of oil, and be sure to allow the oil plenty of time to drain out when you pour it into your drum of fuel. Keep the measuring equipment clean, preferably in a dustproof plastic bag.

Determining how much fuel is in a drum is not easy. The drum may say that it contains 20 litres, but this can vary considerably, even when the drums are factory

Two-stroke performance tuning

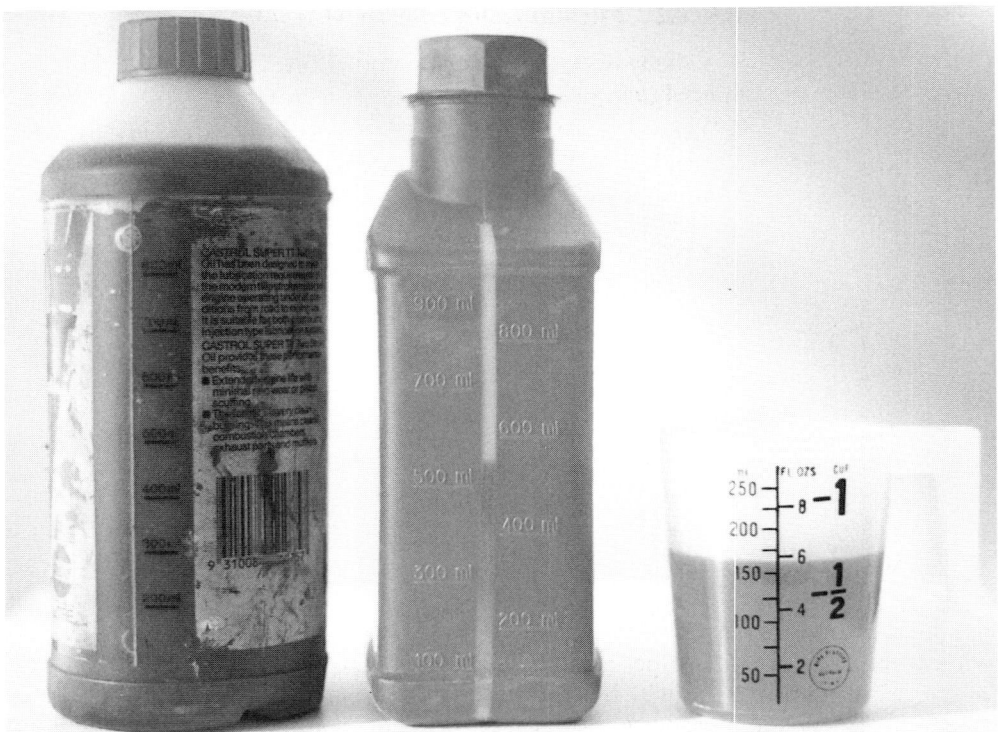

It is essential to measure oil carefully by using a clean and accurate graduated jug (this one was first checked for accuracy using a burette). Do not rely on the scale on the side of oil containers unless you are preparing at least 20 litres of fuel. However if the scale is printed on a plastic wrap, as on the container on the far left, this can move up and down on the container so never *risk using it.*

filled, as in the case of racing fuel. If you mix your own racing fuel the inaccuracy may be even worse, as a 20 litre drum will actually hold 22 litres filled nearly to the top. What I recommend is, assuming that you have brought your fuel in a drum that is supposed to contain 20 litres, drain the fuel from the drum and then refill it with precisely 20 litres measured with a suitable, accurate, 1 or 2 litre measure. Then take a rule and measure how many inches the fuel is from the top of the drum. Next make a gauge from light metal as shown in Figure 8.1, to fit in the neck of the drum and indicate the height of fuel for 20 litres. You can then use your gauge on any other 20 litre drum of fuel that you buy, assuming the drum style doesn't change.

When blending oil and fuel you must be careful not to be confused by volumes which, on the surface, appear similar. In the Imperial system one pint is 20 fluid ounces, whereas in the US system one pint is 16 fluid ounces: considerably less. To assist you with mixing for various fuel/oil ratios refer to Table 8.2.

Incompatibility/insolubility of oil and fuel can mean big trouble, so don't take it for granted that all oils and fuels will mix properly. Mineral and synthetic oils will blend with any of the leaded and unleaded fuels available out of the pump at the local garage, but they may not blend with some 100 octane racing fuels or Avgas 100/130

Table 8.2 Fuel/oil volume for fuel blending

Ratio	Volume of oil for stated amount of fuel				Metric (per 5 litre)
	Imperial (per gallon)		US (per gallon)		
	fl oz	cc	fl oz	cc	cc
12:1	13.3	379	10.7	317	417
14:1	11.4	325	9.1	269	357
16:1	10	284	8	237	313
18:1	8.9	253	7.1	210	278
20:1	8	227	6.4	189	250
22:1	7.3	207	5.8	172	227
25:1	6.4	182	5.1	151	200
27:1	5.9	168	4.7	139	185
30:1	5.3	152	4.3	127	167
32:1	5	142	4	118	156
35:1	4.6	130	3.7	108	143
40:1	4	114	3.2	95	125
45:1	3.6	101	2.8	84	111
50:1	3.2	91	2.6	77	100

Note – 1 Imperial fl oz = 28.4cc, 1 US fl oz = 29.6cc

without the addition of 5–15 per cent of benzol, or toluol (methyl benzine).

Castrol R and A747 will blend with some regular pump fuels, depending on whether they contain a proportion of benzol or toluol. They will also blend with any 100 octane racing fuel or Avgas 100/130 containing 15 per cent toluol (methyl benzine) or benzol.

When methanol is used, it is necessary to mix it with a specially formulated oil.

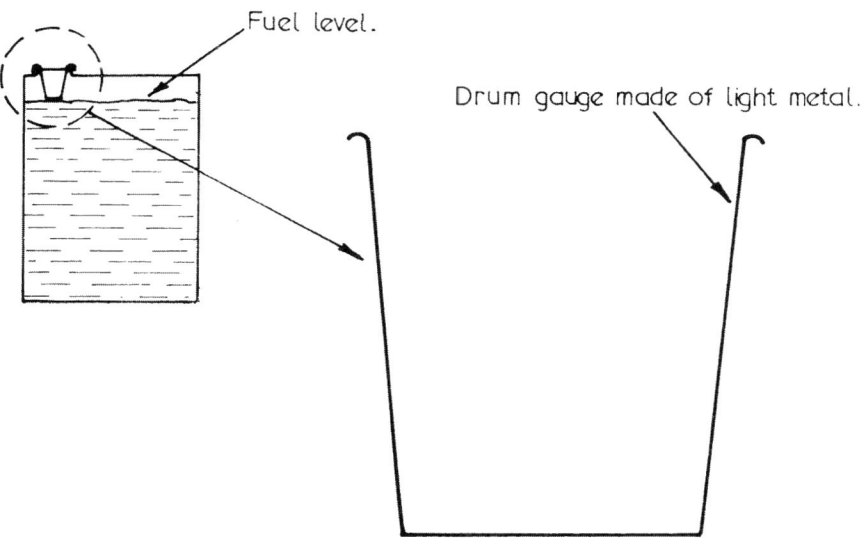

Figure 8.1 A fuel drum gauge is essential for accurate fuel and oil blending.

Two-stroke performance tuning

Castrol M castor oil and Shell Advance M castor oil are both soluble in methanol.

To determine the compatibility of your fuel/oil mix, make up a small sample at the correct ratio in a clear glass bottle. Shake it well, as you should always do when blending oil and fuel. Leave it to stand for 24 hours, and check for separation. If there appears to be some insolubility, try adding 5 per cent toluol, benzol or methyl benzine. In extreme cases you may need up to 15 per cent of these fuels added, to maintain solubility.

At times the oil may not completely separate out of the fuel, but, instead, may form in layers through it. When you find this problem, again try blending 5 per cent methyl benzine, benzol or toluol with your mix.

No matter what type of oil you run, or how well it is blended, you will still end up with premature cylinder/piston wear and possibly even risk seizure if you don't allow the engine to warm up before working it hard. I recommend that you don't ride off until the barrel is getting reasonably warm. This will ensure that piston and bore wear is kept to a minimum. I have seen engines seized by being operated too hard

Check for oil separation in a clear glass jar (left). After mixing the oil and fuel in the correct ratio leave a sample standing for 24 hours. When there is a clear boundary between the oil and fuel, as in this test sample, it is obvious that a blending agent such as acetone or toluol is required. If alcohol fuel is being used such separation indicates that the oil is not compatible with alcohol. When fuel and oil layer like this (right) in a test sample it indicates that a larger amount of acetone or toluol is needed in the blend. The fuel sample clearly shows why the kart from which this fuel was taken blew its engine.

Lubricating and cooling

right after being started. This occurs because the piston gets hot first and expands at a faster rate than the barrel, which takes much longer to warm up and expand the correct amount to provide the proper piston to cylinder working clearance.

We normally refer to two-strokes as being either air-cooled or water-cooled, but in reality water-cooled engines rely indirectly on air to stabilise the temperature of the cylinder head and barrel. The cooling arrangement of every internal combustion engine performs the vital function of dissipating heat in order to maintain normal engine operation.

The two-stroke engine is, in fact, a heat engine in that it relies on the conversion of fuel into heat, and then into mechanical energy to produce power at the crankshaft. Only about 23 per cent of the heat is converted into power, another 33 per cent is lost through the exhaust, and the rest is eliminated through the cooling system.

Lately, a lot has been said about applying a ceramic insulating coating to the combustion chamber and piston crown, to reduce to some extent the heat energy which is lost to the cooling system. It was felt that since it was heat energy, produced by the burning of a fuel, which heated the gases in an engine and caused them to expand and force the piston down, then reducing heat conduction to the cooling system should increase cylinder pressure, and result in more power.

In theory, ceramic coatings to thermally insulate the combustion chamber and piston crown sound a logical way to increase power, but in practice it hasn't worked out. In many instances the reverse has occurred, due to the end gases detonating as a result of increased pressure and temperature within the combustion chamber. I think many tuners realise that a liquid-cooled engine operating at a coolant temperature of 75°C will make significantly more power than if operated at 95°C, even though in the latter case considerably less combustion heat energy is lost to the cooling system. Why is this? Well, charge density will be superior with a cool engine and the combustion process will be more controlled, reducing the incidence of detonation.

Ceramic coatings are, I feel, only beneficial in low-speed engines and engines operated mainly at small throttle openings. Low-speed engines lose much more heat energy to the cooling system than high-speed engines, because each combustion cycle is longer. In the case of engines operated at light throttle openings the combustion process is often retarded, due to excessive dilution of the fuel charge by residual exhaust gas. With a ceramic coating applied to the combustion chamber and piston crown, combustion will be faster and more complete, due to the increase in combustion temperature.

The only other situation in which ceramic coatings may be beneficial is for coating the piston crown only in engines used for desert racing or those burning exotic fuel such as nitro and nitrous oxide. Such engines seem particularly prone to piston burning and, under these circumstances, ceramic coatings appear to offer a degree of protection.

One company which applies ceramic coatings is Heany Industries in America. Using a plasma-spray system, a ceramic coating 0.012–0.014in. thick is applied. The plasma coating process, called Heanium coating, utilises an electric arc device, into which argon gas is injected to generate a plasma stream of high temperature gas (up to 30,000°F). Powdered materials introduced into this plasma stream turn into a molten spray as they are propelled towards the surface to be coated.

As the Heanium coating is 0.012–0.014in. thick, a piston which has had the crown coated will cause an increase in the compression ratio and a decrease in the piston to head (squish) clearance. To overcome both of these problems a thicker head gasket will have to be used.

Heanium coating may also be applied to the exhaust port and the inlet port of motors with relatively straight ports. This will not do much to improve power, except that inlet charge density may increase a little, but cylinder distortion and overheating will be reduced. Cylinder distortion is not such a problem with liquid-cooled barrels, so I feel that you would be wasting your money coating the ports of these engines. Air-cooled barrels definitely will benefit from Heanium coating in the exhaust and inlet ports. Coating the exhaust tract will reduce the amount of heat which the finning on the exhaust side of the barrel has to dissipate. Hence that side of the motor will be cooler and, as a result, the cylinder bore will distort less. Conversely, coating the inlet port will increase the temperature of this side of the barrel because the insulating barrier will prevent the fuel charge from cooling the metal surrounding the inlet tract. The end result will be a lower temperature differential between the exhaust and inlet sides of the cylinder, and less distortion.

It is essential to ensure that your cooling system is working at 100 per cent capacity. Heat radiation from the cooling fins is retarded by the presence of oil and mud, so make sure they are clean. Fins and crankcases painted flat black radiate heat considerably better than shiny silver surfaces. Anything that is obstructing air flow on to the head and barrel should, if possible, be relocated elsewhere. On road bikes, check to see that the horn is not blocking air flow to the head. Also investigate to see if the exhaust can be better located, as the header pipe always seems to be in the way. Every move you make to encourage air flow over the engine will help performance and reliability.

When mounting kart engines it is important that they be wedged at 10° to 15°, with the front of the engine tipped forward. This improves cylinder head cooling in particular by exposing it to a larger flow of cooling air. Additionally this tilting of the engine creates a good deal more air turbulence between the fins on the barrel which increases heat dissipation.

Mud and oil on the fins drastically reduce engine cooling. In this case the engine did not seize but the engine was way down on power due to the massive piston and bore wear caused by the overheating. Piston clearance grew from 0.0025in to 0.0065in.

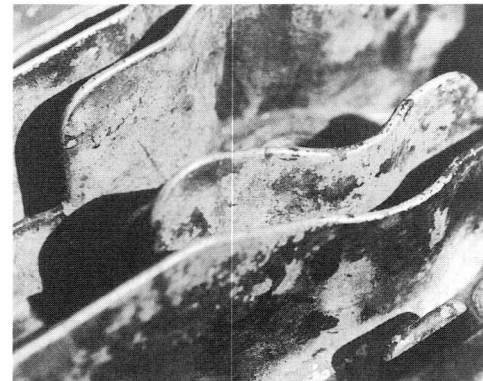

Lubricating and cooling

One difficulty with kart engines is that they are mounted back to front to make expansion chamber fabrication and mounting easier. However this is all wrong from the standpoint of efficient engine cooling. The cool side of the barrel receives good air flow around it but the hot exhaust side by comparison has very little air movement across the fins. This means that the temperature differential between the front and the rear of the barrel is very high, therefore there is more distortion of the bore. Consequently ring seal is reduced, which loses horsepower, and there is more risk of piston seizure.

This problem cannot be eliminated entirely, but that is not to say that the situation cannot be improved upon. Basically we have to get better air flow across the back of the barrel. To achieve that goal it means that some type of air scoop or ducting is required. An air scoop is more effective, but if the engine is buried behind the driver a duct may also be needed. The scoop is in reality just an air deflector. It is a piece of flat aluminium which collects air at the side of the barrel and deflects it across the back of the barrel, thus ensuring that the full area of cooling fins are doing their job of shedding heat. On fast circuits the deflector need only jut out into the airstream 20–25mm, but on slow tracks this will increase to 35–40mm (Figure 8.2).

Water or liquid cooling is now looked on as the answer to two-stroke cooling difficulties. However, liquid cooling is not without problems peculiar to itself. The two major deterrents to proper heat transfer from the combustion chamber and cylinder to the liquid cooling medium are deposits and air in the cooling system.

Metallic oxides twelve thousandths of an inch thick formed in the water passages will cut heat transfer by up to 40 per cent. Therefore, in order to maintain optimum heat transfer, the cooling passages should be cleaned in a special bath that won't attack aluminium. Additionally, the system should contain an inhibitor that will keep coolant passage surfaces clean and free of deposits.

Some tuners mistakenly assumed that this cowling around the barrel (left) was purely to stop the fins vibrating to cut noise levels. It does that, but it also captures all air entering the front of the cowl and forces it to flow over the fins on the hot exhaust side of the barrel. To prevent corrosion and mineral build-up in liquid-cooled engines a mixture of distilled water and inhibitor, or a mixture of distilled water and anti-freeze should be used.

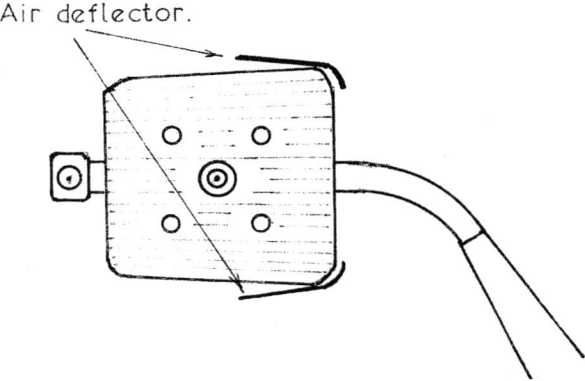

Figure 8.2 An air deflector assists cooling on the exhaust side of kart engines.

There are two basic types of inhibitors: chromates and non-chromates. Sodium chromate and potassium dichromate are two of the best and most commonly used water-cooling system inhibitors. Both are toxic, so handle them with care.

Non-chromate inhibitors (borates, nitrates, nitrites) provide anti-corrosion protection in either water or water and permanent anti-freeze systems. Chromates must not be used with anti-freeze.

If you decide to use a coolant other than water, ethylene glycol is to be recommended. Methyl alcohol-based anti-freeze should not be used because of its very low boiling point and its damaging effect on radiator hoses and water pump seals.

When ethylene glycol anti-freeze is used in concentrations above 30 per cent, additional inhibitor protection against corrosion is not required. I do not recommend the use of cooling solutions composed of more than one-third ethylene glycol and two-thirds water, as heat transfer is adversely affected.

Anti-freeze containing cooling system sealer additives should not be used, as the sealer may plug the radiator core tubes and possibly even coolant passages in the engine. Stop leak or sealer of any description is not to be recommended, except in an emergency to get you home or to finish a race. Then, as soon as possible, it should be cleaned out by a cooling system specialist, using a high pressure air and water flusher.

Petroleum-derived products such as soluble oil, often used as a water pump lubricant and corrosion inhibitor, should never be used. A 2 per cent concentration of soluble oil can raise the cylinder head deck temperature by up to 10 per cent, due to reduced heat transfer efficiency of the coolant. One popular radiator stop-leak contains a high proportion of soluble oil, which is an additional reason for staying clear of radiator sealers. Soluble oil turns water milky when it is added.

The presence of air bubbles in the coolant reduces its heat transfer capacity and the efficiency of the water pump. Air can be sucked into the system through a leaking hose or gasket, and gas bubbles can form in the system due to localised boiling, or due to pump cavitation.

In the first instance air can be kept out by ensuring that the system is free of air or water leaks, and by maintaining the coolant at the correct level. However, a good deal more is involved. At the very least take care to bleed the cooling system carefully whenever it is disturbed in any way to replace a hose, etc. You must ensure that there

Lubricating and cooling

are no pockets of air trapped in the barrel and head. Many engines will self-bleed fairly satisfactorily, while others may have to be bled through a factory-fitted bleed nipple. Others, however, require the addition of a bleeder nipple in a high point in the head.

After letting trapped air escape through the bleeders, do not assume that the cooling system is now free of air. With a race engine running a restrictor, start the engine and throttle it up and down for a couple of minutes. The vibration and water surge should get most air bubbles moving to the high spots, and then the bleeders can be opened to release any air and the radiator topped up.

A road bike is a bit different as we also have to worry about getting air out from under the thermostat. What we need to do after initially bleeding the system is to take the bike for a run for about 15 minutes. The thermostat will open in this time and the combination of engine vibration and water surge in the passages would get any trapped air moving, either into the high point in the head or into the radiator. At the end of the journey, with the engine still running, carefully open the top bleeder to let any air escape. With that done, shut off the engine and allow it to cool. When down to ambient, check the coolant level and top up if necessary.

In the second instance gas bubbles or steam pockets can be prevented by pressurising the system to the degree necessary to prevent the coolant boiling. Many wonder why it is we pressurise the cooling system. The boiling point of water is 100°C (at sea level), so why is it necessary to increase the boiling point by pressurising the system when most motocross, enduro and road bikes operate at around 75°–80°C and most road racers operate at an even lower temperature, usually 60°–70°C?

First, the system is pressurised to prevent boiling after the engine is turned off. Once the coolant stops circulating, its temperature climbs rapidly from its normal temperature to something like 110°C, way past the boiling point of water at sea level pressure (14.7psi). If the water boiled each time the engine was stopped, a considerable amount of coolant would be lost, and if the system was not re-bled an air pocket would form.

Secondly, regardless of what the temperature gauge is reading, the temperature is very high in the water passage around the exhaust port and combustion chamber. Remember that the temperature gauge is only giving a reading of the circulating water temperature, not the temperature of the water around the exhaust port and combustion chamber, where the temperature is well above the boiling point of water. Therefore, to prevent the water boiling and forming a steam pocket, the cooling system has to be pressurised. If the coolant were allowed to boil here, localised heating of the metal would occur, creating thermal stress points that would lead to cracking of the metal.

By pressurising the system using a 14psi radiator cap the boiling point is raised to approximately 125°C at sea level. As well as preventing boiling when the engine is switched off, the radiator pressure cap also serves to stop gas bubbles in a number of other situations. For example, a road engine may be given a lot of throttle at low engine speed, which will give rise to rapid heating of the combustion chamber and exhaust port area. At low engine rpm the water pump is turning slowly so the water flow is limited and the water pump will not be creating any pressure in the barrel and head. It is only the radiator cap that stops this sort of localised boiling.

A similar sort of thing occurs when a race bike pits. The bike is extremely hot because of all the full-throttle running out on the circuit, but now, with the engine idling while adjustments are carried out, the engine would boil without a pressure radiator cap.

However, when water pump speed increases to its peak efficiency speed it is not the radiator cap that stops boiling but water pressure created by the water pump. Even many race engine tuners do not seem to understand that regardless of the radiator cap pressure a water pump spinning at maximum efficiency rpm will produce a pressure head of around 30–40psi in the engine when water flow out of the head is limited by a thermostat or restrictor plate. This pressure packs coolant around the top of the cylinders and around the combustion chambers to carry away combustion heat and stop an insulating blanket of gas bubbles forming in these areas.

An old wives' tale states that if you discard the thermostat or the restrictor the engine is damaged because the water is flowing through it too quickly to draw off excess heat, or it is flowing through the radiator too fast to give up its heat. This is not so; what causes the engine damage is insufficient water pressure because there is no outlet restrictor, so coolant isn't being packed in tight around hot spots in the engine. Then any water that hits these hot spots dances about like water droplets on a sizzling barbeque plate without drawing off any heat. As the water boils off, the size of the hot spot grows and a bigger and bigger steam pocket forms in the high point in the head.

Now if the water outlet going back to the radiator is at the high point in the head it will bleed off some of this steam, and because the cooling system is still functioning to some degree the engine will continue to run. However while these hot spots exist the combustion chamber, the piston crown, and the top of the cylinder (in particular close to the exhaust port), will overheat and distort. At this point the piston may tighten up in the bore and begin to seize. More probably the engine will begin to detonate, horsepower will fall, and if the situation continues, the engine will seize.

The other possibility, which applies to engines with the water outlet in a low point in the head, is even more rapid engine failure. In this situation steam bubbles congregate at the high point in the head, but because the water outlet is lower, this steam pocket grows larger and larger. Very rapidly one side of the combustion chamber ceases to have contact with any coolant and the engine detonates and seizes. (Note – when the water outlet is on the low side of the head a steam vent should be installed at the high point in the head. Connect a 4–6mm diameter hose to direct steam from this area back to the top tank of the radiator.)

In both scenarios the conclusion would probably be reached that the engine damage was the result of the mixture being too lean or the ignition being too far advanced. In truth the engine damage was probably due to lack of water pressure in the engine allowing the formation of hot spots.

However a poorly defined coolant flow path or lazy coolant flow through the engine is even more likely to allow hot spots to develop. In reality we want reasonably rapid water flow through the engine as this tends to reduce the incidence of stagnant high-temperature pools. Additionally the rapid flow will scrub off gas bubbles as they appear in the hottest parts of the engine before they have a chance to congregate into an impenetrable steam pocket. In fact the solution to cooling problems is not so much a matter of moving more water through an engine as moving less more rapidly. Clearly we have to be thinking about accurate delivery of coolant at higher velocity to critical areas where hot spots are most likely to form – around the exhaust port,

Lubricating and cooling

When the water outlet is at the high point in the head air and steam is vented back to the radiator. However there is a problem with this engine – the water outlet is right over the water inlet! Consequently because the combustion chamber does not get much cooling benefit the mixture has to be rich and the spark has to be retarded to avoid detonation.

combustion chamber, spark plug and the top of the barrel. Achieving this will pay large dividends in more horsepower and superior engine reliability.

Unfortunately when water cooling became the norm, competition two-stroke engine tuners breathed a sigh of relief at the thought of increased reliability and more hp, and many never thought of progressing the system a step further. In reality there is much 'free horsepower' available, along with more reliability when cooling deficiencies are corrected.

My first investigation into two-stroke water-cooling was as a result of the early Rotax 250 tandem twin engine. These engines with advanced state of the art porting and rotary valve induction promised much, but out of the box they didn't deliver when compared with the cheaper and simpler Yamaha TZ250. Apart from sloppy assembly the main problem limiting performance was a cooling system which could only be described as poor.

Figure 8.3 illustrates the original system employed on these engines. The water pump had two outlets which fed coolant into the base of each barrel on the left side, in the vicinity of the exhaust port. The water then flowed upward into the single cylinder head and exited through an outlet toward the rear of the head on the right-hand side.

The flow path appears to be reasonable enough but consider what was really occurring. In the case of the front cylinder the water entering the left side of the barrel did a credible job of cooling the left side of the exhaust port. The major portion of coolant then flowed into the head cooling the left side of the combustion chamber. It then passed over the rear cylinder combustion chamber and out of the outlet to return to the radiator. However, there was virtually no flow past the right side of the exhaust port or over the right side or the front of the front combustion chamber. As a

Two-stroke performance tuning

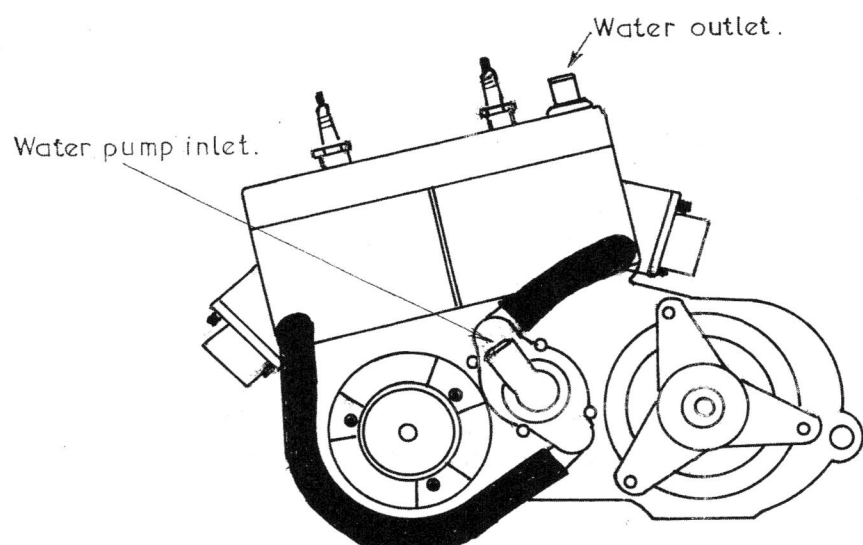

Figure 8.3 The early Rotax 250 tandem cooling arrangement.

consequence the front cylinder had to run reduced compression and spark timing to avoid detonation and even then on long straights the power would drop off due to localised fluid boil in these stagnant high-temperature pools.

The situation in the rear cylinder was no better in fact, but for coolant flow from the front cylinder passing over the rear cylinder combustion chamber the scenario would have been very poor. As for the front cylinder the coolant entered the left side and flowed past the exhaust port, but then without even flowing over the combustion chamber water exited the head through the water outlet located right in line and above the water inlet! Again there was virtually no water flow over the right side of the exhaust port or around the top of the cylinder, which led to localised coolant boil in these areas and resultant loss of horsepower.

The simple solution to these flow problems is shown in Figure 8.4. First the water outlet in the head was blocked off and then a new outlet was located in the middle of the head. (A 4mm steam vent hose was run from the high point at the end of the head back to the radiator header tank. As previously noted this vents steam, but it also allows air to be bled from the system.) Then, to ensure a good supply of coolant into the right side of both barrels a 'Y' piece was installed in both hoses from the water pump and with water inlets installed on the right side of both barrels, in line with their respective exhaust ports, coolant flow was equalised into both sides of the engine. This change enabled a modest boost in compression ratio and ignition timing which lifted power from a mediocre 55–56hp to a reliable 57–58hp. However just as importantly the engine would now hold maximum horsepower load for almost twice as long before power began to drop away from excessive localised heating. Later this awkward external plumbing arrangement was changed to an internal system which comprised a length of copper tube fitted between the left side water inlet and a drilled metering orifice under the exhaust port. The copper tube was bonded to the barrel and sealed around the ends with Silastic. All coolant entering the left side of the barrel

Lubricating and cooling

passed into this tube. Half of the flow passed through under the exhaust port to cool the right side of the exhaust port and cylinder while the other half of the flow erupted from a metering hole in the top of the copper tube to impinge on the left side of the exhaust port and cylinder.

The next important change was directed at ensuring maximum coolant flow over the combustion chambers. As modified it was still possible for a large volume of coolant to short circuit and by-pass the combustion chambers. What was happening was the streams would kiss the exhaust ports and then flow toward the centre of the engine to exit through the water outlet without necessarily flowing up into the head and over the combustion chambers. This meant the combustion chambers were overheating and inducing detonation.

The tack now employed was to bond several pieces of flat aluminium, using Silastic, into the head to block off all except the four water passages at the extreme ends of the head (Figure 8.5). The four open passages were also plated and each plate had a diverter finger, indicated by the dotted line, to curtail flow down the outside of the head. This move was to ensure that coolant flow was concentrated over the centre of the combustion chambers so that there was maximum scrubbing effect to get rid of steam bubbles. Additionally the blocking plates were drilled to allow a steam escape route and to encourage a little coolant movement around the top of the cylinders. These holes also allowed the barrels to be vented of air when the engine was refilled with coolant. We don't want any air at all trapped in the engine.

Figure 8.4 Rotax cooling improved by modifying the location of water inlets and outlet.

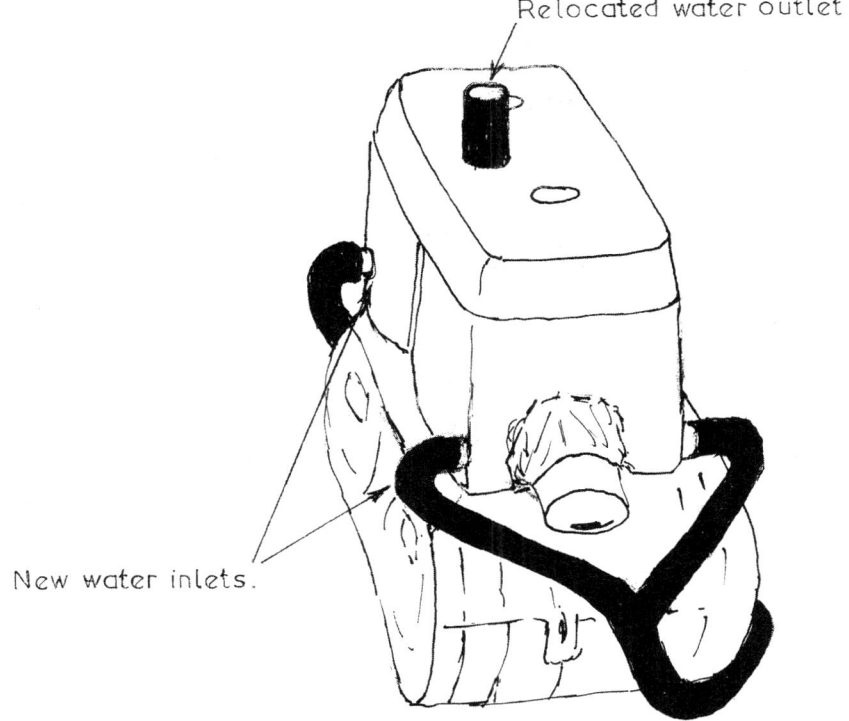

Two-stroke performance tuning

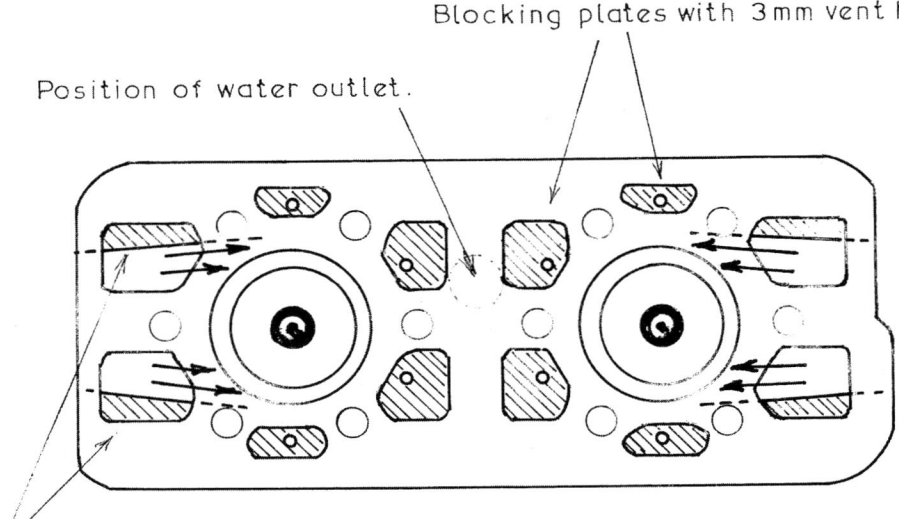

Figure 8.5 These aluminium plates bonded into the head altered the coolant flow path to ensure good even combustion chamber cooling.

With these modifications the engine was able to run more compression and spark advance. Additionally ring seal improved because with more uniform temperatures around the cylinder, there was significantly reduced bore distortion. The carburation was also much easier to get right. In fact, previously we had to run these engines at

The ring on the left has been sealing effectively as shown by the fairly even shine on its lower surface. However the other ring has been jamming due to excessive heat on the exhaust side of the piston.

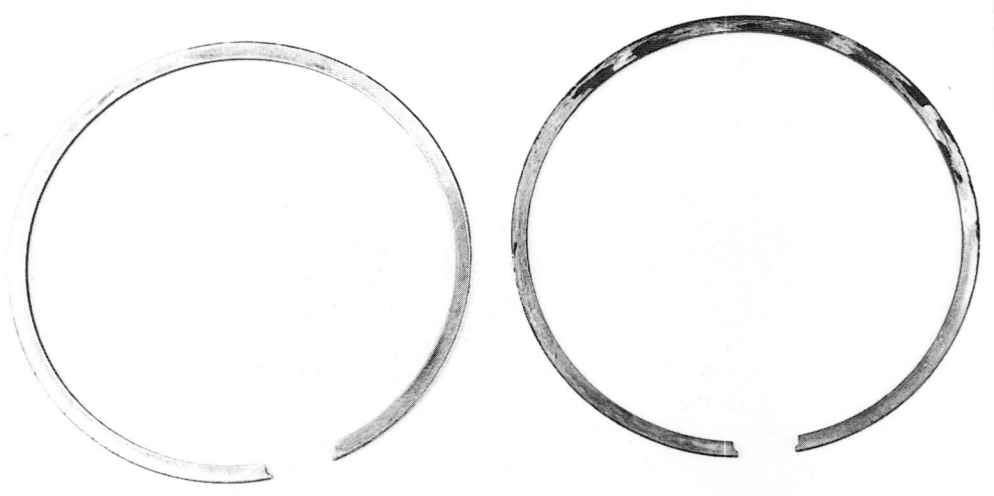

When we closely examine the faces of these rings we can see how the ring on the left has been sealing effectively while the ring which has been jamming has allowed gas to blow by.

70°C which is 10° too hot simply to get sufficiently 'clean' bottom end carburation. The end result was that power was now a reliable 61–62hp, which at the time represented a 10–11 per cent power gain.

My conviction is that we have to equalise as much as possible the temperature around the full circumference of the bore and across the whole area of the combustion chamber. The greater the temperature difference between the exhaust side of the cylinder and the opposite side, then the greater the bore distortion. More distortion means reduced ring seal and less power, but also greater danger of seizure. Also the piston will heat up more on the exhaust side which means the compression ratio and spark timing will have to be reduced to avoid detonation. Again this spells less horsepower.

The situation is similar with the combustion chamber. If coolant flow is such that one part of the chamber is hotter than the rest of the chamber then that hot spot will dictate just how much compression and spark advance the engine will accept before the end gases overheat and kick-off the detonation cycle. Ridding the combustion chamber of that hot spot will permit more compression and/or spark advance, which spells more horsepower.

It should be obvious therefore that we have to concentrate coolant flow around the exhaust port side of the cylinder and over the combustion chamber. Not only that but we should ensure that the cool water flowing from the radiator should course to these hot areas first. When the water circulates in this manner it will gather a little more heat from the hottest areas in the engine and then transfer some of that heat to cooler areas, thus better balancing out temperatures around the bore and across the combustion chamber.

The coolant flow path in many engines is completely wrong to my way of thinking. It makes no sense to have the coolant entering the barrel above the inlet port and then flowing around past the exhaust port, and then moving up into the head, flowing over the combustion chamber and exiting at the rear of the head. Equally it makes little sense to have the water enter the rear of the head and then flow down the inlet side of the barrel, and then across to the exhaust port and up into the head, exiting just in front of the spark plug.

In both the above examples the cool water is flowing first to the cooler inlet side of the cylinder. Some have suggested that manufacturers do this to keep the inlet and transfer ports cooler, thus reducing heat transfer to the fuel/air charge. Others suggest that if the coolant from the radiator were to go first to the hottest areas, the exhaust port and combustion chamber, the sudden chilling of these areas would cause metal fatigue and cracking.

When we consider that the water entering the engine will be less than 10°C (usually closer to 7–8°C) cooler than the water exiting the engine it should be obvious that such explanations are nonsense. In all probability when manufacturers do it in the

Track debris and bent fins (due to being impacted by stones) both restrict air flow through the radiator. Track debris should be picked out and bent fins should be straightened (right) using flat-blade tweezers.

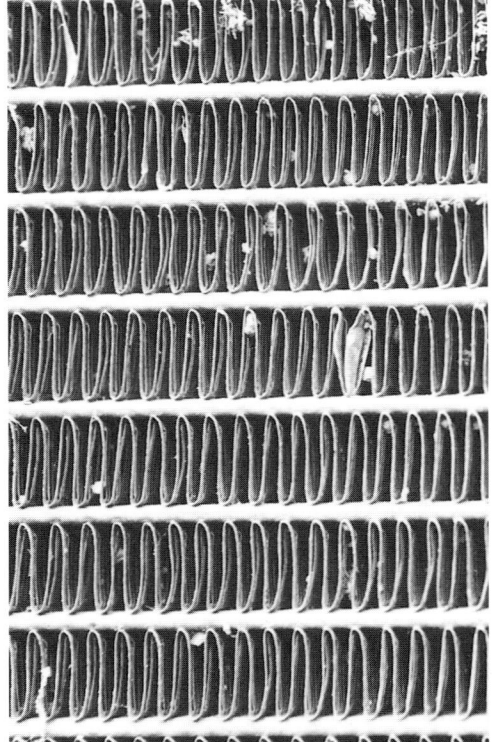

back to front manner described it is to simplify plumbing connections between the engine and the radiator.

As the actual heat exchange between the cooling medium and the air takes place at the radiator, it is essential that it is completely free of mud, bugs or other debris that would restrict air flow and hence reduce cooling efficiency. Any fins which are bent must be straightened, as bent fins present as much impediment to air flow as rubbish trapped in between the fins. The radiator should be painted matt black to provide the best radiating surface, and also to minimise the effects of external corrosion.

When mounting the radiator in a gearbox kart remember that the top tank or auxiliary surge/swirl tank must be positioned higher than the high point of the engine (Figure 8.6). Additionally when filling the system be mindful of locations where air could be trapped, and be sure to instal bleeders or vents in such locations. When it appears that hoses could be holding air try to reposition bends so that they are lower. Alternatively if the hoses are soft progressively squeeze them to expel the trapped air and get it moving back into the radiator to be vented.

It is a mistake to run the cooling system without some form of temperature control. If the engine is over-cooled it will be impossible to get the carb jetting right, and if too much heat is transferred to the coolant, power will be lost. Road bikes should run with a thermostat, but competition bikes require a restrictor in the water outlet in the head (usually with a 9–11mm orifice, depending on water pump capacity, to build water pressure in the engine) and the application of '200mph tape' to block off part of the radiator.

Some people feel that high engine coolant temperature in the range of 80–90°C is good for power. The theory is that anything lower just takes away heat energy which we should be using to push the piston down and produce more horsepower. To some extent this is true, but it must also be remembered that we may have to run at a lower temperature than is theoretically ideal simply to cool the hottest areas in the engine adequately. For example water entering the engine at 50°C will reduce the peak temperature around the exhaust port, the combustion chamber and the top of the cylinder more effectively than coolant entering at 70°C. Then when we get those areas down to a reasonable heat level we can win back much more horsepower because we can now more accurately control the combustion process. With more precise

Figure 8.6 In water-cooled karts the top radiator tank or the surge/swirl pot must be higher than the highest point of the cylinder head.

combustion control we can then explore higher compression ratios and more spark advance. Basically we are in a position to win back more power than the heat energy which we have sacrificed.

Only dyno testing will confirm what is ideal for a particular engine but in my experience I can't remember seeing a two-stroke make more power running higher than 70°C or less than 50°C. Naturally what temperature makes best power has to be balanced against other considerations. For example on a motocross or enduro bike will an additional 4 per cent hp, which you will only be able to take advantage of for around perhaps 30–60 per cent of the circuit, compensate for the extra weight and bulk of large radiators and extra coolant volume? Perhaps it won't which is why many manufacturers are setting their dirt bikes up to run at around 80°C.

From my testing I have found that for every 5°C you lower the coolant temperature the engine will pick up 1–1.5 per cent hp. For example Rotax recommend a coolant temperature of 60°C, but we found on the early 250 tandem engine, with its cooling problems, that we had to run it hotter, at 70°C. After we got the cooling sorted we gained power every time we dropped the temperature. Going from 70°C to 57°C we picked up 3.2–3.5 per cent at every point between 8,500rpm and 12,750rpm, but when we went from 57 down to 50°C the gain was only 0.5–0.7 per cent. This basically confirmed what Rotax had advised, however we also found that the engine would make full load horsepower for considerably longer at 50°C. In practical terms that indicated that if you entered a long straight with the gauge at 50°C you would be making more power after 20 seconds at full throttle in 5th and 6th gears than if you entered the same straight running with 57°C coolant temperature.

With these thoughts in mind it would seem that on a circuit with short straights, where the engine is not spending much time at full load in 5th and 6th gears and where there are many corners allowing deceleration cooling time, then a maximum coolant temperature in the range 60–65°C will not be restricting the engine's horsepower potential. However on circuits with long straights, hills, few corners, headwinds, etc, where the engine is working for long periods in 5th and 6th gears then there is gain in keeping the coolant temperature around 55–60°C. (The water temperature should be measured close to the water outlet in the head, or at the water inlet in the radiator.)

Obviously to achieve these lower coolant temperatures in hot weather may mean that a change of radiator is required. When it comes to radiator size and design there are two things we have to take into consideration; water flow through the tubes and air flow over the fins. If either water volume or air volume is too small there will be insufficient heat exchange to cool the engine adequately.

When it comes to air volume a lot of people have the mistaken idea that the faster the bike is moving, then the more air that will flow through the radiator to provide better cooling. This just isn't so as once speed increases above 40–50mph, depending on radiator thickness and fin density, air flow through the radiator does not increase significantly. This occurs due to air turbulence in the radiator core preventing any more air from flowing through. Excess air then stacks up in front of the radiator and flows out around the bike's fairing without actually contributing anything to the cooling effort.

You might be thinking air turbulence in the radiator core is not a good thing, however without that turbulence very little cooling would in fact take place. If you have a close look at the radiator fins you will note that they are finely louvred across

Lubricating and cooling

Louvres on the fins increase air turbulence across the surface of the fins, thus raising their heat shedding capacity.

their surface. It is these tiny louvres which create the turbulence to enhance the radiating efficiency of the fins.

The amount of turbulence is controlled in two ways. First the louvre design, its height and its angle of attack to the air and the number of louvres on the fin, affect the degree of turbulence created. Obviously the more aggressive the design the greater the cooling potential in radiators which are thin (say one row of 25mm tubes) and which have a low fin count (say 10–12 fins per inch).

The second factor used to control turbulence is what we refer to as the fin count (Figure 8.7). Closely spaced fins at 17 per inch will generate more turbulence and potentially remove more heat than fins spaced at 10 per inch.

However fin count, or density, and louvre style have to be balanced against the thickness of the radiator core. To understand this we have to appreciate that greater turbulence slows air flow through the core, which means that the air gathers up more heat on the way through. With this point in mind we should be able to see that as radiator thickness increases the air moving to the rear of the radiator could potentially become heated up to very nearly the same temperature as the coolant and radiator tube and fin temperature. Consequently the air will not be able to take up any more heat as it moves those last few millimetres through the radiator. In effect the rear of the radiator will not be contributing to the cooling effort.

With this understanding it is evident that we can run a high fin count and aggressive louvres on thin radiators, perhaps one 25mm tube deep. However if the radiator has a depth of one 38mm tube or two 25mm tubes then the fin count may have

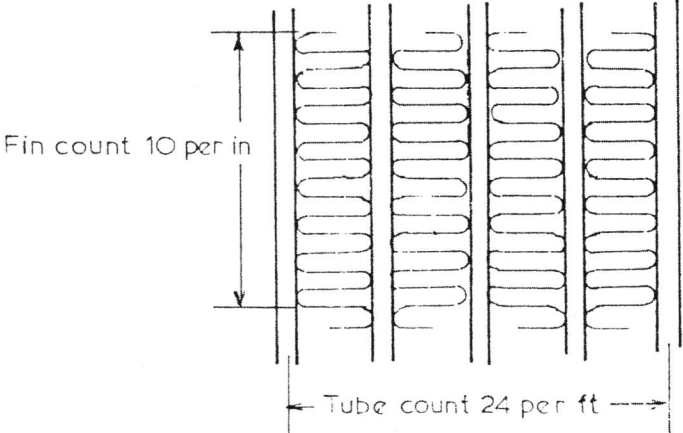

Figure 8.7 Radiator heat shedding ability is influenced by the spacing of both the coolant tubes and the fins.

to come down to 12–14 fins per inch, but this too has to be balanced against how tightly the tubes are bunched.

If you think about it, it is obvious that closely spaced tubes heat the air passing through the core more rapidly, just as would a thick core. Hence while a radiator with tube spacing of 29 tubes per foot (305mm) may work well with narrow 25mm tubes in 5°C winter weather, that spacing may have to be reduced down to 22 tubes per foot when the tube width increases to 38mm and the mercury climbs to 30°C.

Naturally when we increase the number and/or width of the tubes we slow down how quickly the coolant passes through the radiator. For example if coolant flow is 80 litres per minute through 20 tubes the flow in each tube is 4 litres per minute. If we double the number of tubes, while retaining the same tube width, the flow drops to 2 litres per minute. (In reality it will be more than 2 litres per minute as

Figure 8.8 The splitter plate in the side tank and the water inlet and outlet in the same tank identify this as a double pass radiator.

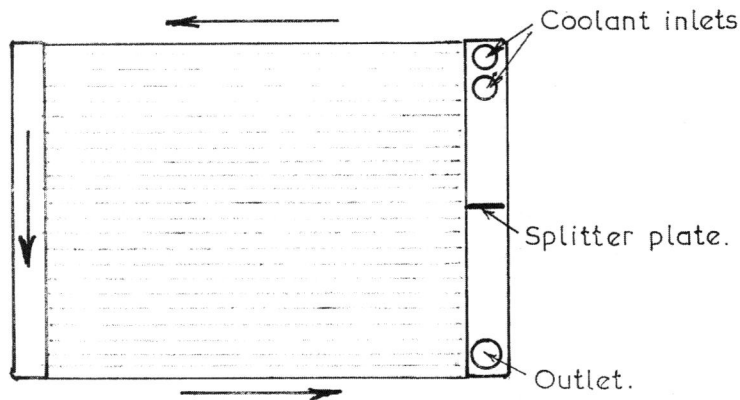

Lubricating and cooling

the reduced frictional losses will allow the water pump to increase its flow rate, perhaps 25 per cent to something like 90 litres per minute.) The effect of this is to slow the rate of coolant flow through the radiator so there is more time for the coolant to transfer heat to the tubes, then on to the fins and finally to the air. Additionally when we increase the size or number of tubes we also increase the radiating surface area of the fins. This too, if we don't go too far in increasing the fin area and subsequently the amount of turbulence, will increase the heat shedding potential of the radiator.

Another way we can increase the heat dumping ability of the radiator is to build it as a double-pass unit (Figure 8.8). This would seem to contradict what I have just stated as it is obvious that when a splitter plate is welded in one tank the effect is to force all the coolant through half the number of tubes, and over twice the distance as the water flows from one side to the other across the top half of the radiator, and then back again through the lower half. The reality is though, providing the water pump is

These water jacket cavities are very wide, allowing water flow to slow or even stagnate around the exhaust port and the top of the cylinder.

up to the task, that a two-pass radiator will dump 15–20 per cent more heat than an otherwise identical conventional single-pass radiator.

This happens in spite of the flow rate through the tubes dropping 25–30 per cent due to increased friction and water turbulence. With the water flow down from say 80 litres per minute to 56 litres per minute the flow rate in half the number of tubes, 10, would seemingly speed flow to 5.6 litres per minute as compared with 4 litres per minute previously. However these 10 tubes are now in effect twice as long so in reality, the water will be giving up heat to the tubes, fins and air for twice as long.

This accounts for part of the increased radiator efficiency but the main gain comes as a result of increased water turbulence in the tubes. At low flow rates the water clinging to the walls of the tubes pretty much stalls, and in effect becomes an insulator which reduces how much heat is shed by the main body of coolant flowing through the centre of the tube. When the flow rate is pushed up water turbulence increases and this has the effect of breaking up this stagnant insulating blanket of coolant clinging to the tube walls, which in turn improves the rate of heat transfer from the coolant to the radiator.

Having progressed to this degree in understanding coolant flow and heat transfer we begin to appreciate that big pools of coolant surrounding the cylinder and combustion chamber do little to cool the engine efficiently and allow it to be tuned to make maximum potential horsepower without damaging detonation. Manufacturers are beginning to narrow down the width of coolant jackets around cylinders, exhaust ports and combustion chambers to achieve rapid, scrubbing, turbulent flow activity. However many earlier two-strokes, and even some more recent models, have water jackets which are far too wide. Such wide passages should be narrowed right down using suitable aluminium stuffers to increase flow activity around the hottest areas of the engine.

Chapter 9

Power measurement and gearing

I am constantly surprised that so few two-stroke tuners ever spend any time testing and developing their engines on a dynamometer. Now don't get me wrong, engine dynos are definitely not the 'be all and end all' of engine tuning. There are clearly definable limits to their usefulness, due to the nature of the load applied to the engine. On the dyno you can't, for instance, check an engine's rideability, or how smoothly it comes on to the power, these factors can only be determined on the race circuit. However with modern computerised dynos an experienced operator can get a 'feel' for how crisply an engine will accelerate, or what its throttle response will be like.

With dyno testing, you are in a good position to see precisely what influence a change in the spark advance will have on the power output. You can determine how much power increase, if any, a larger carburettor will give at maximum rpm and ascertain what the hp loss is at lower speeds. Without the benefits of a dyno work-out, you can only tune according to instinct and/or past experience, and then rely on your lap times or 'feel' through the seat of the pants to indicate if a particular modification is successful or not. This takes a lot of time and is often inconclusive as, unfortunately, most of us can't feel a 5 per cent difference in power.

There are two basic types of dynamometers: the rolling road and the engine dyno. The rolling road is not very popular for two-stroke engine tuning, as its worth is rather limited. With this type of dyno, the bike is tied down with the rear wheel on a heavy roller or a pair of rollers, which are connected to some sort of loading device (the brake). The engine is run and, according to the twisting force applied through the rollers to the brake, the torque and horsepower figures are calculated.

The main problem with the rolling road is that it can never be made sensitive enough for fine engine development work. Because of tyre slip and friction on the rollers, the twisting force being transmitted to the brake is in a state of constant fluctuation. Therefore, to keep the readout fairly steady, it must be heavily damped, by introducing a 'smoothing factor'. As a result of the damping, the dyno becomes unresponsive to small changes in power output, so it would be difficult to check the

Two-stroke performance tuning

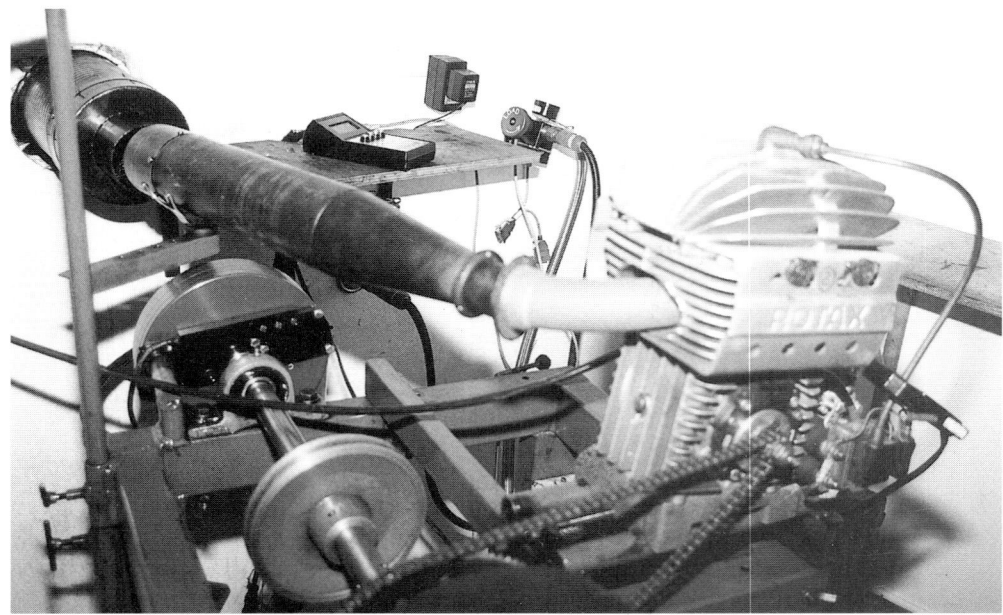

When tuning is carried out on an engine dyno the effects of changes to the expansion chamber, the header length, the spark advance, etc. can be quickly assessed.

effect of a small change in spark advance or main jet size.

The engine dyno, on the other hand, is quite sensitive and will give a clear indication of precisely how much advance or what size main jet the engine prefers for best power. With this type of dyno, the engine is coupled with the brake via a chain, driven from the countershaft sprocket. Therefore there is no slip (unless the clutch is slipping) and the frictional losses in the primary drive, gearbox and secondary drive remain fairly steady.

As I mentioned earlier, you can't check an engine's rideability on the dyno. Usually the areas where you will get caught out are with carburation and expansion chamber testing. For example, you may check three or four different brands and sizes of carburettors and find they give a seemingly identical performance on the dyno. But at the race circuit there will generally be one setup which is superior to the rest, allowing improved lap times, or maybe just a better 'feel' when coming on to the power as you exit a turn.

There are several types of engine dynos in use and, while all do a good job in enabling us to see which way our development work is going, you should not take a lot of notice of the power figures. Variations of up to 10 per cent from one dyno to another are not uncommon. The reason for this is that tuning companies do not have the money to spend on the latest, most accurate dynos. Instead, in many cases, they make do with older, or maybe even new but less sophisticated types. Also dyno manufacturers use various methods of calibration, which tend to give a different power reading from one brand of dyno to another.

This in itself is not such a bad thing, providing all the tuning is done on the same dyno. Otherwise you might try out some new trick part and find it gives you 7 per cent more power on a dyno over the other side of town, when in truth you had lost power. It was just that this dyno is reading higher than the one on which the engine was originally tested (Table 9.1).

Table 9.1 Dyno comparison test of Yamaha YZ250

rpm	Dyno A		Dyno B	
	hp	Torque (lb/ft)	hp	Torque (lb/ft)
4,000	8.6	11.3	8.8	11.5
4,500	9.5	11.1	10.0	11.7
5,000	11.1	11.7	11.6	12.2
5,500	15.4	14.7	17.7	16.9
6,000	21.1	18.5	23.0	20.1
6,500	24.9	20.1	27.1	21.9
7,000	26.4	19.8	30.0	22.5
7,500	25.6	17.9	30.3	21.2
8,000	25.1	16.5	29.1	19.1
8,500	19.9	12.3	22.3	13.8

Note – This motor was not altered in any way between the two tests. The hp and torque figures have been corrected to compensate for changes in air density.

Probably the most important factor in dyno tuning is with regard to the experience and integrity of the operator. For example, how quickly and at what rpm the operator rolls the throttle on during acceleration tests can have a big effect on the shape of the power graph. If in one run he slowly rolls the throttle wide open from 8,000rpm and in the next he slams it open from 7,800rpm the two graphs will look quite different.

In this example the first graph will most likely show the engine has more bottom end grunt because the carburation doesn't have such a problem catching up to moderately increasing air flow. However when the throttle is opened rapidly, particularly from lower rpm, the carburation will be trying to play catch-up for a good portion of the run.

If the operator has erred due to lack of experience this will cause tuning problems because there is no consistency from one run to the next. In effect the changes which are being made to the engine and the resulting power graphs are meaningless. On the other hand if the operator is trying to impress you with some product which he retails by giving the engine the slow roll-on to pad out the power curve you lose out even more. First you paid out good money for a dyno session and for a product of dubious worth, but even worse you now have a piece of paper to prove just how much more powerful the engine is and when you get on to the circuit and the bike doesn't perform you will be left scratching your head and wondering just where the problem lies. The inclination is, because of that vital piece of paper, that you will blame all sorts of other things and you will not suspect that you have been duped.

Something you are sure to have noticed is the variation between the manufacturer's claimed hp and the actual power figure at the countershaft sprocket. The main reason for the difference is that with a few exceptions manufacturers take their power figures at the crankshaft, not at the countershaft. Therefore their hp

Two-stroke performance tuning

numbers will be around 12–15 per cent higher usually as there are no frictional losses due to the primary drive, the gearbox, the secondary chain driving the countershaft to the dyno, and also the clutch churning gear oil in the case of bikes with wet clutches.

In reality these losses can vary considerably from the usually quoted 12–15 per cent. I generally find that engines with a claimed output of 30–45hp at the crank will have about 6–7hp less at the countershaft. More powerful engines have wider gears with correspondingly greater parasitic losses. Thus I expect to see engines claimed to make 55–80hp at the crank actually show 8–9hp less at the back wheel. Correspondingly those in the 90–125hp range will read 10–12hp less at the back wheel. However if the bike has a dry clutch or a dry gearbox (gears not immeresd in oil, rather a pump sprays oil on the gears) the parasitic losses will be reduced from these figures.

Unfortunately, few people realise how little the horsepower figures actually tell about engine performance, or how well an engine has been modified. The true measure of engine performance is the torque and brake mean effective pressure figures. These indicate much more to us and show where we are heading with our modifications.

Torque is a measure of the twisting force at the crankshaft expressed as pounds-force foot (commonly called foot pounds). For example if an engine is producing torque of 20lb ft it means that it will lift a load of 20lb with a lever 1ft long connected to the crankshaft. Now if the engine moves this load through one revolution work is being done; in this instance 126ft/lb (twisting force x revolutions x lever length x 2π). Power is the rate at which this work is being done hence,

$$\text{Power} = \frac{\text{work (torque x revolutions)}}{\text{time}}$$

In the Imperial system, power is measured in lb/ft per minute. However, these units are small, so the unit we know as horsepower (hp) is the one used today. One horsepower equals 33,000lb/ft per minute. This was worked out as a result of experiments done by James Watt, using strong dray horses.

It is obvious, realising power is the rate at which work is done, that two motors both producing 20lb/ft torque could have differing power outputs. In fact, if one motor lifted its 20lb load twice as quickly as the other, then it must be twice as powerful, or have double the horsepower. Engine speed is measured in revolutions per minute, so this is the time unit we use in calculating horsepower, therefore:

$$\text{hp} = \frac{\text{torque x rpm}}{5252}$$

Also

$$\text{hp} = \frac{\text{BMEP x L x rpm}}{6500}$$

where L = engine capacity in litres

Earlier, I mentioned that high horsepower figures can be misleading. We can end up with a big power figure because the motor turns a lot of rpm, which should increase top speed, but unless the engine produces a higher torque output over a wide rpm

Power measurement and gearing

range lap times could be slower due to poorer acceleration or an increase in the number of gear changes required. For this reason we have a measure called brake mean effective pressure (BMEP). This gives a true indication of how effectively the engine is operating regardless of its capacity or its operating rpm. It is, in fact, a measure of the average cylinder pressures generated during both engine strokes. We calculate the BMEP using the formulae:

$$BMEP = \frac{hp \times 6500}{L \times rpm}$$

or

$$BMEP = \frac{lb/ft \times 1.238}{L}$$

L = engine capacity in litres

The highest BMEP will occur at the point of maximum torque, which also happens to be where peak volumetric efficiency occurs.

To give you an indication of how important the BMEP is, and to show how it gives a more meaningful expression of an engine's true performance level and its potential for future development, I have included some dyno figures for the old Yamaha YZ125E and Suzuki RM125C (Table 9.2 and Figure 9.1). As you can see, both engines put out about 22hp, but just look at the differences in the mid-range power and the width of the power band. The Yamaha makes maximum power at 11,000rpm and then the motor proceeds promptly to drop dead, whereas the Suzuki, because it is working much better (as evidenced by the high BMEP pressures), holds within a few per cent of its peak hp over a 2,000rpm range (from 8,500–10,500rpm)

Table 9.2   Dyno test of stock Suzuki RM125C and Yamaha YZ125E

	Suzuki RM125		Yamaha YZ125	
rpm	hp	Torque (lb/ft)	hp	Torque (lb/ft)
6,500	7.8	6.3	7.4	6.0
7,000	10.0	7.5	7.3	5.5
7,500	13.1	9.2	8.7	6.1
8,000	16.1	10.6	12.5	8.2
8,500	19.4	12.0	15.4	9.5
9,000	21.1	12.3	17.5	10.2
9,500	21.7	12.0	17.9	9.9
10,000	21.5	11.3	19.2	10.1
10,500	20.6	10.3	20.4	10.2
11,000	15.1	7.2	21.8	10.4
11,500	9.9	4.5	11.6	5.3

Note – Maximum BMEP for RM125 = 123psi

Maximum BMEP for YZ125 = 104.5psi

Both engines were tested on the same dyno and the figures have been corrected to compensate for differences in air density.

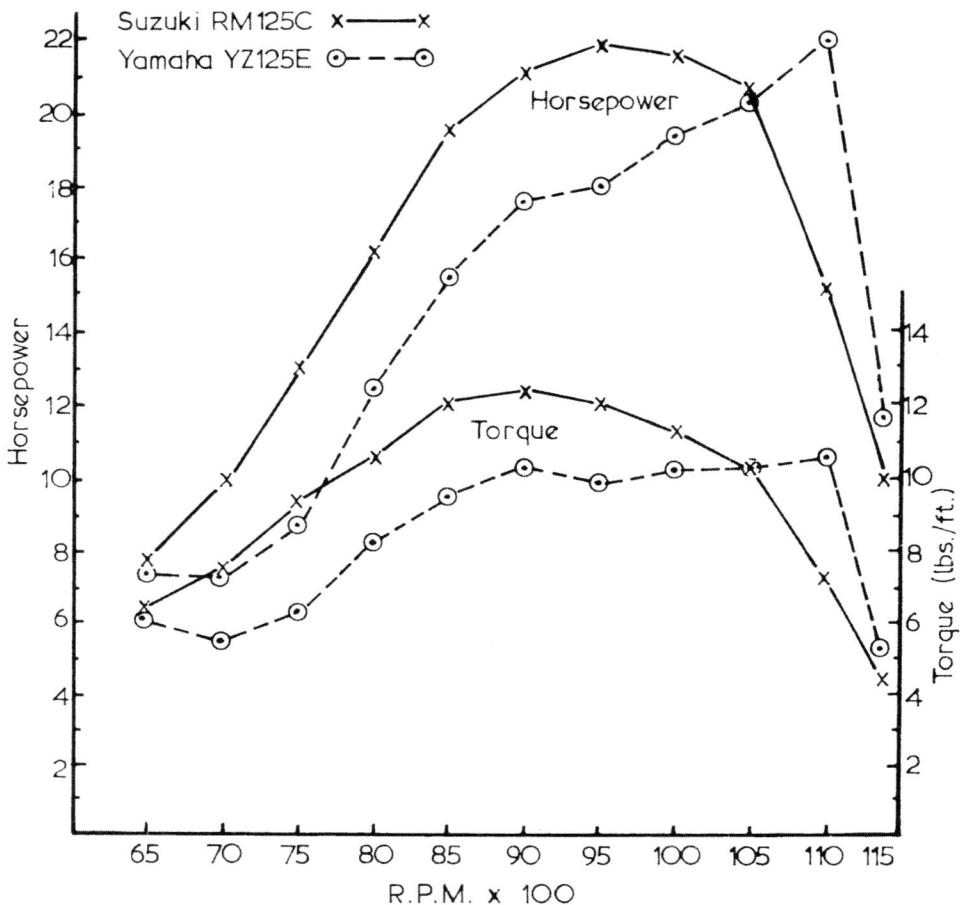

Figure 9.1 Yamaha and Suzuki 125 power curves.

and it is still making good power at 11,000rpm and down at 8,000rpm.

This statement should not, however, be confused with the idea of making more power at high rpm. If this Suzuki could have been re-tuned to make a BMEP of 123psi at 11,000rpm, equivalent to 25.8hp, you can be sure there would have been smiles all around. Actually herein lies the real art of engine tuning where we get the engine running at higher rpm, but without dipping below the manufacturer's original peak BMEP number.

Even more difficult is a striving for a higher BMEP without an increase in engine speed. For example in a sports road bike or in an endurance engine we may not be able to reliably increase engine speed, and in a road racer we may already be running up to the absolute maximum safe engine speed. Hence the only way we can get more horsepower is to increase the BMEP at the peak and from the peak up to maximum rpm, without allowing much of a drop off at the bottom of the power band.

Back in 1980 the works Minarelli 125 twin road racer was operating at a BMEP of 165psi. That benchmark figure was on rotary valves and we all wondered if reed

Power measurement and gearing

valve engines would one day get close to that figure. Today ordinary production 125 motocross engines easily surpass that number with BMEPs close to 200psi (measured at the crank) and even production 250 motocross engines, with cylinders four times as large as the Minarelli racer are pumping BMEPs of 185psi (at the crank).

Two-stroke engine speeds have not increased significantly in that time period, so what accounts for this huge increase in performance? The major advances have been in improved porting and ignition mapping, but there has also been a rethink in the areas of piston crown and combustion chamber shapes. In fact port flow capabilities have improved to such an extent that we are now turning more rpm with shorter exhaust durations and these big horsepower increases have been attained without any increase in carburettor sizes.

What the manufacturers have done should serve as a model for all two-stroke tuners, namely aim to improve flow into and out of the cylinder at current engine speeds thereby bumping up the maximum BMEP. If that is not feasible the alternative aim should be to at least maintain the current BMEP number while striving to push the horsepower higher in the area between peak BMEP rpm and redline rpm.

Figure 9.2 Suzuki RGV250 power curves.

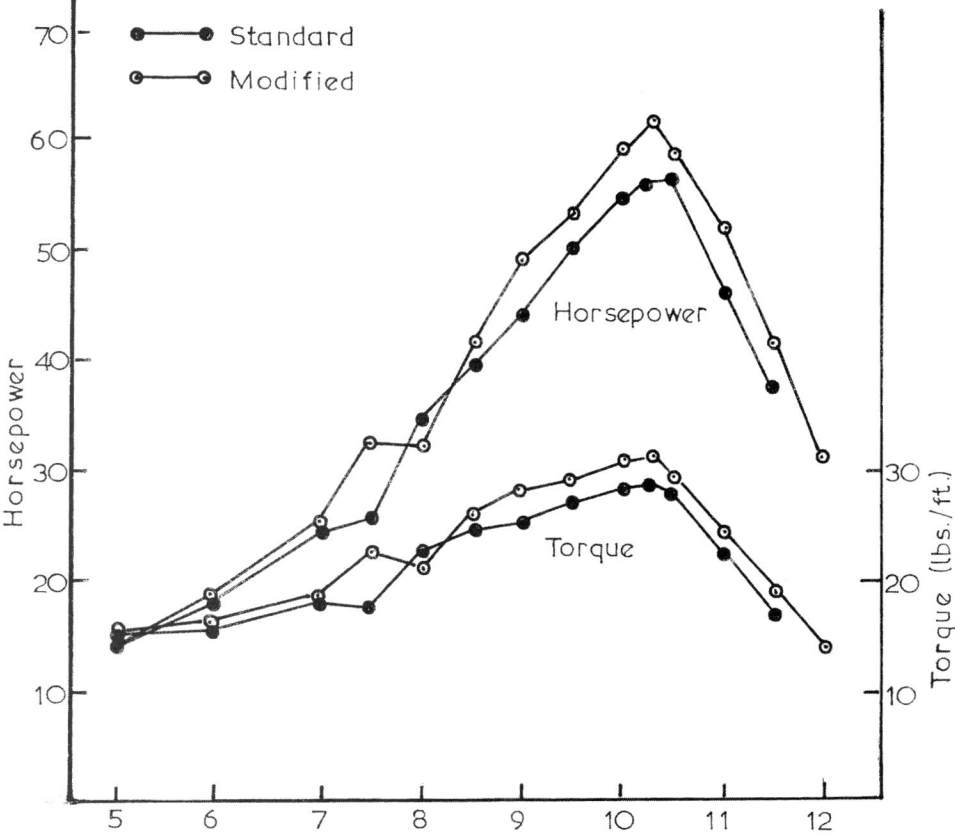

Two-stroke performance tuning

To give an example of what I am getting at, take a look at the Suzuki RGV250 street sports bike. From the factory it is already very highly tuned and to get any sort of engine life it must not be regularly buzzed beyond 10,500rpm. Additionally the gaps between the lower gears are quite wide so any tuning must be directed at not only increasing the peak horsepower and horsepower in the upper power band up to the redline engine speed, but also the horsepower in the lower half of the power band below peak horsepower must not be reduced at all, otherwise the engine will start blubbering between gear changes at wide open throttle.

Figure 9.2 illustrates the power curve of the standard engine, and also the power curve when a new set of expansion chambers were added and the ignition was remapped to suit. These modifications lifted power about 9 per cent, but of equal importance the engine was stronger at all points except over a narrow 450rpm band between 7,800rpm and 8,250rpm. Not so easily visible in this diagram is the fact that the torque has increased from 29.0lb/ft at 10,250rpm to 31.4lb/ft at the same engine speed, peak BMEPs of 143.6psi and 155.5psi respectively, while maximum horsepower has risen from 56.7hp at 10,500rpm to 61.4 at 10,250rpm.

Note that the BMEP of the modified engine has increased from 143.6psi standard to 155.5psi. Such an improvement shows that the engine is working more efficiently. If on the other hand the horsepower had increased to a maximum of 61.4 at 11,500rpm and maximum torque was also at this same speed it would indicate that our modifications while raising the horsepower had actually decreased engine efficiency, as evidenced by the BMEP dropping to 138.8psi.

When a motor is run on older engine dynos, a record is made of its output each 500rpm over its operating range. These output figures must then be converted to tell us what the hp and torque is, as most of these older dynos do not give a direct reading in lb/ft. At the time these calculations are being made, a 'correction factor' is also introduced to keep the output figures standard. If this was not done we would have no way of accurately comparing the engine's power level on another occasion, when the atmospheric conditions are sure to be different.

Throughout the dyno session a check is made of the barometer reading and at frequent intervals wet and dry bulb air temperature readings are taken, as these factors influence the air density. It stands to reason, the cooler the air and the higher the air pressure, then the more oxygen and fuel you can cram into the engine, which in turn gives more power. Conversely, if the temperature is high and the barometric pressure low, the performance will fall. To compensate for this during the test session, and to give a true comparison with earlier and possibly subsequent tests, a correction factor is added to the conversion formula.

The conversion formula to calculate the torque on one particular type of dyno is this:

$$\text{Torque} = \frac{W \times 26.26 \times C/F}{2}$$

where W = readout indicated by the dyno needle, and C/F = correction factor.

For example, if the dyno indicated the twisting force (W) to be 2.18, and the barometric pressure was 30.06in, with a wet and dry bulb temperature of 48°F and

68°F respectively, the engine would be producing the following torque:

$$\text{Torque} = \frac{2.18 \times 26.26 \times 1.016}{2}$$

$$= 29.08 \text{lb/ft}.$$

The correction factor, in this example 1.016, is found from tables or graphs which are readily available.

With modern computerised dynos the test procedure is quite different. Now there is the option of running either steady state tests or transient acceleration tests. The steady state test is much slower, but it is a very useful test in that it allows an engine to be fully loaded at a fixed engine speed for an indefinite period of time. This allows the engine to become fully heated internally so it provides more information on just what is safe in the way of ignition advance, compression ratio, etc.

An acceleration test on the other hand is a quick test, with the engine accelerating at a fixed rate up to maximum engine speed. With this type of test an experienced tuner can get a clear picture fairly quickly with regard to what modifications are working and what are not. Additionally because each test is completed so rapidly the operator can do a huge number of pulls with minimal wear on the engine and little risk of a big blow-up. However because the engine is not fully loaded for an extended time period it is easy to overestimate how high the compression ratio can be pushed or how far the ignition can be advanced.

These modern dynos also speed up testing in that we no longer have to record such things as the barometric pressure and the wet and dry thermometer readings and then manually calculate the torque and horsepower at various engine speeds. Rather the computer receives inputs from the various sources and instantaneously factors in these things before providing a printout of the engine's horsepower and torque.

On this kart dyno a water brake loads the engine.

Unfortunately, because there is no visible 'human factor' in producing this printout, some racers naively believe that somehow such printouts cannot be 'fudged' as they were at times in the past by a few dyno operators not recording the actual dyno needle reading honestly or else fudging the figures by means of a bogus correction factor. Today the figures can also be fudged quite simply by faking inputs to the computer. For example the temperature probe can be moved around to either inflate or deflate the real figures, or the correction factor automatically calculated by the computer can be overridden and another correction factor can be inserted manually. Consequently I become suspicious of printouts which indicate a correction factor of more than 2 per cent, ie less than 0.98 or greater than 1.02 if the dyno is close to sea level, and I expect the factor to change less than 1 per cent during a whole day of testing, eg. in the range of say 1.05 to 1.15.

When the dyno session is over, don't go home and just file the dyno sheet away. What you must do now, on the basis of information which can be gleaned from the dyno printout, is work out your gear change points for maximum acceleration. Obviously if the engine is revved too far past the power peak engine torque will be dropping so rapidly that acceleration will be slowed. Conversely if the engine is changed up into a higher gear too early the engine will bog in the higher gear because engine rpm has dropped out of the power band.

Table 9.3 shows one method some road racers use to determine their gear change rpm in each gear. By calculating the 'step' between gears it is a simple matter to calculate what the engine rpm will be when you slot into the next gear. Then you can determine to what rpm you must run the engine in each gear to keep the engine in the 'meatiest' part of the power band. Looking at the chart you will note that road racers and motocrossers have the most closely spaced gear sets, while sports street bikes and enduro bikes have wider gear spacing. It is also noticeable that the upper gears are closer together than the lower gears. This may indicate that acceleration in the upper gears could be improved by not running the engine to the same high rpm limit necessary to stop the engine bogging down in the 1st to 2nd change, and perhaps also the 2nd to 3rd change in the case of enduro and street bikes.

Relating these thoughts to the Suzuki RGV250 dyno printout in Table 9.4 we can see that while it is beneficial to run the standard engine as fast as possible in 1st gear to avoid the hole in the power band below 8,000rpm on the upchange into 2nd, a similar change up at 11,500rpm from 3rd, 4th or 5th gear would actually produce less than best possible acceleration. For example the rpm drop between 4th and 5th gear is just a bit less than 1,450rpm. Now when we look at the dyno figures we see that the meatiest part of the power band, over a span of approximately 1,500rpm, is from 9,500rpm to 11,000rpm. In that range the engine doesn't dip below 46.1hp and the average power output is 53.0hp (50.4 + 55.1 + 56.6 + 56.7 + 46.1 ÷ 5 = 53.0). If the engine was run up to 11,500rpm in 4th we see that the minimum horsepower has fallen off to 38.6, a loss of 7.5hp, and the average horsepower between 10,000rpm and 11,500rpm is 50.6 (55.2 + 56.6 + 56.7 + 46.1 + 38.6 ÷ 5 = 50.6), or 4½ per cent less.

With the modified engine the story again changes, for while this engine has the ability to run up to 12,000rpm we would be reducing accelerative performance by running the engine to that speed in all but 1st gear. For example the rpm drop on the change from 5th to 6th gear is less than 1,000rpm. Now if we run the engine to

Power measurement and gearing

12,000rpm we see the peak horsepower over that 1,000rpm span is 51.6 at 11,000rpm, while the minimum is 31.3 at 12,000, with the average being 40.6hp (51.6 + 41.7 + 31.6 + 31.3 ÷ 4 = 40.6). In the meatiest 1000rpm segment of the power band, from 9,500rpm to 10,500rpm, the engine doesn't drop lower than 53.6hp and the average is 57.9hp (53.6 + 57 + 61.4 + 59.4 ÷ 4 = 57.9), or 30 per cent more, so accelerative

Table 9.3 Gearbox gear step and rpm step comparison

	Gear ratio	Gear step	\	rpm drop at redline rpm				
			10,600	11,200	11,800	12,400	13,000	13,600
Rotax 250 road racer								
1st	2.308	+21%	2226	2352	2478	2604	2730	2856
2nd	1.824	+16.2%	1717	1814	1912	2009	2106	2203
3rd	1.529	+11.5%	1219	1288	1357	1426	1495	1564
4th	1.353	+9.7%	1028	1086	1145	1203	1261	1319
5th	1.222	+7.5%	795	840	885	930	975	1020
6th	1.130							
Rotax 125 kart								
1st	2.667	+22.3%	2364	2498	2631	2765	2899	3033
2nd	2.071	+20.7%	2194	2318	2443	2567	2691	2815
3rd	1.643	+16.3%	1728	1826	1923	2021	2119	2217
4th	1.375	+14%	1484	1568	1652	1736	1820	1904
5th	1.182	+11.3%	1198	1266	1333	1401	1469	1537
6th	1.048							
Rotax 125 moto-crosser								
1st	2.667	+25%	2650	2800	2950	3100	3250	3400
2nd	2.000	+20.6%	2184	2307	2431	2554	2678	2802
3rd	1.588	+13.9%	1473	1557	1640	1724	1807	1890
4th	1.368	+12.3%	1304	1378	1451	1525	1599	1673
5th	1.200	+8.8%	933	986	1038	1091	1144	1197
6th	1.095							
Rotax 175 enduro								
1st	3.400	+32.1%	3403	3595	3788	3980		
2nd	2.308	+26.9%	2851	3013	3174	3336		
3rd	1.688	+22%	2332	2464	2596	2728		
4th	1.316	+16.8%	1781	1882	1982	2083		
5th	1.095	+16.6%	1760	1859	1959	2058		
6th	0.913							
Suzuki RV250 street sports								
1st	2.454	+33.8%	3583	3786	3988	4191		
2nd	1.625	+24%	2544	2688	2832	2976		
3rd	1.235	+15.4%	1632	1725	1817	1910		
4th	1.045	+12.3%	1304	1378	1451	1525		
5th	0.916	+8.3%	880	930	979	1029		
6th	0.840							

259

Table 9.4 Dyno test of Suzuki RGV250

rpm	Standard hp	Standard torque	Modified hp	Modified torque
5,000	14.6	15.3	14.9	15.7
6,000	18.1	15.8	19.3	16.9
7,000	24.3	18.2	25.2	18.9
7,500	25.6	17.9	32.8	23.0
8,000	34.7	22.8	32.7	21.5
8,500	39.6	24.5	41.7	25.8
9,000	43.9	25.6	49.7	29.0
9,500	50.4	27.9	53.6	29.6
10,000	55.1	28.9	59.6	31.3
10,250	56.6	29.0	61.4	31.4
10,500	56.7	28.4	59.4	29.7
11,000	46.1	22.0	51.6	24.6
11,500	38.6	17.6	41.7	19.0
11,750			37.6	16.8
12,000			31.3	13.7

performance will be superior with a 5th to 6th upchange at 10,500rpm rather than at higher rpm, or at least it would appear to be so when we make judgements based on the figures available to us in Tables 9.3 and 9.4.

While a lot of people use a gear step chart like Table 9.3 to make such assessments I really feel that they do not give an adequately clear picture about the suitability of the gearbox ratios, or the ideal gear change points. Looking at a gear step table from the aspect of remaining in the engine's power band is all right for a bike being run on the road, but in serious competition there are many other considerations.

Racers often only think of performance in terms of power, handling and brakes, but these are not the only factors in the performance equation. We must also include weight, aerodynamics and gearing. For example more horsepower will accelerate a bike faster, but a change in gearing could also provide a similar performance increase. Gearing is all about torque multiplication, so if we have 20lb/ft torque at the countershaft sprocket and 12:1 gearing, the twisting force at the back wheel will be 240lb/ft. That force will enable the bike to accelerate harder than if it produced 10 per cent more torque but was running with 10:1 gearing.

Looking at Tables 9.5 and 9.6 will help you get a clearer picture of how this works in practice. Except at the higher levels of competition we are generally not concerned about the actual road speed of a road racer. However gear charts and dyno torque figures need to be given this third dimension so that they can be related to the real world. Really the only way we can fully comprehend the situation is to consider axle torque (the torque at the centre of the rear wheel) and road speed and engine speed together. Axle torque is calculated by multiplying engine torque by the overall gear ratio, while road speed can be worked out using this formula:

$$\text{Road speed} = \frac{\text{rpm} \times \text{tyre circumference}}{\text{overall gear ratio} \times 1050}$$

Table 9.5 Stock Suzuki RGV250 rear wheel torque and road speed

rpm	1st 20.84 torq	spd	2nd 13.8 torq	spd	3rd 10.49 torq	spd	4th 8.87 torq	spd	5th 7.78 torq	spd	6th 7.13 torq	spd
5,000	319	17	211	26	160	34	136	41	119	46	109	51
6,000	329	21	218	31	166	41	140	49	123	56	113	61
7,000	379	24	251	37	191	48	162	57	142	65	130	71
7,500	373	26	247	39	188	52	159	61	139	69	128	76
8,000	475	28	315	42	239	55	202	65	177	74	163	81
8,500	511	29	338	44	257	58	217	69	191	79	175	86
9,000	534	31	353	47	269	62	227	73	199	83	183	91
9,500	581	33	385	50	293	65	248	77	217	88	199	96
10,000	602	35	399	52	303	69	256	81	225	93	206	101
10,250	604	35	400	54	304	70	257	83	226	95	207	104
10,500	592	36	392	55	298	72	252	85	221	97	203	106
11,000	459	38	304	57	231	76	195	89	171	102	157	111
11,500	367	40	243	60	185	79	156	93	137	107	126	116

Table 9.6 Modified Suzuki RGV250 rear wheel torque and speed

rpm	1st 20.84 torq	spd	2nd 13.8 torq	spd	3rd 10.49 torq	spd	4th 8.87 torq	spd	5th 7.78 torq	spd	6th 7.13 torq	spd
5,000	327	17	217	26	165	34	139	41	122	46	112	51
6,000	352	21	233	31	177	41	150	49	131	56	121	61
7,000	394	24	261	37	198	48	168	57	147	65	135	71
7,500	479	26	317	39	241	52	204	61	179	69	164	76
8,000	448	28	297	42	226	55	191	65	167	74	153	81
8,500	538	29	356	44	271	58	229	69	201	79	184	86
9,000	604	31	400	47	304	62	257	73	226	83	207	91
9,500	617	33	408	50	311	65	263	77	230	88	211	96
10,000	623	35	413	52	314	69	265	81	233	93	213	101
10,250	654	35	433	54	329	70	279	83	244	95	224	104
10,500	619	36	410	55	312	72	264	85	231	97	212	106
11,000	513	38	340	57	258	76	218	89	191	102	176	111
11,500	396	40	262	60	199	79	169	93	148	107	136	116
11,750	350	41	232	61	176	81	149	95	131	109	120	119
12,000	286	42	189	63	144	82	122	97	107	111	98	121

For example the stock RGV250 is rapidly losing torque, which is in reality accelerative power, once past 10,750rpm. Therefore depending on the actual gearbox ratios we would want to be changing up into a higher gear as soon as possible after 10,500rpm rather than running the engine until it 'hits the wall' at 11,500rpm. (Actually this was quite a good engine, some stock RGV250s, drop dead at 11,300–11,400rpm.) At the bottom end of the power band the torque drops away below 8,500rpm, so if possible we would want to keep the engine above about 8,750rpm. With this in mind we can appreciate that at lower road speeds, where there is reduced air resistance and also reduced rolling resistance, the gearbox ratios should allow the engine to work in that 2,250rpm range from 8,500rpm to 10,750rpm. On this score the gap between 1st and 2nd is way too big and the 2nd to 3rd gap is just bordering on being acceptable.

As road speed increases there is a huge increase in wind resistance and at the same time we have reduced accelerative power due to the higher gearing. This means that to keep accelerating at the best possible rate we must now keep the engine in the meaty portion of the power band, which extends from 9,400rpm to 10,600rpm, a range of 1,200rpm. Again we can see that except for the 5th to 6th gear step, the upper gear ratios are too widely spaced to get maximum performance out of the engine.

The problem is that the stock RGV250 has the power band of a 125 motocrosser, mated to the gearbox ratios of a 350 street sports bike. Hence the only practical way to obtain a substantial performance improvement with this bike, while retaining the stock gear cogs, is to broaden the power range from the factory 1,200 rpm to something closer to 1,600rpm. The modified engine achieves this with a 1,750rpm power band extending from 8,900rpm to 10,650rpm.

Another important consideration with potent two-stroke road racers is obtaining a suitable set of gearbox ratios which allow you to ride through all of the important corners around the circuit in a 'friendly' portion of the power band. This isn't a big problem with 125s except when the track is wet, but with 250s it is a concern, and with 500s it is a big problem even when the track is dry. Ideally you want to have a gear for most of the corners, and certainly for all of the high speed corners, which allows you to hold just enough throttle to keep the bike rolling around the corner. If the gear ratio is too low the engine will attempt to suddenly accelerate the rear tyre in the wrong part of the turn, breaking traction and causing the rear end to step out. Conversely if the gear ratio is too high you will find it difficult to maintain momentum through the turn because at some point the engine will fall right away from the power band. Then when you roll on a little more throttle there isn't any response, so you roll on a little more, and by now the engine may be verging on loading up so you have to twist on a good deal of throttle to get any response. At this point the power will come on suddenly, either causing the rear tyre to step out or else the loss of traction will reduce acceleration out of the turn. Either way your speed entering the straight will be way down, with the end result being that at the end of the straight you will be only pulling the sort of top speed being attained by bikes with considerably less powerful engines.

When we are forced to run a stock wide ratio gear set this poses a real problem as you will find that there will always be at least one or two key corners where say, 4th gear is too low and 5th gear is too high. However with a close ratio gearbox it is much easier to find a suitable gear for every corner. Perhaps you may still have to select a compromise gear to get around one part of the track, but the compromise will be much easier to deal with.

The ultimate solution is to run a cartridge type quick change gearbox which allows the individual gear cogs to be changed to provide the best possible combination of gear ratios to suit individual circuits or even changing circuit conditions. For example one circuit may have a long uphill start which requires a low 1st gear and a closely spaced 2nd gear to avoid cooking the clutch and to stop the engine bogging. Changing the sprockets to lower the secondary drive ratio isn't an option because this

Power measurement and gearing

circuit has a high speed main straight. Thus the final gear set would include a very low 1st gear to ensure a good clean start and a high 6th gear to provide maximum speed on the straight, along with suitable intermediate cogs selected to suit key sections of the track.

Another circuit may have a high speed corner or straight which is affected by changing wind direction. Thus when a slight tailwind is blowing a 6th gear just a few percent higher will provide an improved lap speed. Conversely a change to a headwind would see a slightly lower 6th gear being fitted. As shown in Table 9.7, ratio changes as fine as 1 per cent are possible in the upper two cogs of the Rotax gear set, however the Honda NSR500V is more representative of what is more usually available in the way of alternative cogs for most road racers.

Of course, all of these considerations will be wasted if your tachometer is inaccurate. Therefore, it must be checked and recalibrated. Only then can you be sure

Table 9.7 Rotax 125/250 alternative gear ratios

1st	2nd	3rd	4th	5th	6th
3.000	2.071	[23] 1.643	[23] 1.438	[23] 1.286	[23] 1.182
2.667	2.000		[23] 1.400	[23] 1.263	[3] 1.167*
2.583	1.933		[23] 1.375	[3] 1.250*	[23] 1.150
[23] 2.500	1.867		[23] 1.353	[23] 1.236	[2] 1.130**
[23] 2.385	[2] 1.824**		1.316	[2] 1.222**	[23] 1.118
[2] 2.308**	[3] 1.786*	[23] 1.600	[23] 1.438	[23] 1.200	[23] 1.105
[23] 2.250	[23] 1.750		[23] 1.400	[23] 1.182	[23] 1.095
[3] 2.200*	[23] 1.722		[23] 1.375	1.167	[23] 1.083
[23] 2.143	[23] 1.688		[23] 1.353	1.150	1.048
[23] 2.083		[23] 1.571	[23] 1.438	1.130	1.000
			[23] 1.400		
			[23] 1.375		
			[23] 1.353		
			1.316		
		1.529***	[23] 1.400		
			[23] 1.375		
			1.353***		
			[23] 1.316		
		[23] 1.500	[23] 1.375		
			[23] 1.353		
			[23] 1.316		
			[23] 1.286		
		[23] 1.467	[23] 1.316		
			[23] 1.286		

[2] optional 125 gear kit
* standard 125 gear
[3] optional 250 gear kit
** standard 250 gear

Note – *Only the 3rd/4th gear combinations grouped together are possible.*

Honda NSR500V alternative gear ratios

1st	2nd	3rd	4th	5th	6th
2.133	1.765	1.556	1.350	1.190	1.091
2.067	1.706		1.300	1.143	1.045
2.000*	1.647		1.238	1.091*	1.000
1.938	1.611		1.190	1.045	0.957*
1.875	1.556*		1.143	1.000	0.917
	1.500	1.500	1.350		
			1.300		
			1.238		
			1.190		
			1.143		
		1.474	1.350		
			1.300		
			1.238		
			1.190		
			1.143		
		1.421	1.350		
			1.300		
			1.238		
			1.190		
			1.143		
		1.350	1.238		
			1.190		
			1.143		
		1.300*	1.238		
			1.190*		
			1.143		

* Standard NSR500V gear

Note – Only the 3rd/4th gear combinations grouped together are possible.

that you are getting the best performance from the engine, changing up at the right rpm and keeping within the power band. Unfortunately, many believe that the tacho reading is infallible and beyond question. However, I have not found this to be so; mechanical units are typically 400–700rpm fast and electronic digital readout tachos can be all over the place, depending on the ambient temperature and the quality of the unit.

What we need in a race tacho first of all is that it is accurate and secondly, it must have a needle that swings around at lightning speed without overshooting the mark. For example, we do not want the needle flying around to 13,600rpm and then dropping back to a correct 13,200rpm. Neither do we want it lagging behind true engine speed. In the past racers regularly had to guess gear changes in the lower gears because of needle lag. Thus if the rev limit was 12,000rpm the rider would change up at 11,300 into second gear and at 11,700 going into third.

With cheaper tachos manufacturers reduce needle lag by reducing damping, but this of course increases needle flicker as the bike bumps around the circuit.

Additionally with a lightly damped needle, needle bounce and overshooting the true rpm and then dropping back are common problems. Expensive high quality race tachos will have these kinds of problems engineered out. Such units may use a stepper motor to drive the needle and they may incorporate a shift light.

Extensive tuning on the race track is necessary, even after a full tuning session on the dyno. This is because the load applied on the dyno may have been just a flash loading, ie just long enough to get a power reading. Such a brief run at full load may not allow sufficient heat build-up in the piston crown, combustion chamber or crankcase, to cause detonation. Under normal racing conditions, where full load may be applied for a much longer period, detonation might occur due to the mixture being a fraction lean, and in some instances the ignition advance may be a little early.

Throughout the testing you should make one change at a time. This is the only way that you are going to find out to what the engine is responding. With just one variable introduced for each test it is often difficult to know just what step to take next, so you will appreciate that the introduction of two or three changes will make it virtually impossible to know where you are heading with your tuning.

At the start, race track tuning can be extremely frustrating, because you seem to keep going up so many dead end streets. However, if you stick with it, and go about your tuning in a systematic way, you are sure to have the engine responding better and making more power than before you started. The thing you must do is make just one change at a time, keep accurate notes of any changes, and make sure that you have got a good system which can be relied upon to time each lap accurately.

Appendix

Table of useful equivalents

1 inch = 25.4mm
1 cubic inch = 16.387cc
1 horsepower = 0.7457 kilowatts
        or 1.0139 PS
1 pound foot torque = 1.3558 Newton metres
        or 0.13824 kg/m
1 pound inch torque = 0.11298 Newton metres
1 psi = 6.89476 kilopascals
    or 68.95 millibars
    or 2.0345 inches of Mercury (Hg)
    or 27.67 inches of water

$°F = \left(\frac{9}{5} \times °C\right) + 32$

1 gallon (Imperial) = 160 fluid oz
        or 4.546 litres
1 gallon (US) = 128 fluid oz
        or 3.785 litres

1mm = 0.03937in
1 litre = 61.024cu in or 1000cc
1 kilowatt = 1.341hp
1 PS = 0.9863 horsepower
1 Newton metre = 0.7376 pound foot
1 kg/m = 7.2336 pound foot

$°C = (°F - 32) \times \frac{5}{9}$

1 fluid oz (Imperial) = 28.4cc

1 fluid oz (US) = 29.57cc

Index

Acetone 155, 162, 230
Advance-spark 12, 15, 27, 145, 159, 162, 164, 167–176, 238, 239, 241, 244, 249, 257
Air:
   bleed screw 122, 137, 140
   cleaner 146–148
   cooling 12, 222, 231
   density 142–145, 256–258
   flow 60–62, 65–67, 71, 73, 77, 79, 83–87
   – fuel ratio 12, 59, 72, 109, 117, 120, 125, 130, 131, 141, 143, 145, 160, 251
Alcohol 122, 155, 159–162, 222, 225, 229, 230
Anti-freeze 234
Aprilia 111
Avgas 155, 156, 158, 228

Balancing 8, 197, 198
Bearings 192–200, 204, 205, 208–210, 219
Benzol 155, 229
Big end bearing 193, 195–200, 226
Bing carburettor 73, 148
Blueprinting 7, 8, 192
Boost ports 33, 34, 50–58, 74–76, 217

Bore-cylinder 203, 215, 216, 232, 239, 241
Boyesen reed petals 73, 81, 82
Brake mean effective pressure 252–254
Bultaco 9, 23, 55, 56, 73

Capacitor discharge ignition 165, 167, 168, 184, 188
Carburettor: 120–141, 148–153, 251
   air screw 122, 137, 140
   basic operation 122–130, 150, 151
   Bing 73, 148
   boring oversize 133, 134
   E.I. 128, 129
   float 120–122, 152
   float level 121, 136, 139, 154
   Keihin Power Jet 128–130, 226
   Lectron 128, 129
   McCulloch 148, 151
   main jet 124, 134, 136, 141, 160, 163
   Mikuni 124, 128, 129, 149
   Mikuni Powerjet 125, 126
   needle 124, 135–141
   needle jet 124, 134–136
   needle and seat 120–122, 160
   pilot jet 122, 136, 137, 140, 162
   pulse type 148–152, 226

size 9, 124, 128, 131–133, 146, 147, 249, 250, 255
slide 125–129, 134, 136, 138, 153, 162
Walbro 149
waterproofing 148
Castor oil 222–225, 230
Ceramic coating 231, 232
Clearance:
  big end 193, 196, 203
  little end 208
  crank end-float 192–194
  piston 8, 15, 207, 208, 219, 220, 223, 231
  piston pin 207
  piston ring groove 211
  piston to head 17, 232
  rotary valve disc 72
  squish 8, 16–19
Coil 164–166, 179, 188, 189
Combustion chamber: 12–21, 169, 174, 255
  squish area 15, 18, 20, 232
  volume 23–26, 223
Combustion rate: 12, 13, 15, 20, 22, 59, 72, 132, 145, 162, 169, 175, 186, 231, 243
  abnormal 12, 13, 15, 20, 59, 142, 145, 157, 159, 162, 170, 231, 236, 238, 239
Compression ratio: 9, 12, 21–28, 159, 162, 169, 175, 223, 232, 238, 239, 241, 244, 257
  calculation 22–27
  permissible 21, 24, 27, 28, 163
Condenser 164–166
Con-rod 8, 17, 195–197, 199, 202, 203, 207, 219
Cooling:
  air 12, 222, 231–233
  water 16, 222, 231, 233–248
Crankcase: 203–205
  compression 48–50
  reed valve 56, 63, 72–86
  seals 194, 198, 204, 205
Crankpin 195–197, 200
Crankshaft: 7, 107, 194–203, 207
  balancing 8, 197

run-out 194, 203
Crankwheel 8, 66, 194, 200–203
Cylinder: 7, 8, 17, 203, 215–219
  bore accuracy 203, 215, 216, 232, 233, 239, 241
  honing 216, 217
  plating materials 218, 219
Cylinder head: 12–20, 232
  fin pattern 12
  gasket 17, 27, 28
  sealant 28, 30
  squish clearance 8, 16–19
  tension 29, 30

Detonation 12, 13, 15, 20, 142, 145, 157, 159, 162, 167, 170, 174, 176, 180–182, 184, 223, 231, 236, 238, 239, 265
DKW 31
Dynamometer 174, 175, 249–251, 255–258

Efficiency-volumetric 22, 130, 169
Electronic ignition 165–172, 176
End gases 13, 182, 231, 241
Ethanol 122, 159–162, 222, 225
Exhaust:
  flow 43–48, 106, 255
  port 7, 33, 36–48, 217, 219, 232
  port timing 10, 36, 40–43, 255
  pulse waves 44, 46, 88–90, 94, 96–99, 106, 117, 175
  sealant 117, 118
Expansion chamber: 22, 26, 36, 41, 43, 89–119, 130, 174, 176, 250
  baffle cone 99–101
  decoking 118, 119
  diffuser cone 95–99
  fabrication 102–107, 116
  header pipe 90–95, 110, 111
  muffler 114, 115, 174
  stinger 101, 102

Filter-air 146–148
Float-carburettor 120–122, 136, 139, 152, 154
Flywheel: 9, 194, 200–202

Index

concentricity 8, 194
  lightening 200–202
Fuel: 154–163
  octane rating 145, 154–159
  oil mixing 160, 162, 219, 223–230
  pump 152

Gasket:
  head 17, 27, 28, 232
  sealant 28, 30
Gasoline 154–156
Gearing 8, 130, 255, 258–264
Gudgeon pin 197, 207, 219

Header pipe 90–95, 110, 111
Head gasket 17, 27, 28, 232
Hone-cylinder 216, 217
Honda 38, 50, 75, 89, 90, 129, 132, 176, 218, 263, 264
Horsepower 251, 252
Humidity 143–145
Hylomar sealant 28

Idle jet 122, 136, 137, 140, 162
Ignition: 164–181
  abnormal 12, 13, 15, 20, 22, 59, 72, 132, 145, 162, 170, 231, 236, 238, 239
  advance 12, 15, 27, 145, 159, 162, 164, 167–176, 238, 239, 241, 244, 249, 257
  capacitor discharge 165, 167, 168, 184, 188
  coil 164–166, 179, 188, 189
  condenser 164–166
  leads 167, 190
  magneto 165–167, 179
  points 164–166, 179, 180
  rotor 9, 10, 201
  timing 12, 15, 27, 145, 159, 162, 163, 176
  waterproofing 190, 191
Inhibitor-cooling system 16, 233, 234
Inlet:
  manifold 79, 83–86, 154
  port 7, 58–62, 65–67, 71, 77, 83–86, 217, 232

port timing 10, 58–60, 64–66, 70, 73, 76
pulse waves 58

Jet:
  idle 122, 136, 137, 140, 162
  main 124, 134, 136, 141, 160
  needle 124, 135–141
  power 125, 126, 129–131

Kawasaki 36, 38, 90, 107, 219, 220

Little end bearing 55, 56, 86, 197, 208, 209, 219
Lubrication 55, 60, 193, 198, 199, 208–210, 216, 219, 222–230

McCulloch 28, 148, 151, 177
Magneto 165–167, 179
Main bearing 194, 198, 199, 204, 205
Main jet 124, 134, 136, 141, 160, 163
Methanol 122, 159–162, 222, 225
Methyl benzine 155, 156, 159, 229
Minarelli 67, 255
Mineral oil 222, 228
Morbidelli 10, 42, 67, 219
Muffler 114, 115
MZ 33, 50

Needle 124, 135–141
Needle jet 124, 134–136
Needle and seat 120–122, 160
Nikasil 218
Nitromethane 122, 155, 162, 163
Nitropropane 155, 163

Octane rating 145, 154–159
Oil: 222–230
  castor 222–225, 230
  – fuel ratios 160, 162, 219, 223–230
  mineral 222, 228
  synthetic 222–224, 228
  viscosity 222, 223

Permatex sealant 28, 118
Petrol 154–156
Piston: 7, 17, 53–55, 60, 107, 130, 198, 203, 205–212, 219, 220, 231

circlips 205–207, 219
clearance 8, 15, 207, 208, 219, 220, 223, 231
lubrication 60, 208, 219, 226
pin 197, 207, 219
rings 60, 208, 210–215, 222, 226, 241

Points:
ignition 164, 179, 180
magneto 165, 179, 180

Port:
boost 33, 34, 50–58, 74–76, 217
exhaust 7, 10, 33, 35–48, 217, 219, 232, 255
inlet 7, 10, 58–62, 65–67, 77, 82–85, 217, 232, 255
transfer 31–34, 48–58, 77–79, 217, 255

Power:
band 7–9, 22, 35, 36, 45, 49, 109, 253–255, 258–262
jet 125, 126, 129–131
measurement 249–251, 255–258
valve 9, 33, 35, 36, 44, 45, 90, 223, 224

Pre-ignition 180, 182, 184, 187

Pressure:
cooling system 235, 236
fuel 152, 153

Propylene oxide 155, 162

Pulse waves:
exhaust 26, 44, 46, 88–90, 94, 96–99, 106, 117
inlet 58, 65

Pump – fuel 152

Radiator: 242–248
inhibitor 16, 233, 234
mounting height 243
sealant 234

Reboring:
carburettor 133, 134
cylinder 203, 215, 216, 232, 233, 239, 241

Reed valve 56, 63, 72–87
Relative air density 142–145, 256–258
Ring-piston: 60, 208, 210–215, 240, 241
flutter 210, 211
gap 211

Rotary valve 10, 62–72
Rotax 33, 35, 55, 69, 71, 77, 112, 132, 168, 176, 219, 237–241, 244, 263
RPM limit 7, 107, 172, 200–202, 221, 254, 255
Running-in 219

Sealant:
cooling system 234
exhaust flange 118
head gasket 28, 30
Hylomar 28
Permatex 28, 118
Silastic 30, 118, 191, 238, 239

Seals-crankcase 194, 199, 204, 205
Shot peening 202
Silastic 30, 118, 191
Silencer 114, 115
Soluble oil 234
Spark advance 12, 15, 27, 145, 159, 162, 163, 238, 239, 241, 244, 249, 257

Spark plug: 182–190
cap 190
cleaning 189, 190
electrode style 184, 186
gap 187, 188
heat range 182–185
leads 167, 190
reading 141, 142, 183, 184

Squish 8, 12–21, 232
Suzuki 26, 33, 38, 48, 49, 59, 76–78, 87, 110, 146, 225, 226, 253–256, 258–262

Tachometer 264, 265
Temperature:
cylinder head 174, 175
water 235, 241, 243, 244

Tension – cylinder head 29, 30
Thermocouple washer 174, 175, 187
Thermostat 234, 235, 243
Timing:
exhaust port 10, 36, 40–43
ignition 12, 15, 27, 145, 159, 162, 163, 176
inlet port 10, 58–60, 73, 76

    marks 168, 177, 179
    rotary valve 10, 64–66, 68–72
    transfer port 10, 41, 48–53
Toluol 155, 156, 159, 229
Torque:
    engine 252, 256, 257
    rear wheel 260–262
Transfer ports 10, 31–34, 48–58, 77–79, 217

Volumetric efficiency 22, 130, 169

Water cooling 16, 222, 231, 233–248
Waterproofing:
    carburettor 148
    ignition 190, 191

Yamaha: 7, 17, 20, 33–36, 48, 49, 59, 72, 73, 77, 78, 82, 86, 107, 125, 132, 146, 149, 170, 172, 174–176, 181, 192, 200, 204, 217, 219, 237, 251, 253
    power valve 9, 33–36, 44, 45, 90, 223, 224